Analysis and Strategies for Sustainable Regional Planning

Juanjo Galan

Analysis and Strategies for Sustainable Regional Planning

Sierra Calderona Strategic Plan, Spain

Serra Náquera Olocau Marines Gátova | Plan Territorial Estratégico Sierra Calderona

Springer

Juanjo Galan
Department of Architecture
Aalto University
Espoo, Etelä-Suomi, Finland

ISBN 978-3-030-07205-6 ISBN 978-3-319-96508-6 (eBook)
https://doi.org/10.1007/978-3-319-96508-6

This Springer imprint is published by the registered company Springer Nature Switzerland AG
The registered company address is: Gewerbestrasse 11, 6330 Cham, Switzerland

Foreword I

This book presents a comprehensive regional model that is the result of exhaustive multidisciplinary research. It features a sustainability-orientated integrated plan for the south-central area of the Calderona Mountain Range (Sierra Calderona). The Plan exceeds strict municipal demarcations and covers an area of incomparable heritage, natural and cultural wealth.

The Plan for the south-central area of Sierra Calderona was jointly developed by the Universitat Politècnica de València and the municipalities of Serra, Náquera, Marines, Olocau and Gátova, with the invaluable assistance of their fellow citizens. It offers an exceptional example of the possibilities of sustainable planning and of the effective management of natural, cultural and social resources. The implementation of this comprehensive plan will certainly contribute to improving the quality of life of the inhabitants of the Sierra Calderona.

It is essential to treat both the Sierra Calderona and its habitats with due respect so that future generations will be able to discover and enjoy its privileged and rich environment. That is why an effort was needed to forge a strong complicity between nature and society, based on an integrative approach and on a balanced combination of layers and interests. Development paths that used to run in parallel converge now harmoniously in the Strategic Plan, and the combination of different disciplines leads to intelligent and sustainable growth.

Valencia, Spain

Francisco José Mora Mas
Rector

Foreword II

The traditional urban-rural dichotomy and the classical systems of planning are being challenged by new ways of living, working, producing, consuming and enjoying free time. At the same time, science is providing us with clear evidence of the deep connections between social and ecological systems.

In this context, we try to navigate amongst the legitimate aspiration of each generation to live better and the urgent necessity to understand local and global implications of our individual and collective actions.

We are more people than ever before, but at the same time—and following the emergence of new sustainability frameworks and new social and governance paradigms—we might be better equipped to integrate complexity, to share the planet in a smarter way, and to look for synergies instead of conflicting interests. Somehow, we do not have another moral and ethical alternative.

This book is a pioneering exploration of what can be achieved when we choose the path of integration instead of the highway of division.

Espoo, Finland

Anna Valtonen
Dean of the School of Arts, Design and Architecture

Espoo, Finland

Pekka Heikkinen
Head of the Department of Architecture

Preface

The main reason for writing the book was to show, through a case study, the potential synergies provided by integrative planning and explore new possibilities to link all the disciplines involved in regional planning. The book shows how, in planning, the whole can be greater than the sum of the parts and that the sustainability framework can be an ideal platform for multi-, inter- and trans-disciplinarity. This book is aiming at a wide range of professions that will find themselves represented in the Sierra Calderona Strategic Plan and will get new insights into how their specific work can contribute to a bigger picture.

In particular, this publication includes the analysis and diagnosis of different territorial layers in the Sierra Calderona (Valencia, Spain) and the definition of interconnected regional and sectoral strategies. The whole work incorporates new methodologies and graphical systems for regional and local planning and constitutes an example of multidisciplinary planning based on principles of sustainability, system thinking, participatory governance and local adaptation.

Espoo, Finland Juanjo Galan

Acknowledgements

For their collaboration and support in the preparation of the Sierra Calderona Strategic Plan:

- Alfonso Fernández Morote
- Gemma Peiro Frías
- Ignacio Diez Torrijos
- Antonio Fresneda Colomer
- Ana Gascón Hernández
- Anna Ruiz Pitarch
- Miguel Portolés Castaño
- Lourdes Canos Darós
- Juan Pablo Perales Marhuenda
- Pablo Lambertos Escudero
- Guadalupe Martín Pardo
- Cristina Santandreu Mascarell

For their collaboration and support in the translation of the book:

- Lucía Martínez Estefanía
- Paula Cardells Mosteiro
- Stephen Venn

For their advice and support in the Polytechnic University of Valencia (UPV):

- Vicente Caballer Mellado (Departament of Economics and Social Sciences, UPV)
- Mª del Carmen Blasco Sánchez (Departament of Regional and Urban Planning, UPV)
- Francisco Martínez Pérez (Departament of Regional and Urban Planning, UPV)
- José Vicente Colomer Ferrándis (Departament of Transport Engineering, UPV)
- Martín Rodríguez García (Departament of Life Sciences, UPV)
- Alberto San Bautista Primo (Departament of Agronomic Sciences, UPV)
- Francisco Galiana Galán (Departament of Rural Engineering, UPV)

For their advice and support in the Aalto University:

- Pekka Heikkinen (Head of the Department of Architecture, Aalto University)

For their advice and support in the municipalities included in the Sierra Calderona Strategic Plan:

- Javier Arnal Gimeno (Mayor of Serra)
- Damián Ibáñez Navarro (Mayor of Náquera)
- M. Dolores Celda Lluesma (Mayor of Marines)
- Antonio Ropero Morales (Mayor of Olocau)
- Leopoldo Romero Llima (Mayor of Gátova)
- Fernando Alfaro Ibáñez (Town Councillor of Planning, Náquera)

Contents

About the Author

Juanjo Galan is an Associate Professor of Landscape Architecture at the Department of Architecture, Aalto University. His research focuses on landscape planning, landscape design, sustainable development, regional and urban planning and, on a more general level, on the intersections between social and ecological systems.

After finishing his master studies in Landscape Architecture at the Heriot-Watt University of Edinburgh, he received his Ph.D. in Landscape Planning in 2011 in the Department of Urban Planning at the Polytechnic University of Valencia, where he spent 10 years as teacher, researcher and as founder–coordinator of the Master in Landscape and Garden Design. During this period, he also developed an intense institutional and professional career as president of the Spanish Association of Landscape Architects, as external advisor of the Department of Regional Planning of the Valencian Autonomous Region and as director of the Galan Vivas landscape architecture practice.

Introduction

1

Abstract

The introduction of the book explains the main objectives and the basic structure of the Sierra Calderona Strategic Plan. The plan provides an example of integral planning at a regional and local scale and is informed and driven, by social, cultural, ecological, perceptual, sustainability and economic factors that are supported by a participative process. The elaboration of the Sierra Calderona Strategic Plan was divided into the following four phases that, in turn, structured this publication in two volumes. Phase 1 (*Territorial Context*) includes the gathering and processing of different layers of environmental, urban, demographic, economic and socio-cultural information that were grouped into a single documentary and cartographic database. Phase 2 (*Regional Analysis and Diagnosis*) produced the specific and necessary layers of information for the development of the planning proposals of the Sierra Calderona Strategic Plan. Phase 3 included the definition of *Regional Objectives and Strategies*. Finally, Phase 4 consisted of ten sectoral plans and eighteen pilot projects aimed to define the evolution of the Sierra Calderona, as well as to facilitate decision-making, external investment and the future development of local planning. The three first phases are presented in this book.

1.1 Goal of the Sierra Calderona Strategic Plan

The Sierra Calderona Strategic Plan was developed for the municipalities of the south-central area of Sierra Calderona in order to foster harmonised planning and sustainable development from an environmental, cultural, social, economic, infrastructural and landscape perspective.

The Strategic Plan takes advantage of the synergies and potentials of supramunicipal planning and provides clear and practical guidance for local planning, decision-making and public participation. It also facilitates the communication to third parties of the possibilities that the area offers for both investment and public use.

Additionally, the Strategic Plan aims to establish the necessary link between the main principles of the Territorial Strategy of the Valencian Region and the municipal planning system, by specifying and adapting the principles of the former to a specific area and by expanding the perspectives and possibilities of the later in their local areas of competence.

A first study of the area showed its outstanding value, which is the result of a contrasting physiography, a particularly benign climate and an exceptional natural, cultural and landscape heritage.

The area so defined is characterised by the geographical encounter of a coastal and a mountain system. It presents a rich mosaic of agroforestry in which towns and villages, housing estates, industrial land and infrastructures are integrated with different levels of success. In addition, the area presents a strong social fabric with participative communities and with its own identity, enriched with the contributions of seasonal and occasional visitors.

The Calderona Mountain Range (Sierra Calderona) in general and its south-central area in particular have been historically bonded to the Turia River Valley, and it is currently a strategic enclave at the northern limit of the metropolitan area of Valencia, where it is expected to remain as a place for nature, culture and services. The Calderona Mountain Range aims to open to the outside without resigning to its own history and character. That is the motivation behind this Strategic Plan, which is expected to help its five partner municipalities to better address the opportunities and problems they share.

1.2 Structure of the Sierra Calderona Strategic Plan

The drafting of the Strategic Plan for the municipalities of the south-central area of Sierra Calderona was sequenced in the following four phases that, in turn, structured this

J. Galan, *Analysis and Strategies for Sustainable Regional Planning*,
https://doi.org/10.1007/978-3-319-96508-6_1

publication in two volumes. The results of the phases 1, 2 and 3 are included in this book. A public consultation process that was supported with a specific website (www.tusierracalderona.es) accompanied the development of these phases.

Phase 1: Territorial Context: The elaboration of the plan started with the gathering and processing of different layers of environmental, urban, demographic, economic and socio-cultural information that were grouped into a single documentary and cartographic database. This base allowed a homogeneous treatment and definition of proposals on the subsequent phases of the Plan.

Phase 2: Regional Analysis and Diagnosis: The analysis and diagnosis aimed at generating the specific and necessary layers of information for the development of the Sierra Calderona Strategic Plan, both with regard to the current state of the area and to its socioeconomic and environmental processes. For that purpose, the starting point was the already existing documentation, as well as that internally produced and the results of the first phase of the public consultation.

Phase 3: Regional Objectives and Strategies: The definition of regional objectives and strategies was the preliminary step for the development of plans, actions and concrete proposals. This phase was informed by the second phase of the Public Participation Plan.

Phase 4: Sectoral Plans and Pilot Projects: The ten sectoral plans and the eighteen pilot projects included in the last part of the Strategic Plan defined the environmental, cultural, social, economic and urban planning frameworks for the evolution of the Sierra Calderona. In addition, these sectoral plans and pilots were prepared to facilitate decision-making, external investment and future local planning. The list of sectoral plans included: land-use and landscape planning, forestry, hunting, agriculture, stockbreeding, urban planning, heritage, tourism and public use, socioeconomic development, sustainability, and finally, governance. On the other hand, the eighteen pilots —prepared in the form of projects, studies or programmes— developed in detail some of the most relevant or complex proposals included in the sectoral plans. This last phase was informed by the third phase of the Public Participation Plan.

Public Participation Plan: The Sierra Calderona Strategic aims to respond to the expectations, needs and concerns of both local residents and visitors. For this purpose, a Public Participation Plan was developed and structured in three phases that were progressively activated during the preparation of the Plan and that mainly addressed the following questions: Which territory? Which objectives? Which actions?

1.3 Collaborators

The team of collaborators included landscape architects, architects, agronomists, forest engineers and experts in tourism, many of whom were linked to the academia (professors and final year students) although there was also a significant and essential presence of external professionals (see acknowledgments).

Additionally, a team of supervisors was periodically informed of the work progress and their impressions and suggestions collected. The presence of those experts and decision-makers from different governmental agencies such as the Diputación de Valencia (Provincial Council of Valencia) and the five municipalities promoting the Strategic Plan was crucial for the integration of their particular visions of planning, as well as their diverse experiences and knowledge. The team of supervisors was completed with professors from the Universitat Politècnica de València with specific expertise in some of the topics addressed in the Sierra Calderona Strategic Plan.

2 Territorial Context

Abstract

The *Sierra Calderona Strategic Plan* aims to integrate environmental, cultural, social, economic and urban planning criteria, as well as to guide decision-making and local planning. The Plan was informed and driven by social, cultural, sustainability and economic factors and was supported by a participative process. This chapter gathers and displays the existing territorial information of the south-central area of Sierra Calderona. This information was used to generate new specific cartographies for the Strategic Territorial Plan. The Sierra Calderona Strategic Plan was developed during the years 2013–2014 in five municipalities (Serra, Náquera, Olocau, Marines y Gátova), located at the outskirts of the metropolitan area of Valencia. The Strategic Plan covers a surface of 200 km^2 with a significant portion included in the Sierra Calderona Natural Park and its buffer area. Due to its proximity to the metropolitan area of Valencia (1.5 million inhabitants) and to its natural, cultural and recreational values, the area is exposed to a strong pressure for transformation and public use. Interestingly, the Sierra Calderona displays most of the challenges and potentials affecting other areas located at the limits of urban agglomerations, giving to the whole work and to the proposed solutions a general interest.

2.1 Introduction

This chapter gathers the existing territorial information of the south-central area of Sierra Calderona. In the following chapters, this information will be used to generate new specific cartographies for the Sierra Calderona Strategic Plan.

The thematic and urban cartographies facilitated by both "Sistema de Información Territorial" and "Instituto Cartográfico" of the Valencian Region for the municipalities of Náquera, Serra, Olocau, Marines and Gátova have been merged into one single cartographic database from which the Plan's proposals have been developed.

2.2 Territorial Context

2.2.1 Location and Scope

The Sierra Calderona Strategic Plan covers a total area of approximately 200 km^2 (20,000 ha) and includes the municipalities of Serra, Náquera, Gátova, Olocau and Marines, all located within the Province of Valencia, at the northern limit of the metropolitan area of the city of Valencia.

These five municipalities share a common past, present and future, and via this Plan, they aim to gain a common tool that will support local planning, decision-making and public participation, while showing third parties the enormous potential that Sierra Calderona has to offer as a space for economic investment within an exemplary sustainable development framework.

The evolution of the studied area is inextricably linked to the Sierra Calderona Natural Park ("Parque Natural de la Sierra Calderona"), created in 2002, since around 50% of its surface area is located within the Natural Park and 30% in its buffer zone. Therefore, the whole area is highly affected by the Natural Park regulations: "Plan Rector de Usos y Gestión" or "PRUGSC" (in English: "Use and Management Governing Plan") and the "Plan de Ordenación de los Recursos Naturales" or "PORNSC" (in English: "Natural Resources Management Plan").

The area is connected to two major transport infrastructures: on the east to the motorway A7, also known as "Autovía del Mediterráneo", and the V21 road, and on the south to the CV35, popularly known as "Pista de Ademuz". Railway connections are also present in the area since the neighbouring villages of Bétera and Líria have the terminal train stations of two regional railway lines from the city of Valencia (Figs. 2.1 and 2.2).

J. Galan, *Analysis and Strategies for Sustainable Regional Planning*,
https://doi.org/10.1007/978-3-319-96508-6_2

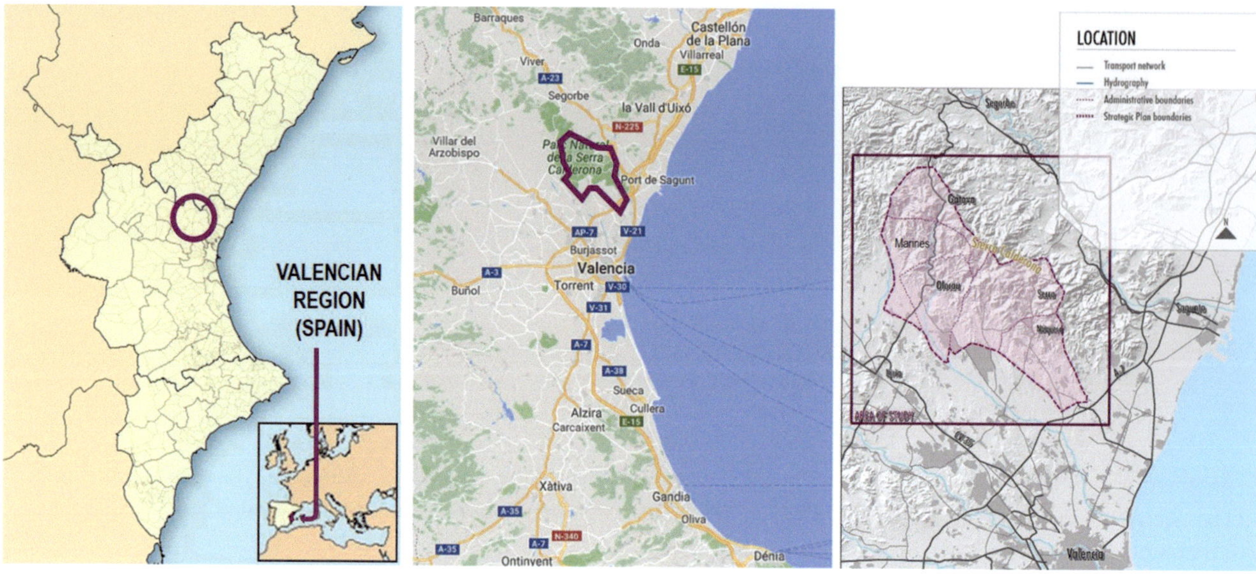

Fig. 2.1 Location of the scope covered by the strategic plan. *Source* Own elaboration

Fig. 2.2 Scope of the strategic plan. *Source* Own elaboration

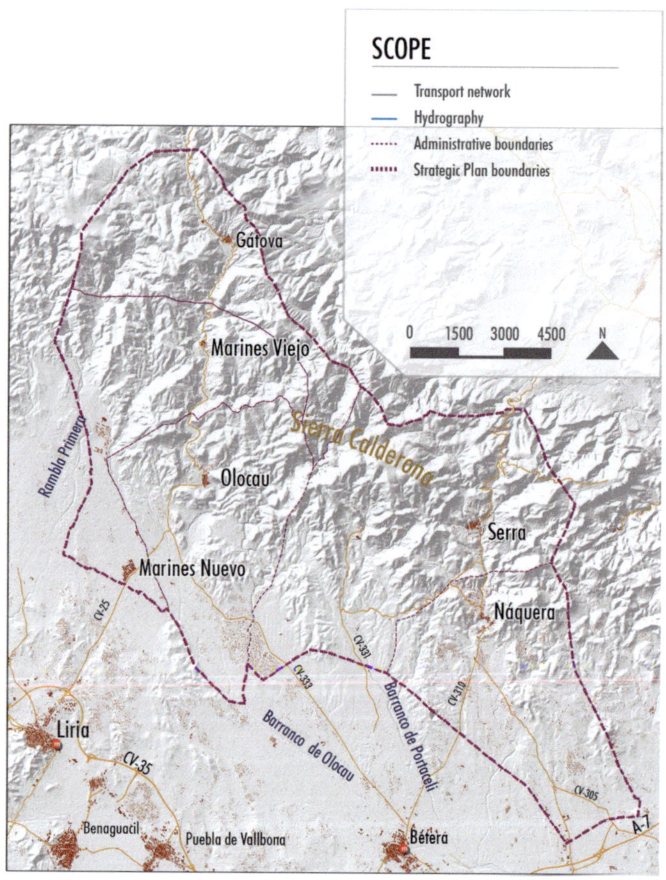

2.2.2 History

As numerous archaeological sites document, the Sierra Calderona was inhabited since prehistoric times, having hosted since then Iberian and Roman settlements that were strategically located between the Palancia Valley in the north, and the Valencian Plain in the south.

The area was strongly influenced by the Muslims, who founded the settlements today known as Serra and Olocau. However, it was after the Christian Reconquest that the settlements of Gátova, Marines and Náquera were consolidated.

From the seventeenth century onwards, the area focused its economy on forestry (charcoal, wood, firewood, esparto), agriculture, horticulture (cereals, vines, olive, fig, almond and cherry trees) and stockbreeding. Mining, quarrying and water and ice extraction also became common activities.

As defence constructions reveal, in the course of the nineteenth and early twentieth centuries, the strategic importance of the corridors crossing Sierra Calderona became significant in different military conflicts. During this period, villas and secondary residences were established in Náquera and Serra, foreseeing the progressive importance that, during the second half of the twentieth century, the construction and service sectors would acquire in the region. It was during the second half of the twentieth century when citrus trees were strongly introduced into the plains of Camp del Turia and when the new town of Marines (Marines Nuevo) was built after the devastation of the old town (Marines Viejo) in the 1957 flood.

Due to their significance in regional planning, it is important to mention two large military zones, one located in Bétera, in the vicinity of the area covered by the Strategic Plan, and a second one in Marines, as well as the creation of a vast Natural Park, which affects, to a greater or lesser extent, all five municipalities.

2.2.3 Climatology

The studied area has a Mediterranean climate with a moderate temperature fluctuation, the average annual temperature being 17 °C. Rainfall is irregular with values ranging between 350 and 600 mm. Peak values are registered in autumn, when torrential rain occurs. Summers are very dry. Occasionally, winter snowfalls can occur at high altitudes.

In general, the flat low-lying areas present a particularly mild climate, promoted by the south-facing slopes, while in the mountainous areas and interior valleys, there is a more marked temperature fluctuation.

2.2.4 Physiography

Physiographically, the area can be divided into three different areas:

- The southern fertile low-lying plains
- The foothills of Sierra Calderona
- The rugged mountainous interior areas, whose major peaks are Gorgo (907 m), Águila (878 m), Rebalsadors (802 m), Oronet (742 m) and Garbi (600 m).

From a lithological perspective, the soil in the lowlands is made up of pebbles, gravels, silts, clays and conglomerates; while in the mountainous areas dolomites, sandstones and clays define a siliceous substrate with a characteristic reddish colour (Figs. 2.3, 2.4, 2.5 and 2.6).

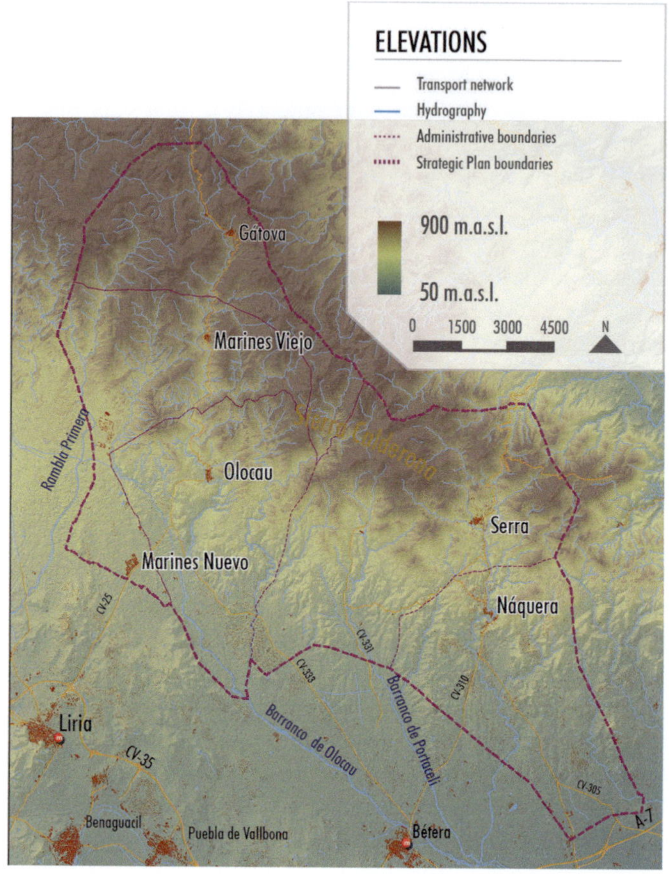

Fig. 2.3 Elevations. *Source* CITMA

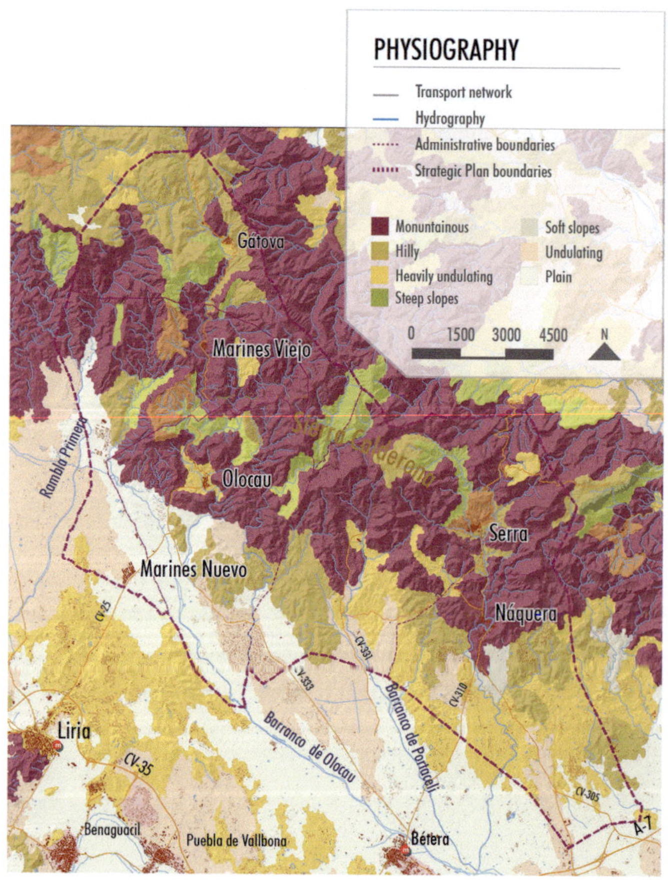

Fig. 2.4 Physiography. *Source* CITMA

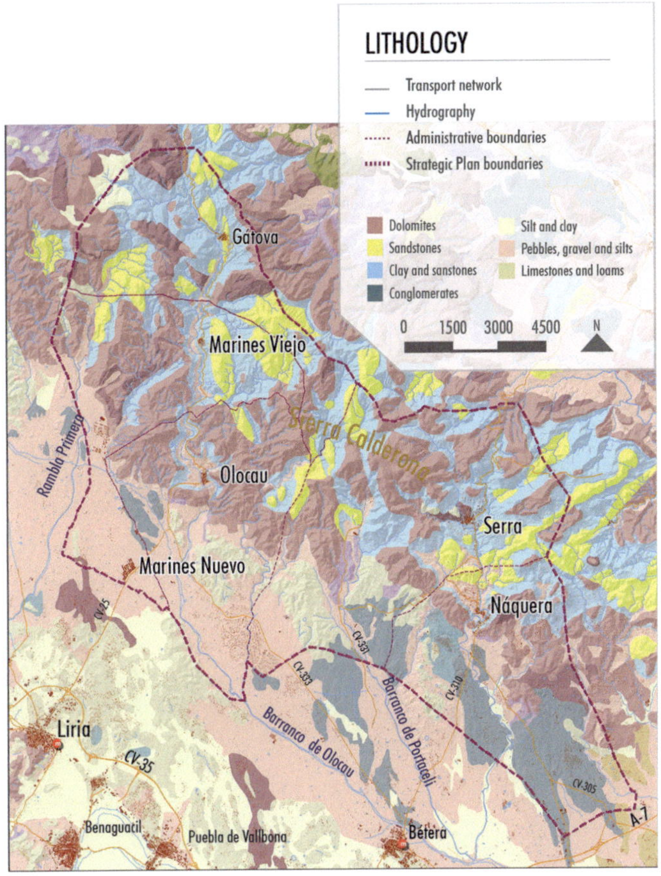

Fig. 2.5 Lithology. *Source* CITMA

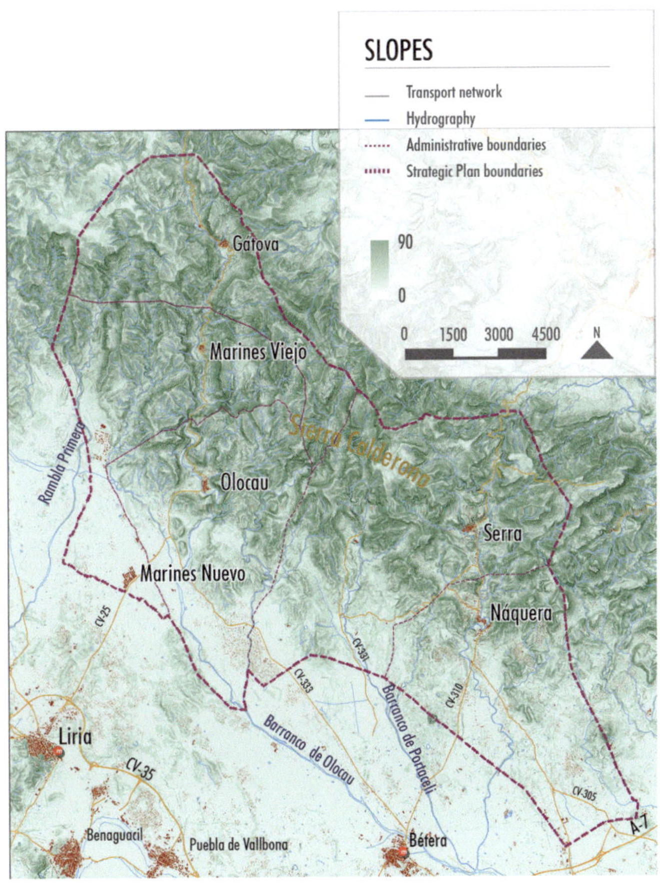

Fig. 2.6 Slopes. *Source* CITMA

2.2.5 Hydrography

The entire study area drains into the Valencian Plain through a system of ravines and gullies that can temporarily transport large flows. The ravines of Olocau, Portaceli and Náquera are the longest and most important in the area; their valleys host the historic urban settlements and the roads crossing the mountain range from south to north. These ravines, after their confluence, become the "Barranco de Carraixet".

The whole area has numerous springs of high-quality water. The artificial hydrographic system includes the ancient networks, ponds and channels in the interior valleys, as well as the channel of Campo del Turia, which crosses the southern part of the Marines municipality, where several water tanks of great capacity are located (Fig. 2.7).

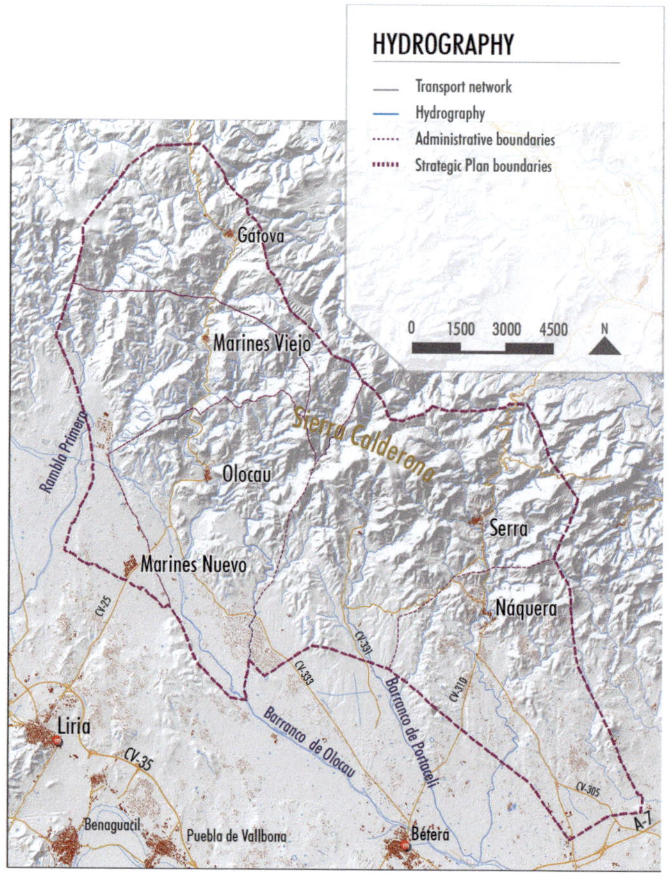

Fig. 2.7 Hydrography. *Source* CITMA

2.2.6 Land Uses

The area covered by the Sierra Calderona Strategic Plan presents a predominant matrix of agricultural and forest land. Amidst this matrix, urban areas have been inserted in the form of ancient towns, new low- and medium-density settlements and, as a singularity, the new town centre of Marines (Marines Nuevo). Industrial areas are scarce and mostly concentrated on the south-eastern corner of the area. According to this pattern, the area can be divided into three zones:

(1) Southern plain and Sierra Calderona first foothills: it presents irrigated fields, mainly citrus trees, and sparse low-density housing estates in the south-eastern part (Náquera). The largest housing estate, Pedralbilla-Torre de Portaceli, is located south of Olocau and Serra. The village of Marines Nuevo is also included within this area.

(2) Interior valleys and Sierra Calderona foothills: this includes the old town centres of Náquera, Serra, Olocau, Gátova and Marines Viejo, as well as adjacent agricultural land, in which both irrigated (vegetable gardens, orchards and citrus tree plantations) and non-irrigated crops (olive, carob, and almond trees) coexist.

(3) Forest area: this stretches over the mountainous areas and is composed mainly of thermo-mediterranean scrub accompanied, especially in the eastern part, by a canopy of pine trees. Occasionally, localised masses of cork oak trees can be found. Integrated in the forest matrix, mainly in the surroundings of small villages and abandoned rural constructions, rainfed orchards with almond and olive trees can be also found (Figs. 2.8, 2.9, 2.10, 2.11 and 2.12) (Table 2.1).

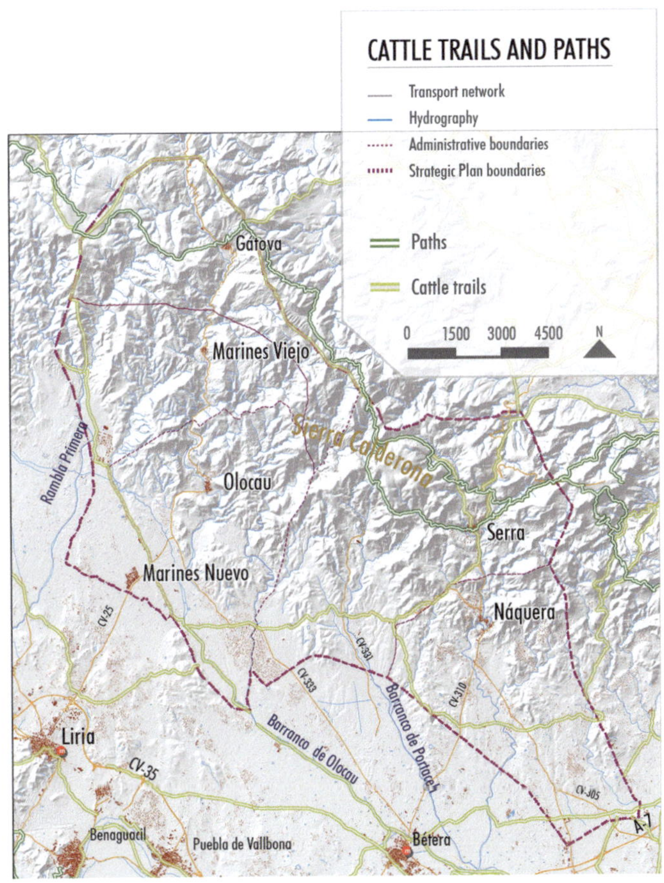

Fig. 2.8 Cattle trails. *Source* CITMA

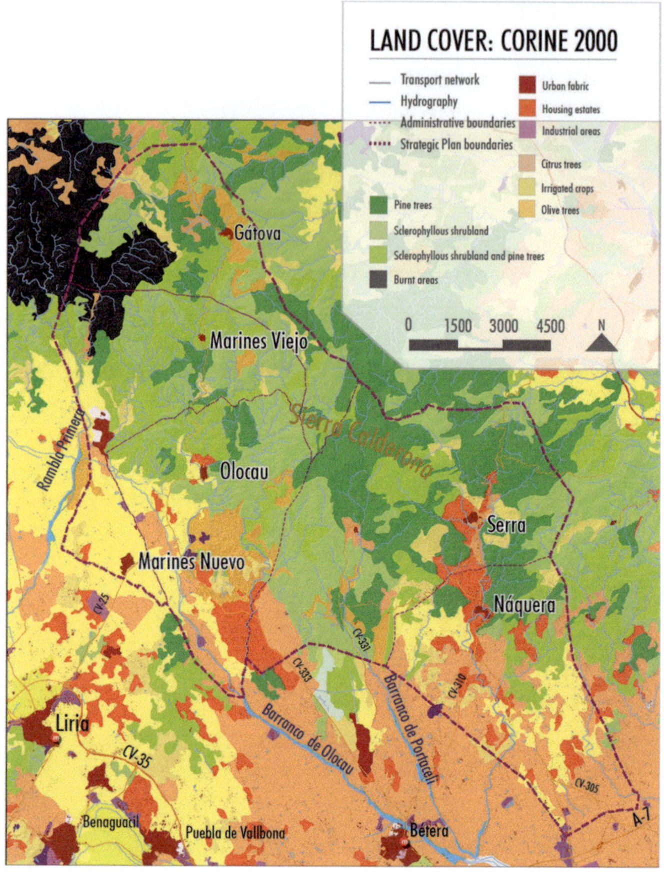

Fig. 2.9 Land cover CORINE 2.000. *Source* CITMA

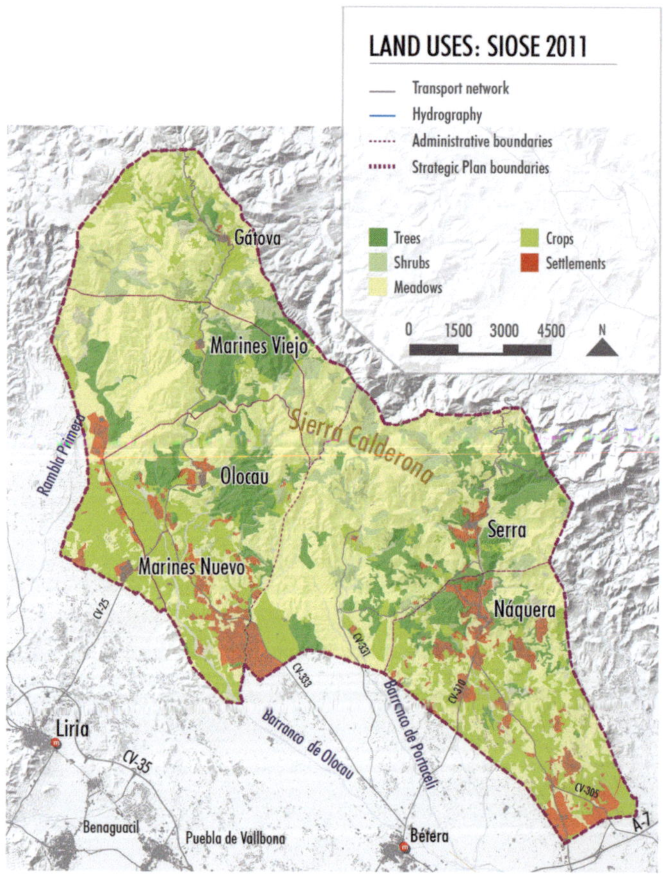

Fig. 2.10 Land uses SIOSE 2.011. *Source* CITMA

Fig. 2.11 Soil fertility classification. *Source* CITMA

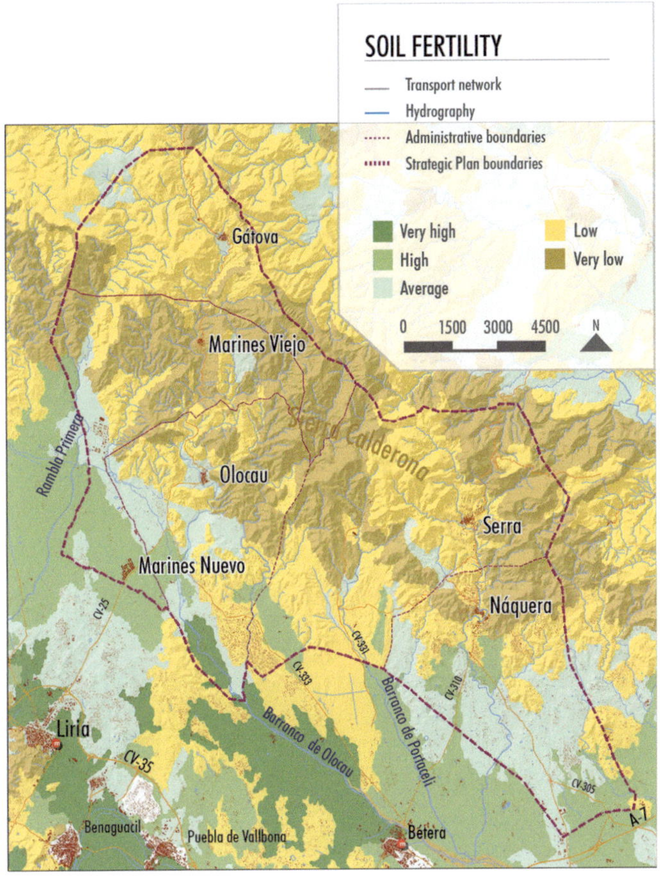

Fig. 2.12 Common and strategic forest land. *Source* PATFOR

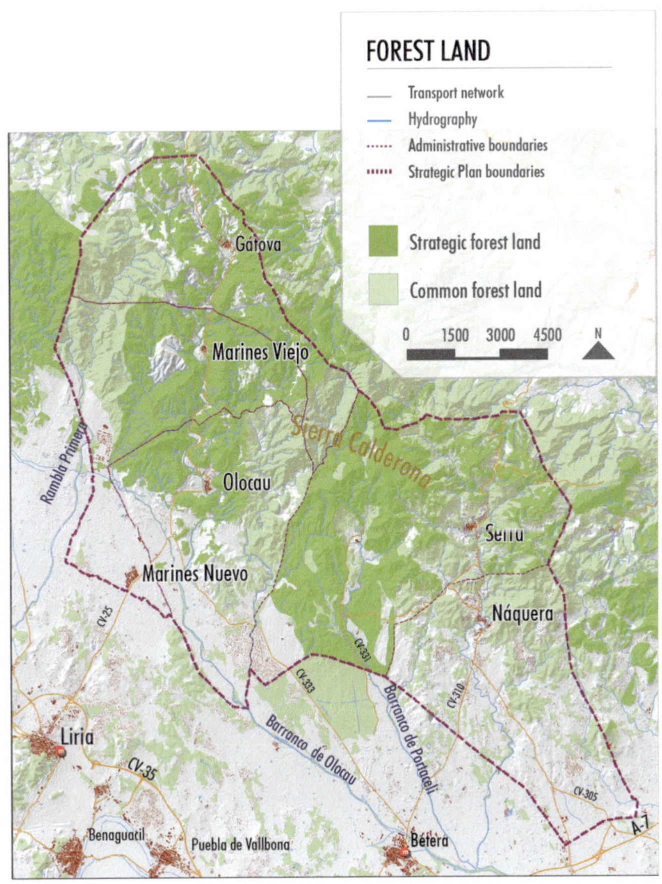

Table 2.1 Percentage of land take by use (CORINE 2000). *Source* Own elaboration

Land take by use (CORINE year 2000)	% total area Strategic Plan	% area SERRA	% area NÁQUERA	% area MARINES	% area OLOCAU	% area GÁTOVA
Continuous urban fabric	1	<1	<1	2	<1	<1
Discontinuous urban fabric	6	6	13	<1	8	<1
Industrial and retail areas	<1	<1	<1	<1	<1	<1
Mining areas	<1	<1	<1	<1	1	<1
Areas in construction	<1	<1	<1	<1	<1	<1
Sport and recreational areas	<1	<1	<1	1	<1	<1
Fruit trees	11	6	36	2	9	3
Olive trees	5	3	<1	2	14	9
Mixed vegetable gardens	11	<1	21	21	18	<1
Partially wild agricultural land	5	5	5	5	2	7
Coniferous forests	20	43	9	7	6	21
Natural meadows	<1	<1	<1	<1	<1	<1
Sclerophyllous scrubs	20	26	3	22	22	26
Transition scrubs	17	11	13	32	20	17
Burnt areas	4	<1	<1	6	<1	17
Total	100.00	100.00	100.00	100.00	100.00	100.00
Total area (km^2)	199.60	57.32	38.72	35.73	37.41	30.42

2.2.7 Urban Areas

It is possible to distinguish the following types of urban areas:

Towns and Villages: They are compact and densely inhabited urban areas with a morphology that follows the landform. They correspond to the historical town centres of Náquera, Serra, Gátova, Olocau and Marines Viejo. They integrate old and new constructions, which forms a dense urban mosaic with varying qualities and degrees of integration. The town of Marines Nuevo also belongs to this typology although it was built in 1967 following the traditional pattern of colonisation villages (settlements fostered by the national government after the Spanish Civil War to repopulate certain rural areas of the country).

Housing estates attached to towns: They are areas of low and medium density that are located next to towns or that have been absorbed by them and are now part of their fabric. They lack the traditional diversity of functions that historic town centres generally have (retail, services, etc.).

Isolated housing estates: They are scattered throughout the territory, and they are either the consequence of programmed developments or the result of the legalisation of some pre-existent unplanned settlement. Their use is mainly seasonal-residential although progressively the population is becoming permanent, especially in those settlements that are well communicated.

Due to their physical separation from towns and their need of public facilities and services, the municipalities where these kinds of developments exist need specific policies to respond to a polynuclear and seasonal urban pattern. Within this group, two new housing estates with either approval pending or under-construction were included.

Dispersed residential buildings: This type includes isolated residential buildings lacking adequate sewage, water supply, etc.

Industrial areas: They are localised in five sectors. The two largest industrial estates are the one in Marines and "Los Vientos" in the southeast of Náquera. The other three industrial estates are located in the south of Náquera, in Serra and next to the centre of Gátova. Some isolated and minor industrial sites can also be found in the area (Fig. 2.13).

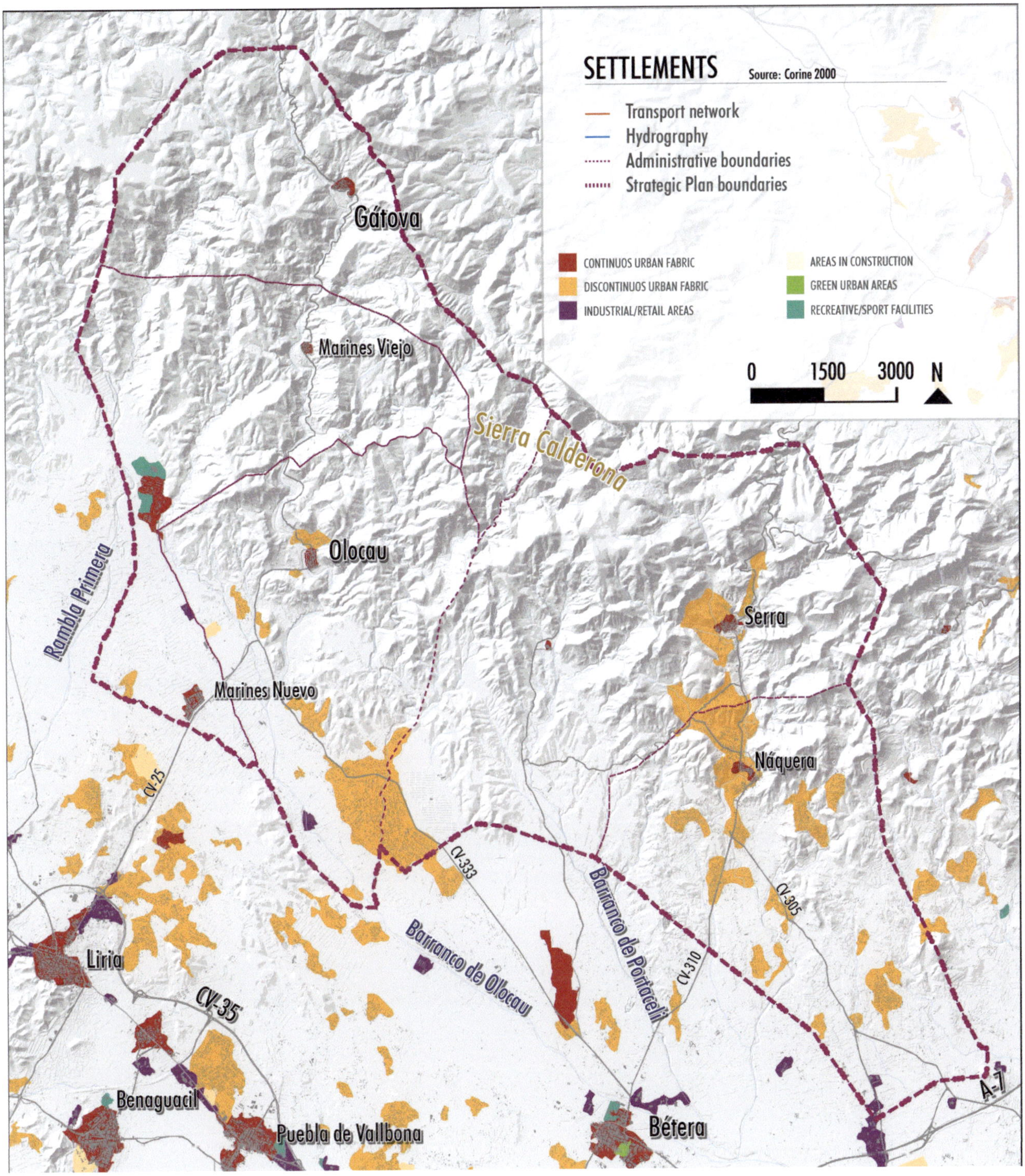

Fig. 2.13 Settlements and urban areas. *Source* CITMA

2.2.8 Transport Infrastructure

The south-central area of Sierra Calderona is surrounded by two major transportation corridors: to the east the A7 motorway and the V21 road, and to the south the CV35 road, also known as *Pista de Ademuz*.

From the east, the road CV305 allows quick access to Náquera. The CV35 has two main exits towards the area: one to the east, connecting with the A7 and CV310, and another one to the west, via Liria, connecting with the CV25.

The roads CV310 and CV25 cross the Sierra Calderona in north–south direction. The CV310 connects Bétera with Náquera, Serra and the Palancia valley. The CV35 connects Liria with Marines Nuevo, Olocau, Marines Viejo, Gátova and Segorbe. Both roads are interlinked by the road CV333, which also connects Bétera with a point between Marines Nuevo and Olocau. This last road also provides access to some other important settlements, mainly housing estates, as well as to the road CV33 leading to "Cartuja de Portaceli" (a Carthusian monastery) and to the CV328 road, which terminates north of the town of Náquera.

In the mountainous interior, there are numerous forest tracks, country roads and trails, as well as access roads to small housing estates. Because of its intensive use, the road CV334–CV3360 connecting the north of Serra with the Garbi peak is also worthy of mention.

The study area is also connected to the railway network via two train stations, in the towns of Bétera and Liria. Both stations are the terminal stops of a metro line from the city of Valencia (Fig. 2.14).

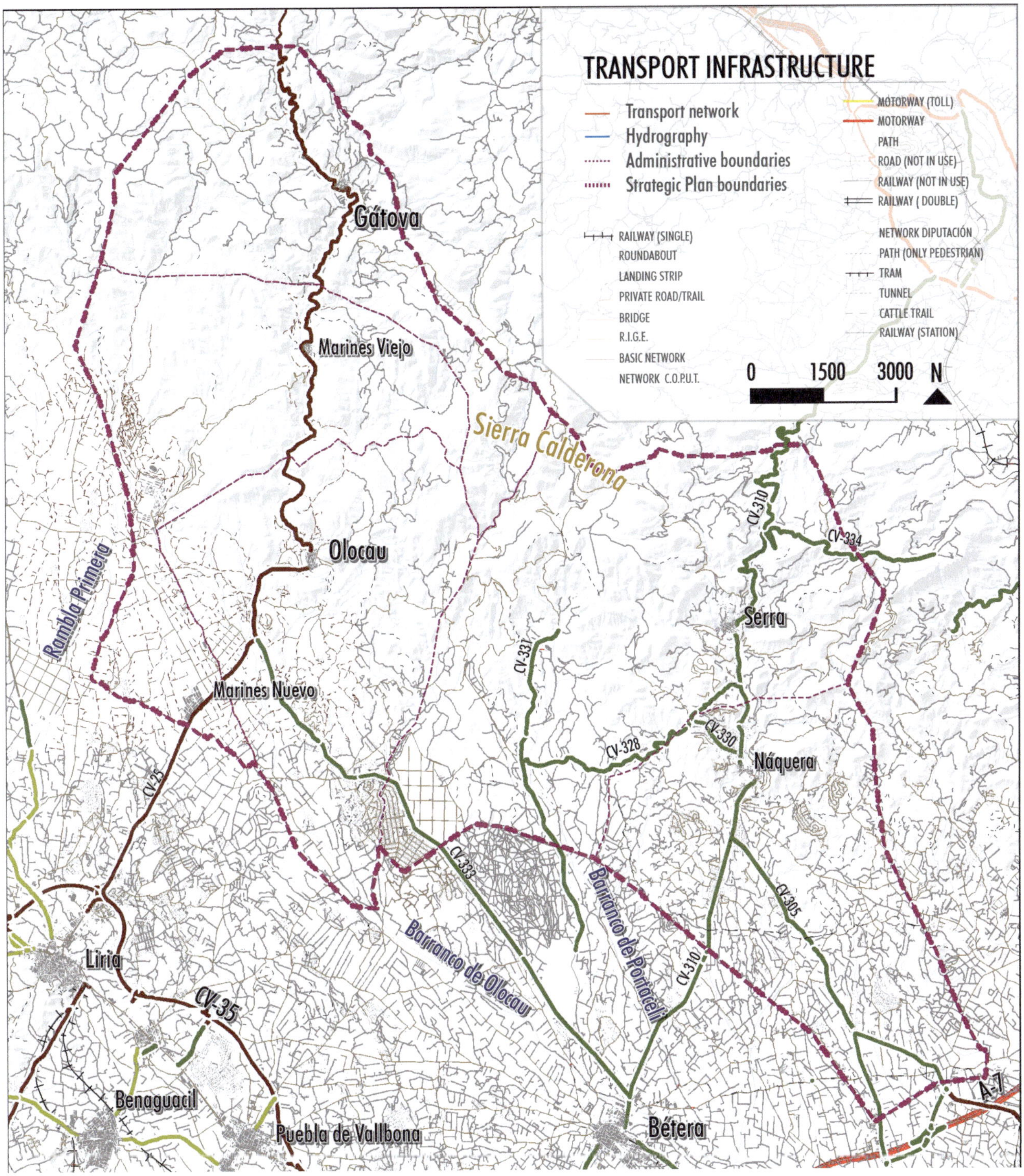

Fig. 2.14 Transport infrastructures. *Source* CITMA

2.2.9 Natural Park: Protected Area

The Sierra Calderona gained the status of Natural Park in 2002. It has a total area of 18,019 hectares, located in fourteen different municipalities and covering 49.33% of the total area included in the Sierra Calderona Strategic Plan, partially or totally affecting its five municipalities.

The PRUGSC (Use and Management Governing Plan), distinguishes five kinds of areas within the Park (Fig. 2.15, 2.16 and 2.17):

- *Áreas de Especial Protección* or AEP (special Protection Areas)
- *Áreas de protección Ecológica* or APE (ecological protection areas)
- Áreas *Urbanizadas* or AU (urbanised areas)
- *Áreas de Actuación Preferente* or AAP (areas with priority of action) including:
  - *Áreas recreativas* or AR (recreational areas)
  - *Áreas afectadas por incendios forestales* or AI (areas affected by forest fires)
  - *Áreas afectadas por actividades extractivas* or AE (areas affected by extraction activities)
  - *Áreas para la Regeneración de Hábitats Naturales* or ARH (areas for the regeneration of natural habitats)
- *Equipamientos* or EQ (public facilities).

Table 2.2 displays the proportion of each of the above-mentioned kinds of areas in relation to the total surface of the Sierra Calderona Strategic Plan, and in relation to the surface affected by the PRUGSC (Use and Management Governing Plan).

Additionally, the area covered by the Natural Park and its buffer is regulated by the Park's PORNSC (Natural Resources Management Plan) (Fig. 2.18).

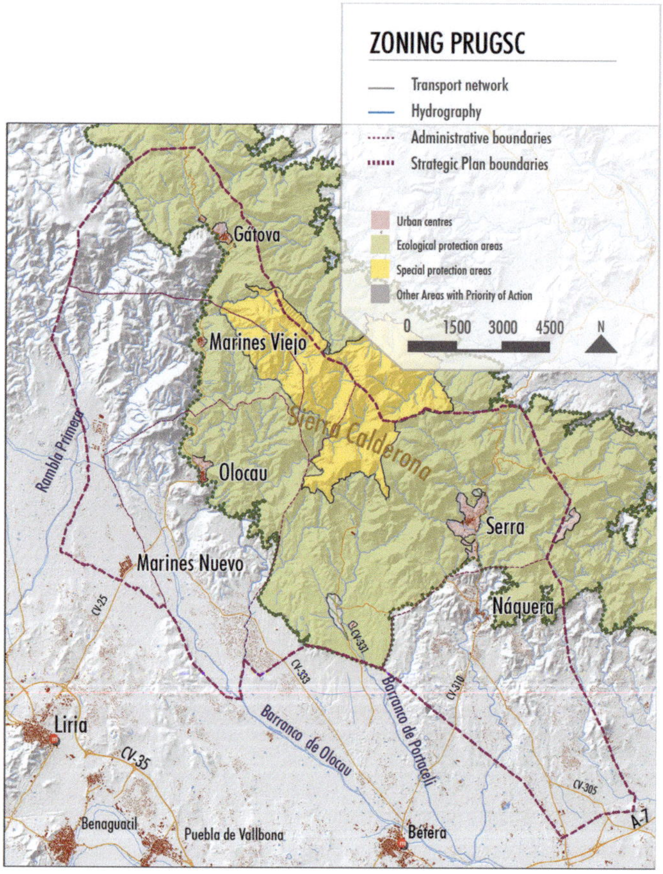

Fig. 2.15 Zoning in the "Use and Management Governing Plan" of the Sierra Calderona Natural Park (PRUGSC). *Source* CITMA

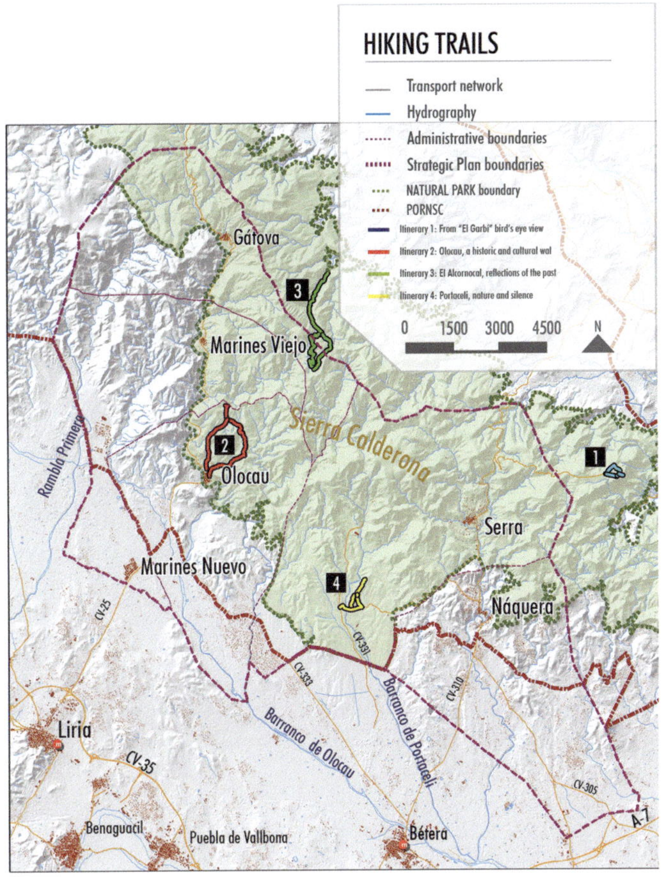

Fig. 2.16 Main hiking trails in the Sierra Calderona Natural Park. *Source* CITMA

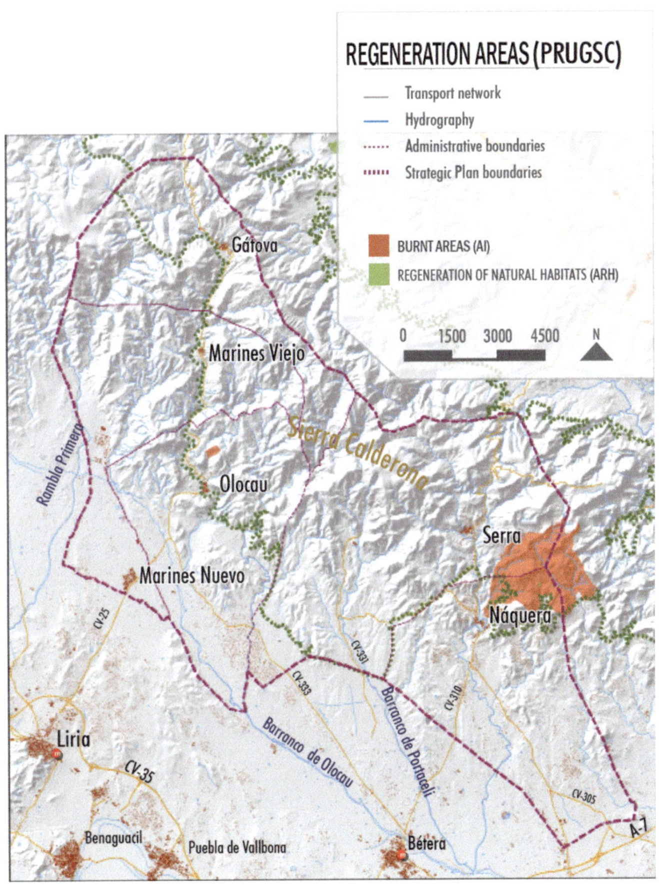

Fig. 2.17 Areas affected by forest fires (AI) and areas for the regeneration of natural habitats (ARH). *Source* CITMA

Table 2.2 Sierra Calderona Natural Park as part of the Sierra Calderona Strategic Plan. *Source* CITMA

Zoning PRUGSC	Surface (has)	% surface in relation to the Plan's area	% surface in relation to the PRUGSC within the Plan's area
Ecological protection areas	8,125.21	40.71	82.51
Special protection areas	1,480.77	7.42	15.04
Recreational areas	46.74	0.23	0.47
Other areas with priority of action	54.61	0.00	0.01
Urbanised areas	194.43	0.97	1.97
Total	9,901.76	49.33	100.00

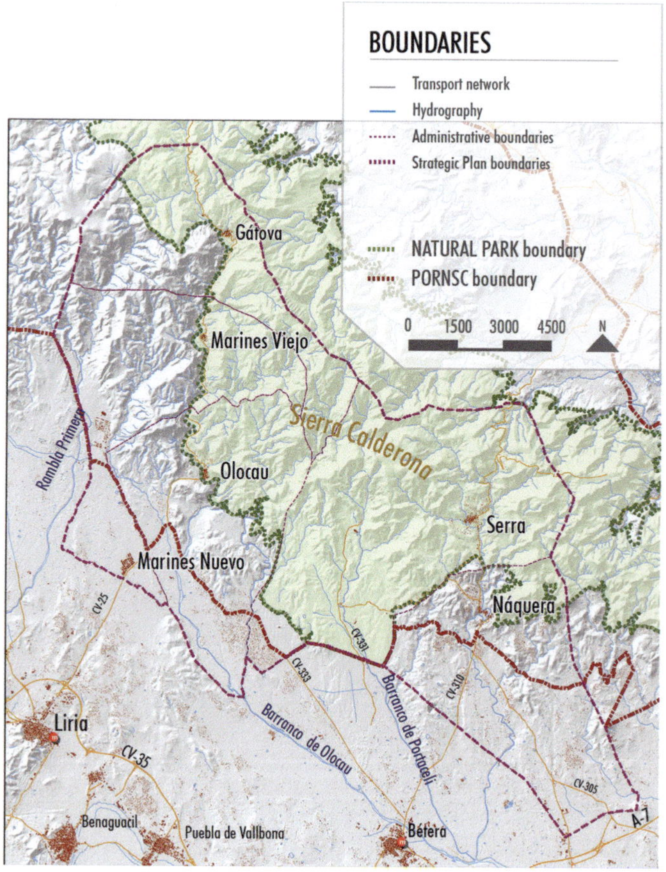

Fig. 2.18 Natural Park of Sierra Calderona. Protected area regulated by the "Use and Management Governing Plan" (PRUGSC) and by the "natural resources management plan" (PORNSC). *Source* CITMA

2.2.10 Natural Park: Buffer Zone

The Natural Resources Management Plan (PORNSC) covers the area of the Natural Park (protected area) and its border (buffer zone), representing 79.60% of the total area of the Sierra Calderona Strategic Plan. The buffer zone includes seven different kinds of areas:

- *Área de protección ecológica* or APE (ecological protection areas)
- *Área de protección paisajística* (landscape protection areas)
- *Área de predominio forestal* or APF (predominantly forest areas)
- *Área de predominio agrícola* or AAG (predominantly agricultural areas)
- *Áreas de Regeneración* or RG (regeneration areas)
- *Áreas Antropizadas* (anthropised areas)
- *Áreas Urbanizadas* or AU (urbanised areas).

Table 2.3 shows the proportions for each of the above-mentioned kinds of areas in relation to the total area of the Sierra Calderona Strategic Plan and in relation to the surface affected by the Natural Resources Management Plan (PORNSC).

The Natural Resources Management Plan also defines the area of socio-economic influence of the Natural Park, a legal and administrative framework for the design, programming and implementation of activities that will promote and sustainably develop the economic and social activity in the area of Sierra Calderona.

Since the area of socio-economic influence of the Natural Park comprises completely the fourteen municipalities affected by the Natural Park, it also includes the entire area covered by the Sierra Calderona Strategic Plan (Fig. 2.19).

Table 2.3 Natural resources management plan (PORNSC) of the Sierra Calderona Natural Park as part of the Sierra Calderona Strategic Plan. *Source* CITMA

ZONING PORNSC	Surface (has)	% surface in relation to the Strategic Plan's area	% surface in relation to the PORN within the Strategic Plan's area
Anthropised areas	242.83	1.22	1.53
Predominantly agricultural areas (AAG)	1,315.70	6.59	8.28
Predominantly forest areas (APF)	1,268.17	6.35	7.98
Ecological protection areas (APE)	4,870.14	24.40	30.65
Landscape protection areas	4,734.39	23.72	29.80
Regeneration areas (RG)	3,046.96	15.27	19.18
Urbanised areas (AU)	410.08	2.05	2.58
TOTAL	15,888.27	79.60	100.00

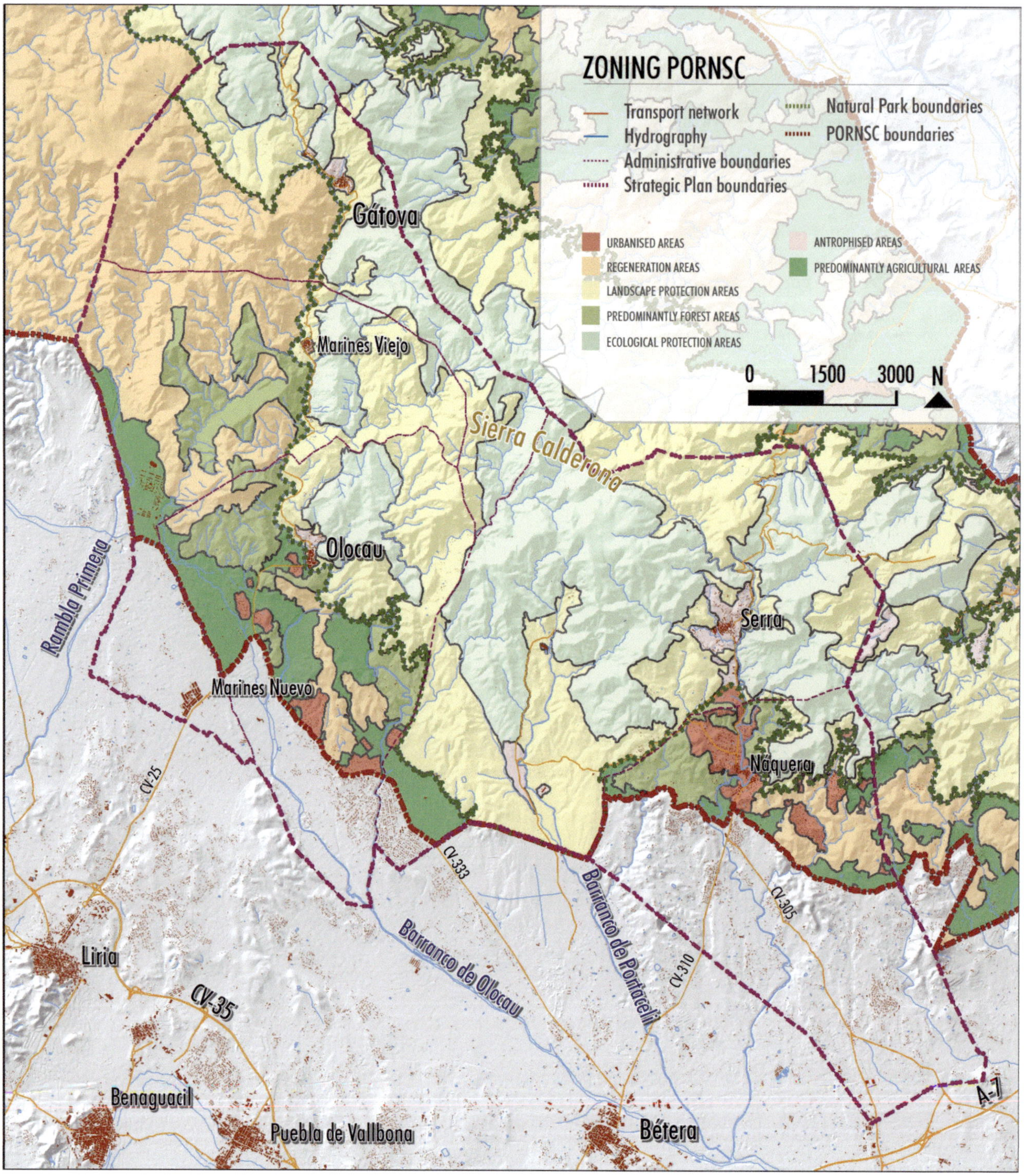

Fig. 2.19 Zoning according to the natural resources management plan (PORNSC). *Source* CITMA

2.2.11 Land-Use Planning: Land-Use Classification

The current municipal planning ordinances (PGOU or subsidiary regulations) in all five municipalities show a concentration of urbanised land in the axis Náquera-Serra, in the "Pedralbilla-Torre Portaceli" housing estates and in a system of isolated settlements in the municipality of Náquera. Developable land is concentrated in those same areas, with the two most extensive pockets located next to the industrial estate "Los Vientos" (southeast of Náquera) and north of the "Pedralbilla-Torre Portaceli" housing estates.

The western part of the area covered by the Plan includes smaller urbanised areas and less developable land. Non-developable land predominates in the mountainous areas, the military base of Marines and in the southern hillside of the central area, which despite its good access has successfully managed to avoid the proliferation of isolated housing estates.

It should be noted that, after the declaration of the Natural Park status, the zoning referred to in the PRUGSC and the PORNSC restricts municipal planning. This implies that, with the exception of urban areas, non-developable land included in the Natural Park becomes "land designated as non-developable land with special protection" ("Suelo No Urbanizable de Especial Protección" or "SNUEP") in municipal plans, with the exception of agricultural plots, which remain as "common non-developable land" ("suelo no urbanizable común" or "SNUC"). All the constructions that were illegally built before the declaration of the Natural Park remain illegal (Fig. 2.20).

Likewise, in the Park's buffer zone, the land designated as non-developable and included in the category of ecological protection ("Protección Ecológica" or "PE") or landscape protection ("Protección Paisajística" or "PP") when applied to forest areas, will have a special protection status for urbanisation purposes. At the same time, the land designated as non-developable and included in areas of forest predominance when applied to forest areas or mountains, will also qualify as "special protection land" for urbanisation purposes. The remaining areas designated as not for building (e.g. predominantly agricultural areas, anthropised areas, areas under natural regeneration areas) will retain their status as "common non-developable land".

Table 2.4 shows the proportion of each land-use class within the whole scope of the Sierra Calderona Strategic Plan and in each municipality. Additionally, in order to emphasise the importance that the Natural Park has in local planning, Table 2.5 shows the proportion of municipal land and of the entire Strategic Plan affected by the Natural Park regulations (PRUGSC and PORNSC).

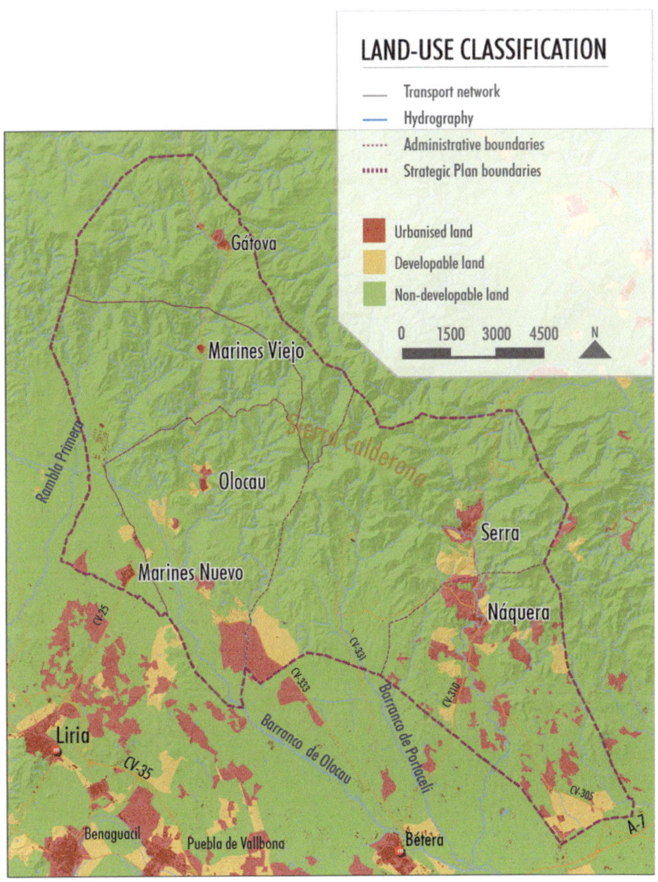

Fig. 2.20 Land-use classification. *Source* CITMA

Table 2.4 Percentage of land affected by the regulations of the Sierra Calderona Natural Park

Land classification	Area Strategic Plan (%)	SERRA (%)	NÁQUERA (%)	MARINES (%)	OLOCAU (%)	GÁTOVA (%)
Non-developable land	90.18	91.42	73.33	97.55	91.55	98.99
Developable land	4.02	4.75	9.75	0.58	3.50	0.00
Urbanised land	5.72	3.83	16.92	1.87	4.68	0.83
Historic urbanised land	0.08	0.00	0.00	0.00	0.27	0.18
Total (%)	100.00	100.00	100.00	100.00	100.00	100.00
Total area (km^2)	199.60	57.32	38.72	35.73	37.41	30.42

Table 2.5 Land-use classification in the area covered by the Sierra Calderona Strategic Plan

Influence Natural Park in municipalities	SERRA (%)	NÁQUERA (%)	MARINES (%)	OLOCAU (%)	GÁTOVA (%)
Area included in the Natural Park (see PRUGSC)	93.96	8.73	25.84	29.27	63.00
Area included in the perimeter protection zone (see PORNSC)	3.92	27.29	55.55	50.40	37.00
Area included in the buffer zone (Natural Park + perimeter protection zone). (See PORNSC)	97.88	36.02	81.39	79.67	100.00
Municipal areas not regulated by the PORNSC or PRUGSC	2.12	63.98	18.61	20.33	0.00
Total area (km^2)	57.32	38.72	35.73	37.41	30.42

2.2.12 Land-Use Planning: Land-Use Zoning

Zoning enables a more detailed definition of the functions and levels of protection for the three land classes described in the previous section: urbanised land, developable land and non-developable land (Fig. 2.21). The following aspects should be emphasised:

- Existence of pockets of non-developable land around urbanised areas: as it can be appreciated after looking at the zoning plan of the Natural Resources Management Plan (PORNSC), an important portion of these areas will have the same regulations as in predominantly agricultural or forest areas, regeneration areas and landscape protection areas.
- Existence of pockets of public land: including the military bases in Bétera (in the border of the Sierra Calderona Strategic Plan) and Marines, as well as a golf course southeast of Náquera (next to the industrial estate "Los Vientos").

For planning purposes, and as outlined in the previous section, it is essential to indicate that the zoning shown in Fig. 2.21 and Table 2.6 does not incorporate the regulations of the Natural Park's PORNSC and PRUGSC. These regulations could increase the level of protection of important land pockets such as some forestland included in the PORNSC as ecological protection area (PE) or as landscape protection area (PP) and some non-developable pockets included in forest predominance areas (APF) when affecting forest and mountains.

Table 2.6 shows the proportion of each land use in the entire scope of the Sierra Calderona Strategic Plan and in each municipality. Table 2.7 shows, in the same way, the proportion of land with military use (military zones and their security buffer zone).

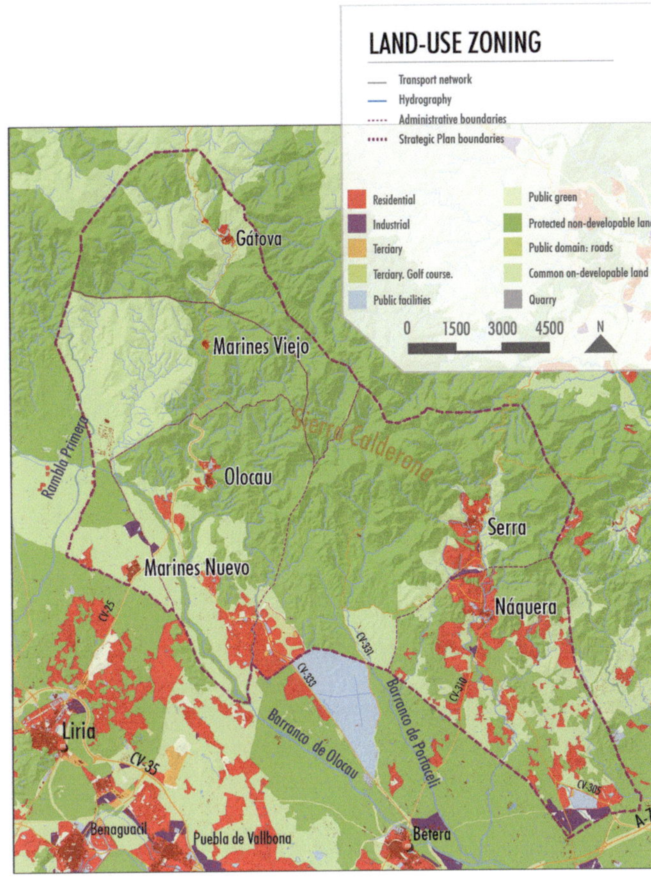

Fig. 2.21 Land-use zoning. *Source* CITMA

Table 2.6 Land-use zoning (2013)

Zoning	Area Strategic Plan (%)	SERRA (%)	NÁQUERA (%)	MARINES (%)	OLOCAU (%)	GÁTOVA (%)
Residential	7.98	6.55	22.37	1.37	7.48	0.76
Industrial	0.70	0.23	2.45	0.71	0.18	0.00
Tertiary	0.07	0.12	0.10	0.01	0.01	0.06
Tertiary golf	0.29	1.00	1.01	0.00	0.00	0.00
Public facilities	0.34	0.20	0.33	0.09	0.20	0.04
Facilities—green areas	0.41	0.48	0.50	0.26	0.61	0.15
Protected areas	62.76	79.93	72.58	38.06	63.68	76.33
Military areas with special ordinances	8.55	0.00	0.00	47.78	0.00	0.00
Public domain—roads	0.75	0.37	0.66	0.16	2.59	0.79
Common land designated as not for building	18.15	11.12	0.00	11.56	25.25	21.87
Total (%)	100.00	100.00	100.00	100.00	100.00	100.00
Total area (km^2)	199.60	57.32	38.72	35.73	37.41	30.42

Table 2.7 Military areas affected by the Sierra Calderona Strategic Plan

Influence of military bases	Area Strategic Plan (%)	SERRA (%)	NÁQUERA (%)	MARINES (%)	OLOCAU (%)	GÁTOVA (%)
Military base	13.22	0.00	0.00	47.17	5.08	25.12
Security perimeter	2.68	0.00	0.00	14.99	0.00	0.00
Total area (%)	15.90	0.00	0.00	62.16	5.08	25.12

2.2.13 Population and Services

The permanent population living in the area covered by the Sierra Calderona Strategic Plan amounts to 13,058 inhabitants (year 2013), 5,900 of which correspond to Náquera, 3,349 to Serra, 1,832 to Marines, 1,517 to Olocau and 460 to Gátova. The total population increases noticeably on the weekends and during the summer, when it increases to as much as 35,000–40,000 inhabitants. Lately, the permanent population has grown in well-connected housing estates as people, mostly young families, have established their first residence there.

A first demographic analysis shows that the population pyramids are similar to the ones in the whole Valencian Region. Gátova's municipality, with a clearly ageing population, is the only exception.

In the absence of more detailed studies on public services, facilities, infrastructure and tourism infrastructure, there seems to be a fairly balanced provision of cultural and sports facilities, and an uneven availability of tourism infrastructures regarding both accommodation and dining. The initiatives developed by the municipality of Olocau to expand and diversify its cultural and tourist facilities deserve a special mention.

2.2.14 Local Economy

Local economy has evolved from an agricultural-stockbreeding-based economy (non-irrigated crops of olive, carob and almond trees in the interior areas, irrigated crops of citrus trees in the southern plains and to a lesser extent horticultural crops and stockbreeding) into an economy of services associated with residential uses (permanent and temporary) and tourism.

Industrial activity has little relevance in the area, with the exception of the industrial estates in the southeast of Náquera and in Marines, where a major paint factory is located. Metal and woodwork industries, traditionally associated with the building sector (previously thriving, now in decline), can occasionally be found. In Náquera, a biomass plant uses forest residues.

Given this context, it is not surprising that the population living in the area works in towns or industrial estates nearby (Valencia, Bétera, Pobla de Vallbona, Massamagrell, Liria, etc.). It should also be noted that the presence of important military bases in the area has not produced any considerable effect on the local economy.

Despite the possibilities that the Park´s Natural Resources Management Plan (PORNSC) envisioned to channelise the socio-economic development of the area from an integrated perspective, virtually none of them have either been articulated or defined by the corresponding sectoral plans (forestry, agricultural-stockbreeding, hunting, tourism, etc.).

The Sierra Calderona Strategic Plan should therefore support the economic development foreseen in the Natural Resources Management Plan (PORNSC) paying attention to both residents and visitors. The primary sector (forestry, stockbreeding, agriculture) should be developed in order to generate local wealth and employment through products with high added value, and complementary activities should help to create a more diversified economy. Currently, there is little profit generated by the public use of the area by the population of the metropolitan area of Valencia in local economies, in spite of the high levels of management and maintenance that they impose on both municipal authorities and local populations. Therefore, strategies and plans are needed to promote and foster activities that can have a positive and effective impact on the population and local entrepreneurs and which facilitate a better and more sustainable public enjoyment of the whole area.

2.2.15 Natural Heritage

Apart from the acknowledged environmental values of the Sierra Calderona Natural Park (geomorphological, ecological, flora and fauna), the area covered by the Sierra Calderona Strategic Plan includes also other protected spaces or elements, such as Special Protection Areas for Birds (*Zonas de Especial Protección para las Aves* or "ZEPA"), sites of community importance (*Lugares de Interés Comunitario* or "LIC"), micro-reserves, monumental trees and listed caves (Figs. 2.22, 2.23, 2.24 and 2.25).

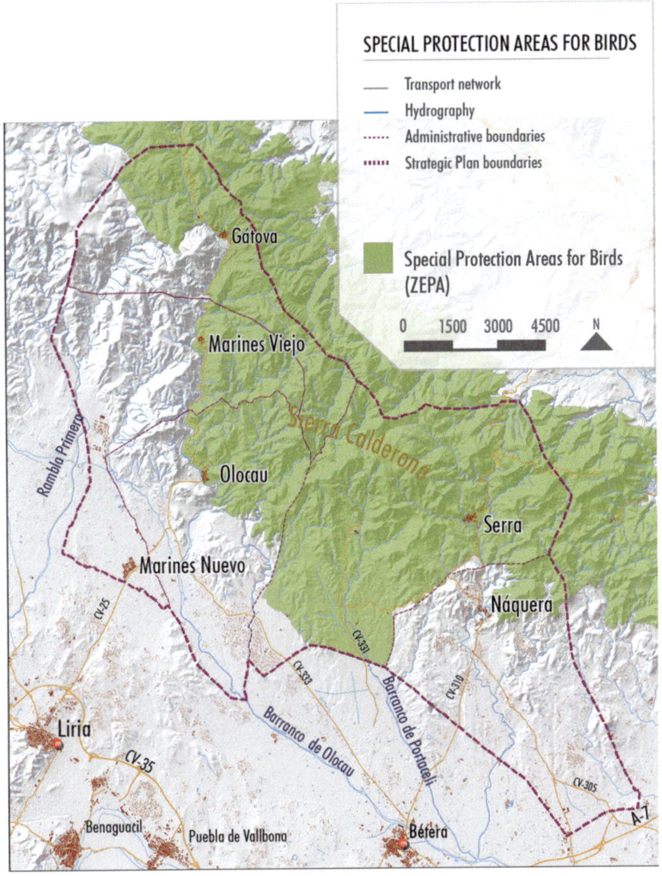

Fig. 2.22 Special Protection Areas for Birds (ZEPA). *Source* CITMA

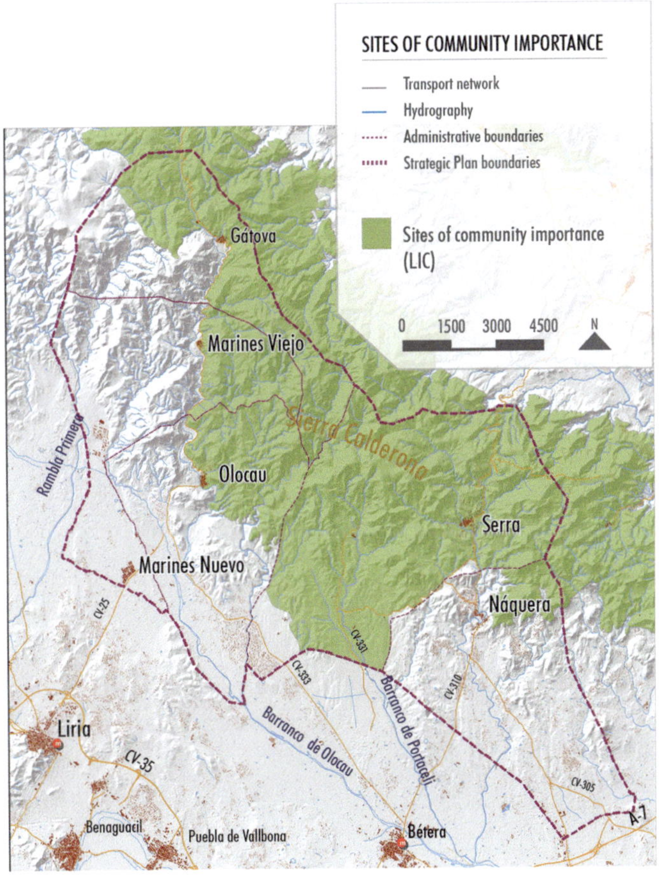

Fig. 2.23 Sites of Community Importance (LIC). *Source* CITMA

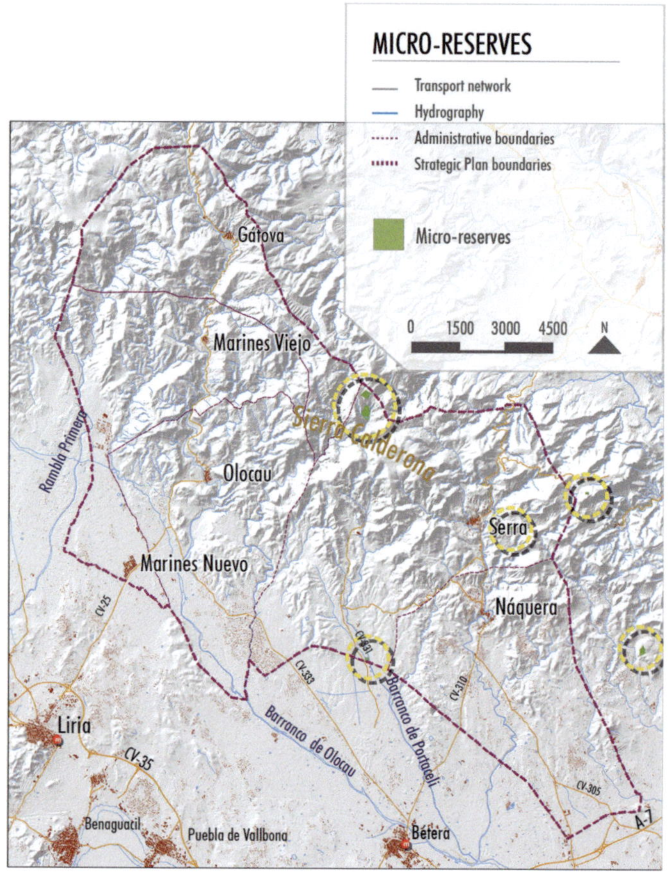

Fig. 2.24 Micro-reserves (Tancat de Portaceli, Peñas Altas, Sagaratillo Ravine and Castell de Serra). *Source* CITMA

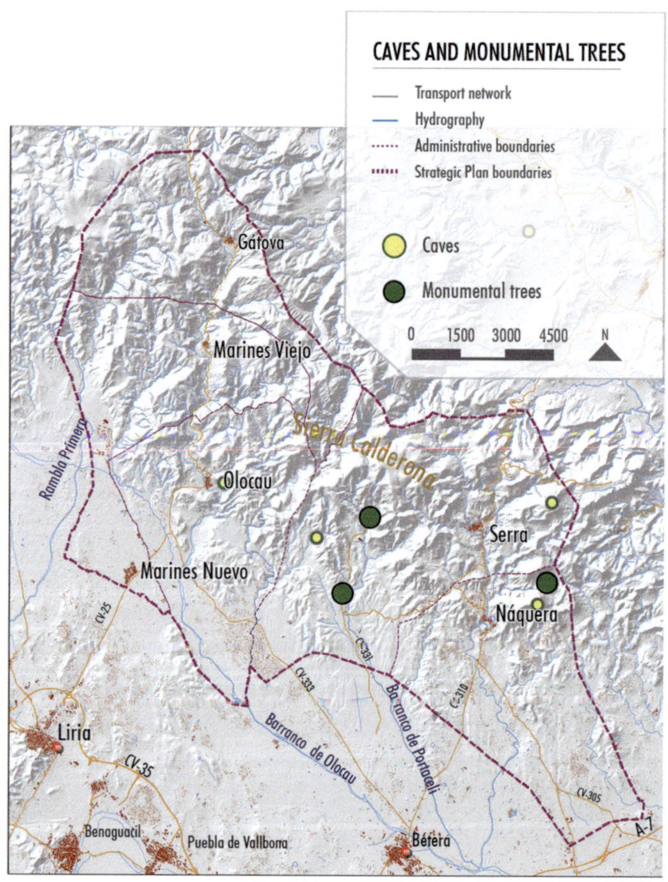

Fig. 2.25 Caves (Sotaranya, Caballo, Puntal de Mateu) and monumental trees (Pi de la Bassa in Portaceli-Serra and Pi de Salt in Náquera). *Source* CITMA

2.2.16 Cultural Heritage

Human presence over a long duration and the particular idiosyncrasy of the area has generated a rich and unique cultural heritage including:

- *Ethnographic heritage*: agricultural and crafting techniques (basketry, stonework and masonry)
- *Architectural heritage*: renowned examples of religious, civil and military architecture
- *Hydraulic heritage*: dams, aqueducts, ponds, springs, drinking troughs, watermills and washing places
- *Urban planning heritage*: Marines Nuevo as an example of a rationalist colonisation village
- *Archaeological heritage*: important Iberian archaeological sites
- *Gastronomical and festive heritage*: traditional meals and celebrations throughout the year, including unique festivities like "Fiesta de la Cereza (Cherry Fair)" in Serra, "Moros y Cristianos" and pig-slaughter (Fig. 2.26 and 2.27).

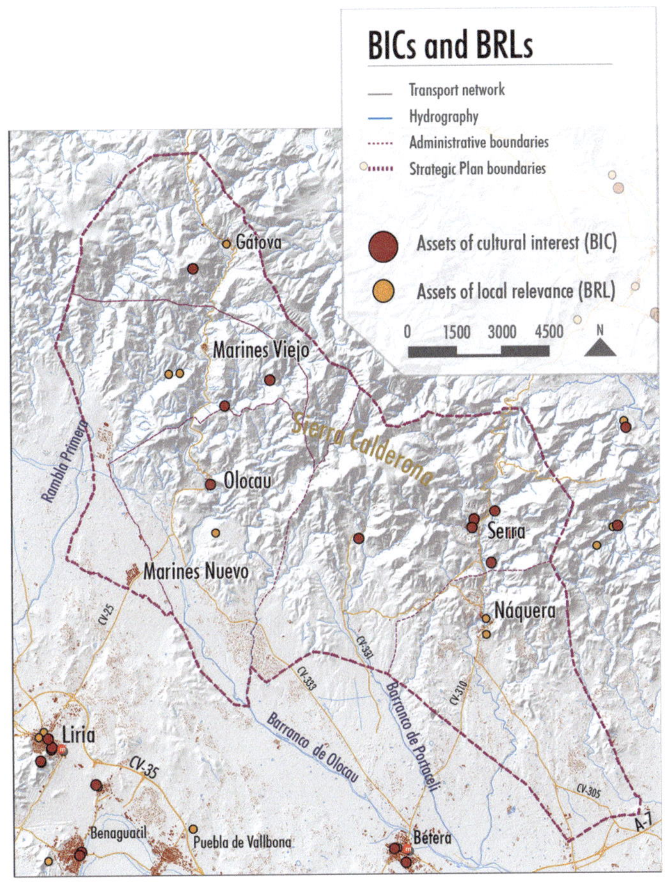

Fig. 2.26 Assets of cultural interest (BIC) and assets of local relevance (BRL). *Source* CITMA

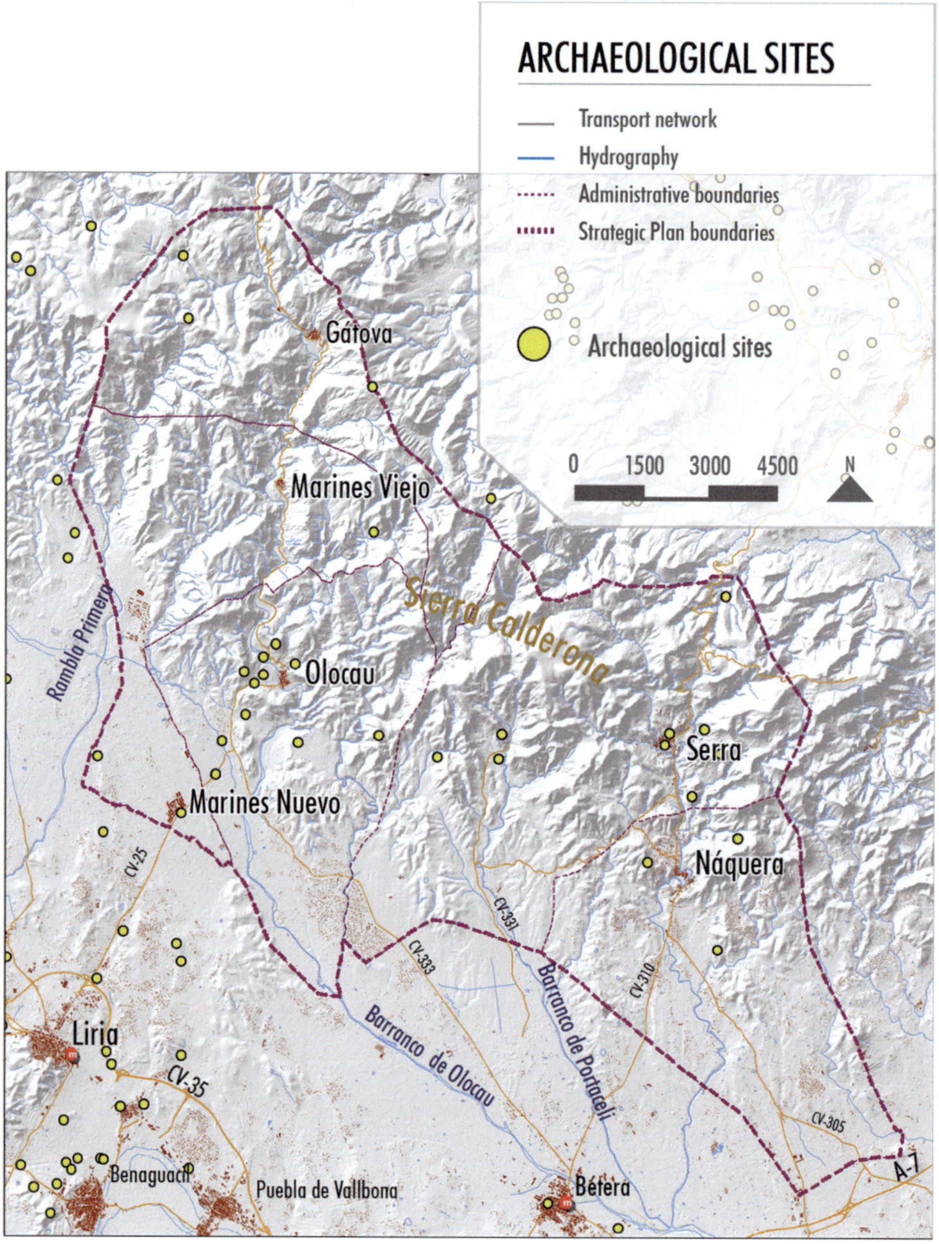

Fig. 2.27 Archaeological sites. *Source* CITMA

2.2.17 Natural Hazards: Physiography and Hydrology

An important area of the Sierra Calderona Strategic Plan is at high risk of erosion due to the steep slopes of Sierra Calderona. It is necessary to ensure an adequate and continuous layer of vegetation in the forest areas. Similarly, the abrupt topography, coupled with specific geological conditions, has generated areas at risk of landslides and rockfalls in the upper section of the Olocau Ravine and in the north of Serra.

The area covered by the Sierra Calderona Strategic Plan does not present flooding risks, except on the system of ravines that drain into the Valencian plain. The aquifers located in the central area of Náquera, as well as in specific areas north of Gátova and east of Serra, are highly vulnerable, while the central area of the Strategic Plan has a medium aquifer vulnerability. This natural risk should therefore be evaluated in relation to the performance of the sewage systems on the urbanised areas (Figs. 2.28, 2.29, 2.30 and 2.31).

Fig. 2.28 Potential erosion risk. *Source* CITMA

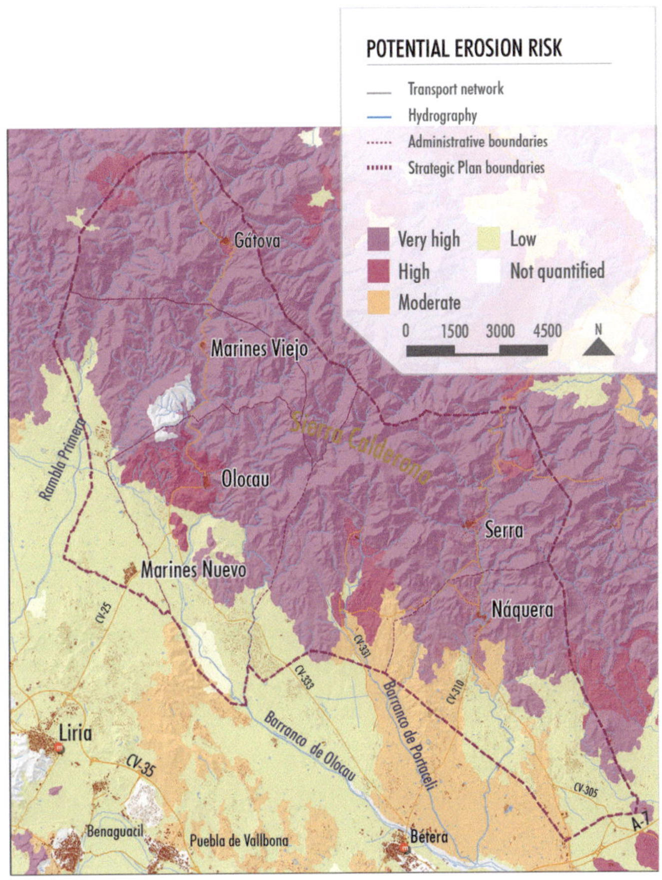

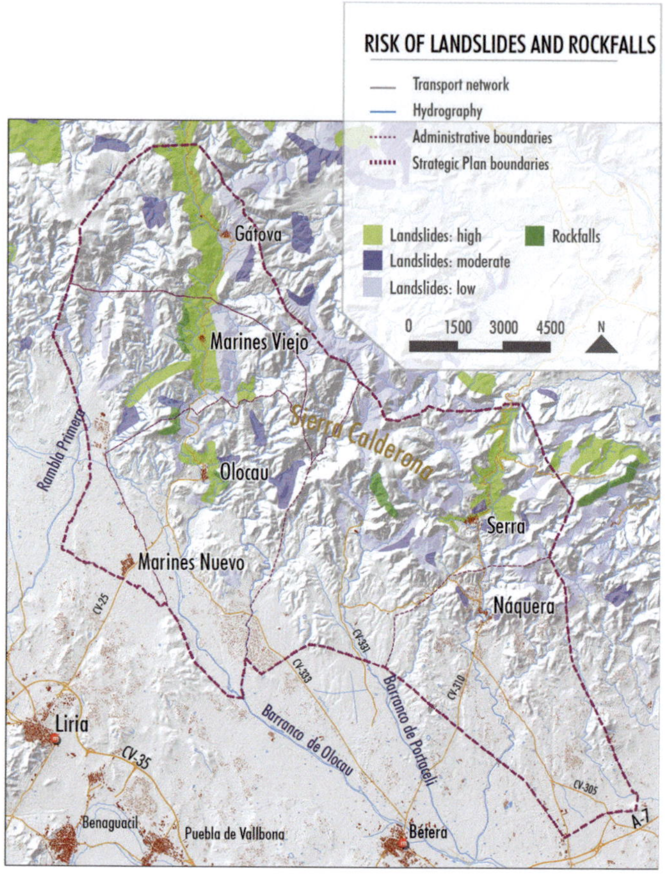

Fig. 2.29 Risk of landslides and falling rocks. *Source* CITMA

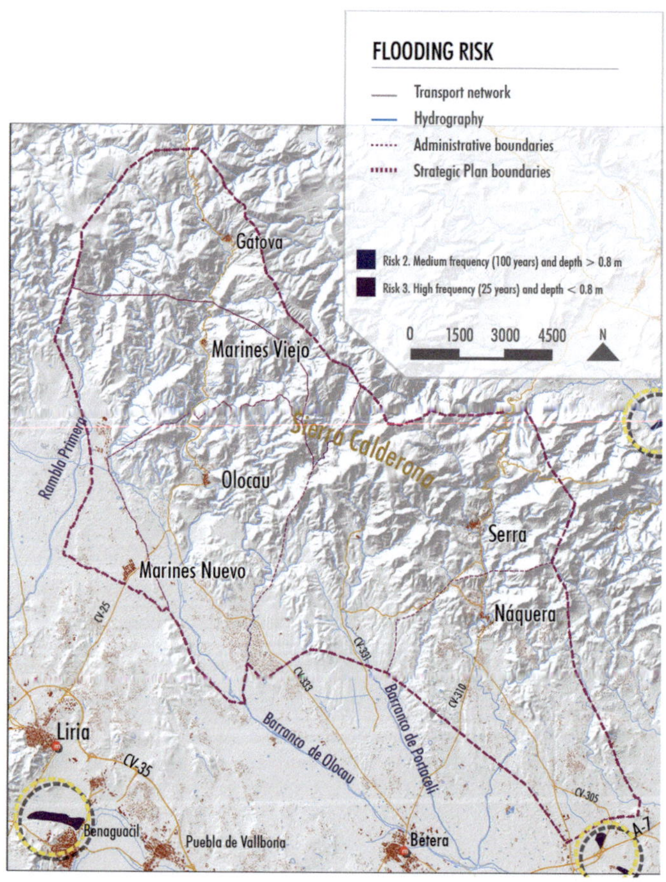

Fig. 2.30 Flooding risk. *Source* CITMA

Fig. 2.31 Aquifer vulnerability. *Source* CITMA

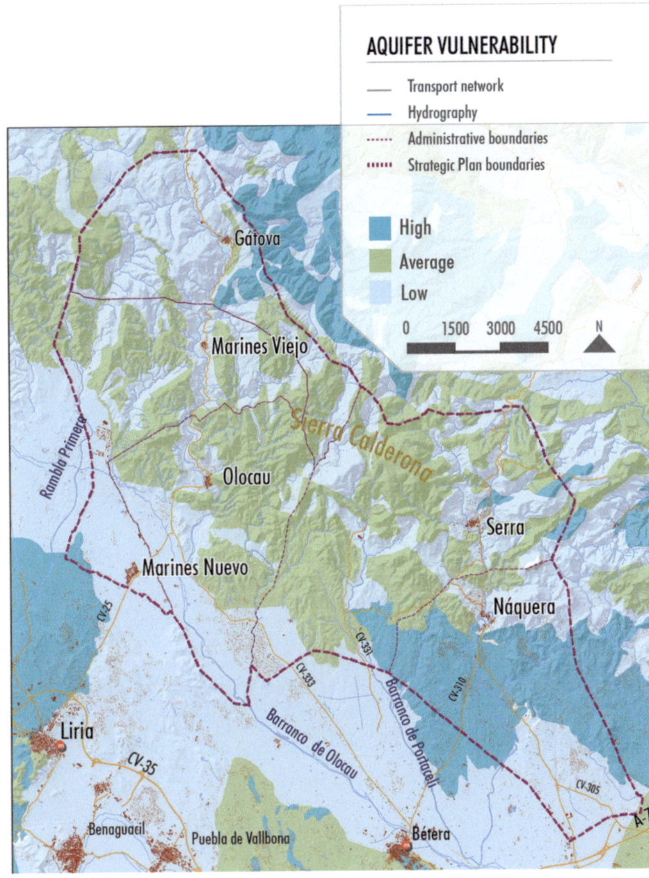

Regional Analysis and Diagnosis 3

Abstract
This chapter presents the current situation of the south-central area of Sierra Calderona from an environmental, cultural, social, urban and economic perspective. In this context, a number of topics have been defined that, through a systematic, proactive and cross-sectional approach, are intended to establish a solid base from which to propose both objectives and strategies for the sustainable development of the municipalities located in that region. The list of topics include: natural environment, agriculture and livestock farming, urban environment and well-being, infrastructures and mobility, cultural heritage, tourism and public use, landscape and territorial models, regional sustainability, socio-demography and local economy. The whole procedure of analysis and diagnosis was supported by a public consultation process, involving both residents and visitors, in which participants were asked to indicate the major problems and opportunities they perceive in the above-mentioned thematic areas. Within the addressed thematic areas, the Plan focused on the aspects most relevant for the socio-economic and sustainable development of the five municipalities covered by the Sierra Calderona Strategic Plan (Serra, Náquera, Olocau, Marines and Gátova).

3.1 Introduction and Methodology

The aim of this chapter is to determine the current situation of the south-central area of Sierra Calderona from an environmental, cultural, social, urban, landscape, touristic and economic perspective.

To achieve this goal, a number of topics have been defined that, through a systematic, proactive and cross-sectional approach, are intended to establish a solid base from which to propose both objectives and strategies for the sustainable development of the municipalities located in that region.

In order to develop this analysis, information from different sources has been collected. That information has been transformed into cartographies and data tables that were subsequently used to draft the above-mentioned regional objectives and strategies.

The present chapter has been structured into ten thematic areas or subchapters that, at all times, have been interrelated in order to offer a cross-cutting profile of the region.

The whole analysis and diagnosis were supported by a public consultation process involving both residents and visitors, in which participants were asked to indicate the major problems and opportunities they perceive in the proposed thematic areas. In order to make the participatory process representative, the consultation was conducted through surveys of predefined samples of the population, offering at the same time the possibility to any citizen to participate via a website specifically created for the drafting and monitoring of the Plan (www.tusierracalderona.es).

Within the addressed thematic areas, the study focused on those aspects more relevant for the socio-economic and sustainable development of the five municipalities covered by the Sierra Calderona Strategic Plan.

The necessity to generate regional models and to define territorial patterns coupled with the unavailability or difficulty of accessing some data implied the need of making occasionally some approximations, extrapolations or estimates. These estimates have been supported in all cases by data from other works or reports of recognised solvency and authorship.

J. Galan, *Analysis and Strategies for Sustainable Regional Planning*,
https://doi.org/10.1007/978-3-319-96508-6_3

3.2 Natural Environment

3.2.1 Generalities

The analysis of the forestland in the area covered by the Sierra Calderona Strategic Plan reveals the following major facts:

- 68% of its surface is forest land, with a remarkable increase of this proportion in the municipalities of Gátova and Serra.
- According to the determinations of the PATFOR (in English: "Forestry Strategic Action Plan of the Valencian Region", 56% of the forest land in the studied area of the Sierra Calderona is strategic and the remaining 44% is common. The common forest land includes dispersed areas in the southern foothills of Sierra Calderona with a major landscape and environmental role in the settlements located on the foothills of Náquera and Olocau.
- Forest land is mostly privately owned (71%) although important pockets of public land, belonging to the municipality or the regional government, also exist within the municipality of Serra.
- The Calderona Natural Park is mainly constituted by a matrix of forest land, with an internal network of rainfed agricultural land and small urbanised area. The Natural Park covers 49% of the area included in the Sierra Calderona Strategic Plan and accounts for 64.5% of its total forest land.
- The area regulated by the "Plan de Ordenación de Recursos Naturales (PORNSC)" of the Natural Park affects 80% of the area covered by the Sierra Calderona Strategic Plan, whereas the area of socio-economic influence of the Natural Park affects its entirety (100%). This implies that the regulations of the Natural Park have a strong influence in most of the area included in the Strategic Plan (Figs. 3.1, 3.2 and 3.3, Table 3.1).

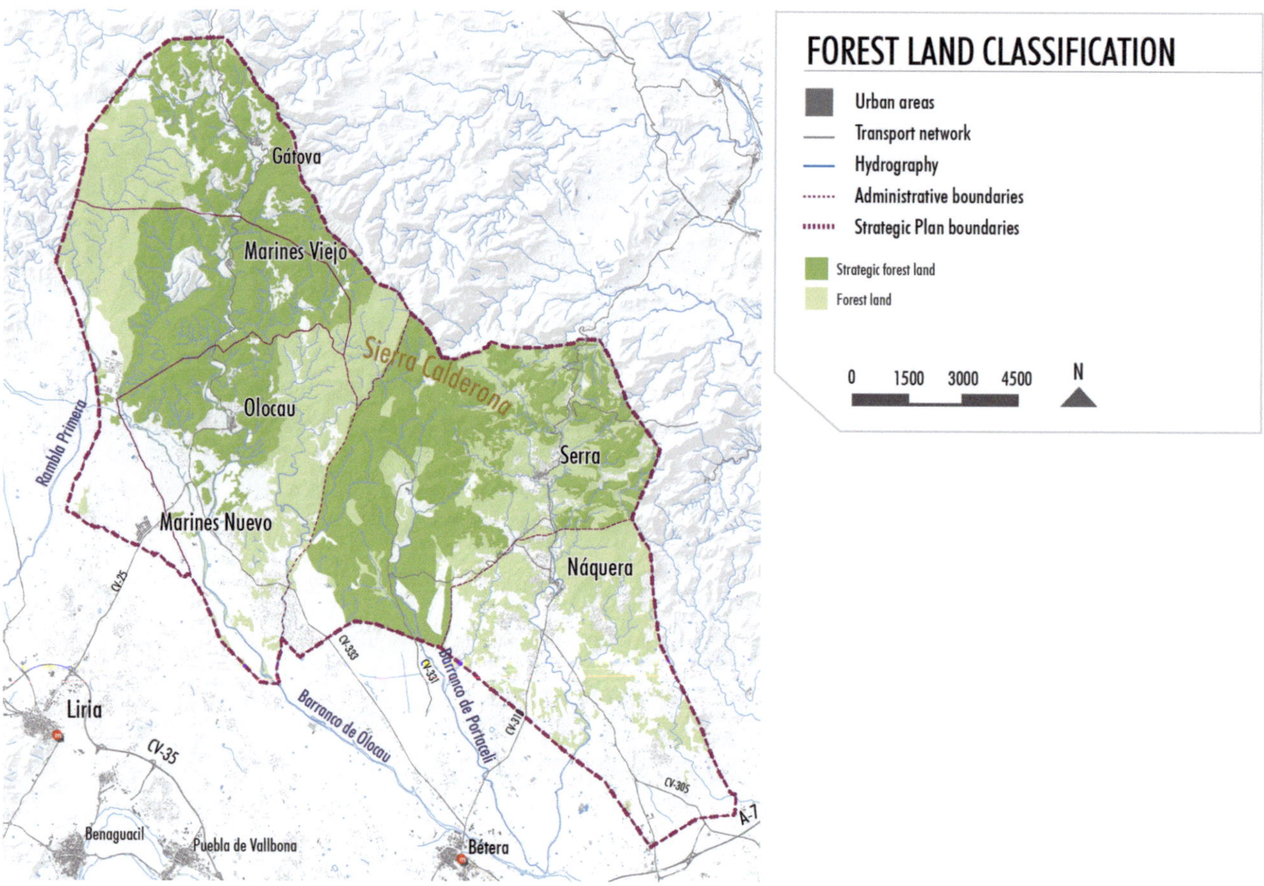

Fig. 3.1 Forest land classification. *Source* PATFOR, Vaersa

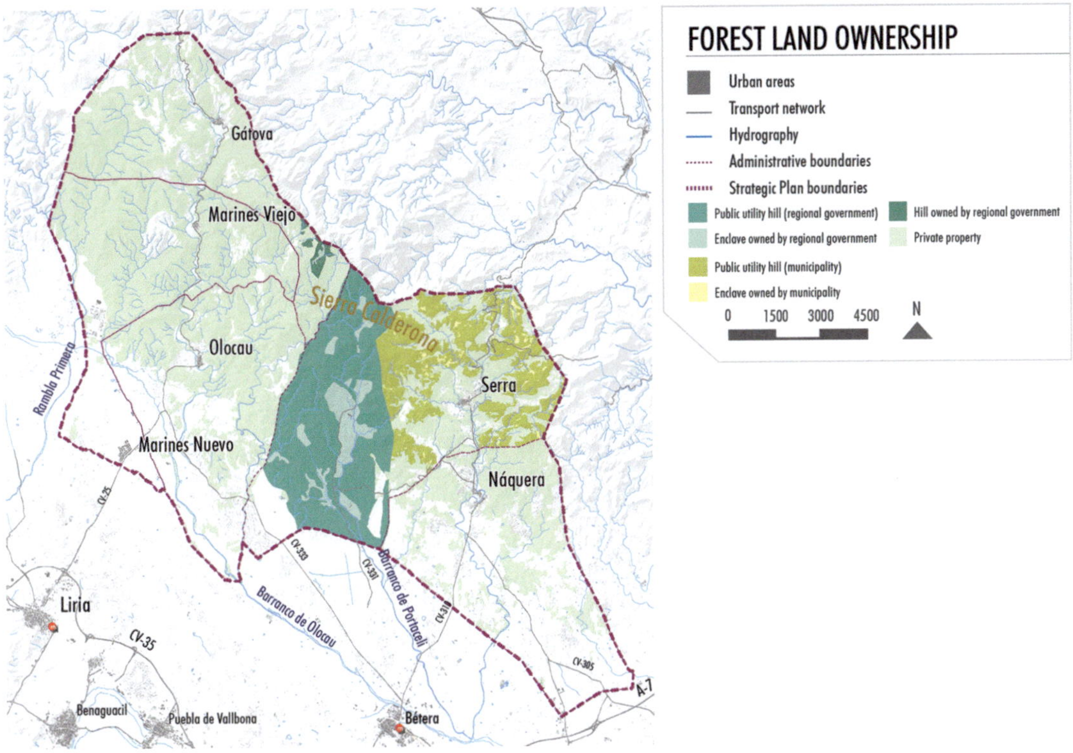

Fig. 3.2 Forest land ownership. *Source* PATFOR, Vaersa

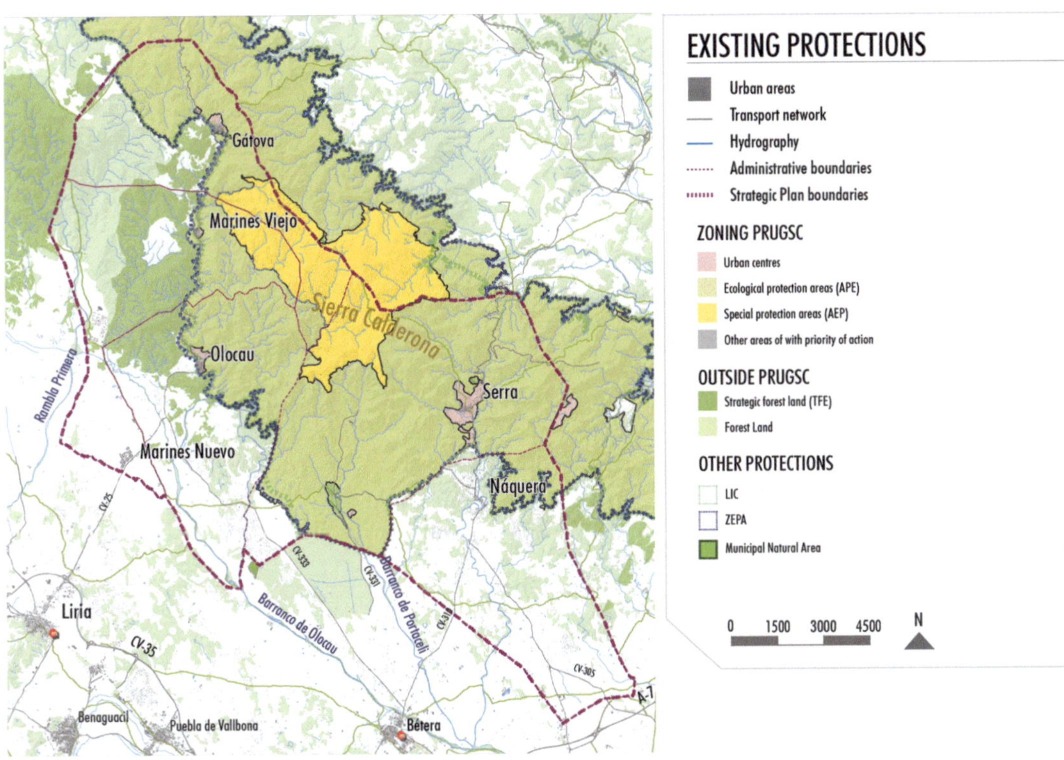

Fig. 3.3 Existing protections. "Plan de Ordenación de Recursos Naturales, PORNSC" and "Plan Rector de Uso y Gestión, PRUGSC" of the Sierra Calderona Natural Park. *Source* CITMA

Table 3.1 Forest land by municipalities and ownership. *Source* Own elaboration based on cartography from PATFOR, Vaersa

	Total	Public property	Public property	Private property	Private property
	Area (ha)	Area (ha)	(%)	Area (ha)	(%)
TOTAL PTE	13,619.23	3,945.14	28.97	9,674.09	71.03
NÁQUERA	1,444.08	4.96	0.34	1,439.13	99.66
SERRA	5,098.04	3,853.90	75.60	1,244.14	24.40
MARINES	2,430.47	0.00	0.00	2,430.47	100.00
OLOCAU	2,183.77	7.49	0.34	2,176.29	99.66
GÁTOVA	2,462.86	78.79	3.20	2,384.06	96.80

3.2.2 Economic Activity

The existing and potential economic activities of the forest land in the area covered by the Sierra Calderona Strategic Plan have been carefully studied, as they are considered strategic for the socio-economic development of the municipalities, where the existing environmental protection restricts the implementation of other productive activities, such as intensive agriculture or industry. The conclusions reached are presented below.

3.2.2.1 Timber Harvesting

Current and potential situation: Currently, timber harvesting is rare. While this activity would be administratively feasible in large parts of the area regulated by the Natural Park, with the exception of the Special Protection Areas, the productivity would be medium and mostly restricted to pine trees.

Timber harvesting areas: 5,442.52 ha would be suitable for use for this activity. 70% is privately owned, and 30% is public property (Fig. 3.4).

Production: It is estimated at 4,013.02 m^3/year.

Economic impact: As will be described in the local economy subchapter of this chapter, the annual revenue is estimated to be 120,390.56 €/year.

Necessary facilities: In order to develop these activities, forest roads (there is already an extensive network) and hauling roads would be necessary. A network of timber processing companies already exists; these companies being mainly located in Serra, Gátova and Náquera.

Environmental impact: Silvicultural management for controlled timber production would increase fire safety in the mountainous areas and improve the pine forest phytosanitary condition. However, the elimination of the forest understory would reduce its biodiversity and ecological quality. It is assumed at this point that the masses of holm and cork oak trees, either already present or growing spontaneously, would not be included in logged areas.

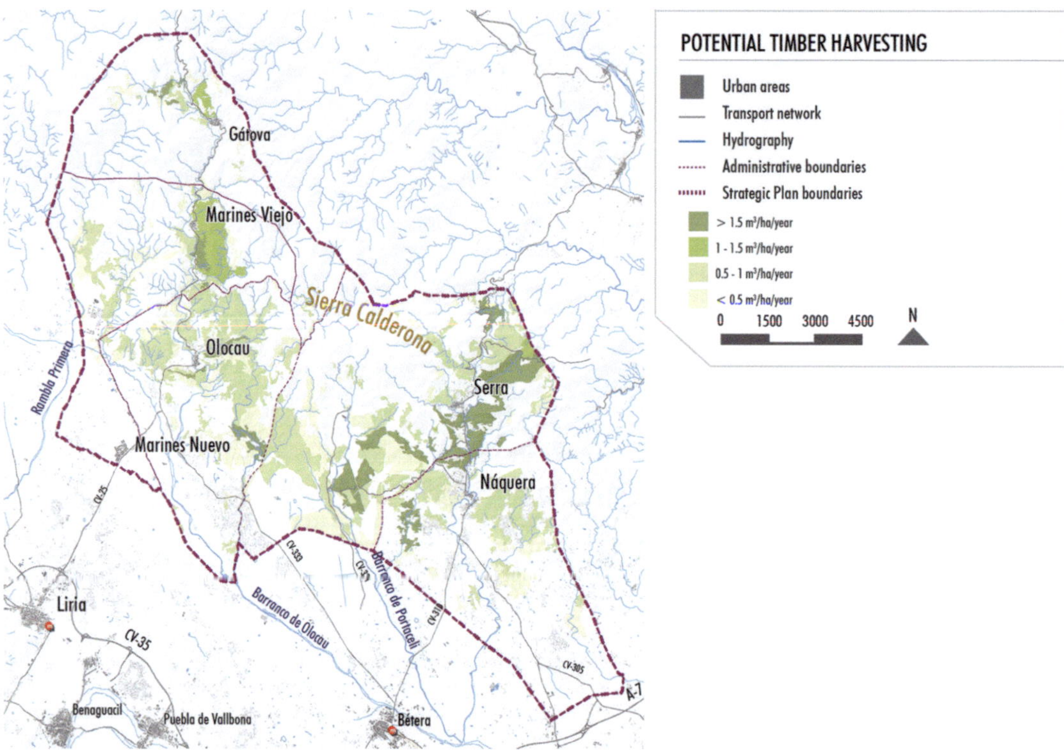

Fig. 3.4 Timber harvesting. *Source* Own elaboration based on cartography from PATFOR, Vaersa

3.2.2.2 Biomass Harvesting

Current and potential situation: Currently, biomass harvesting is not common and could be reinforced by using pruned material and material from agricultural activities. While the activity would be legally possible in a large part of the area regulated by the Natural Park, with the exception of the Special Protection Areas, the potential productivity is estimated to be mainly low.

Harvesting areas and production: 5,441.95 ha would be suitable for the production of chips and pellets. Regarding ownership, 70% is privately owned and 30% public property. Both area and activity are compatible with timber harvesting, although for logging there are more restrictions in steep areas (Fig. 3.5).

Production: It is estimated that 1,374 m^3/year of chips, or 852 tons/year of pellets, could be produced each year.

Economic impact: This would depend on the quantity of biomass used to produce pellets or chips. As it will be described in the local economy subchapter of this chapter, the annual revenue is estimated to be 87,859.79 €/year for producing pellets and 34,059.80 €/year for producing chips. It must be noted that when producing pellets, it would be necessary to increase the quantity of material to be processed up to 6,000 tons/year and that would only be possible by collecting agricultural and forestry residual materials from other areas. Regarding grants and subsidies, the "Agencia Valenciana de la Energía" (AVEN) subsidises 45% of the eligible costs for biomass installation projects and "Conselleria de Infraestructuras, Territorio y Medio Ambiente" subsidises 51€/ton of the forest biomass coming from silvicultural work.

Necessary facilities: The production of chips requires simpler and cheaper facilities than the production of pellets does. The production of pellets, to be viable, requires a larger volume of raw material to be processed. A highly versatile plant producing biofuel from organic products already exists in Náquera.

Environmental impact: The same advantages and disadvantages as for timber harvesting would apply.

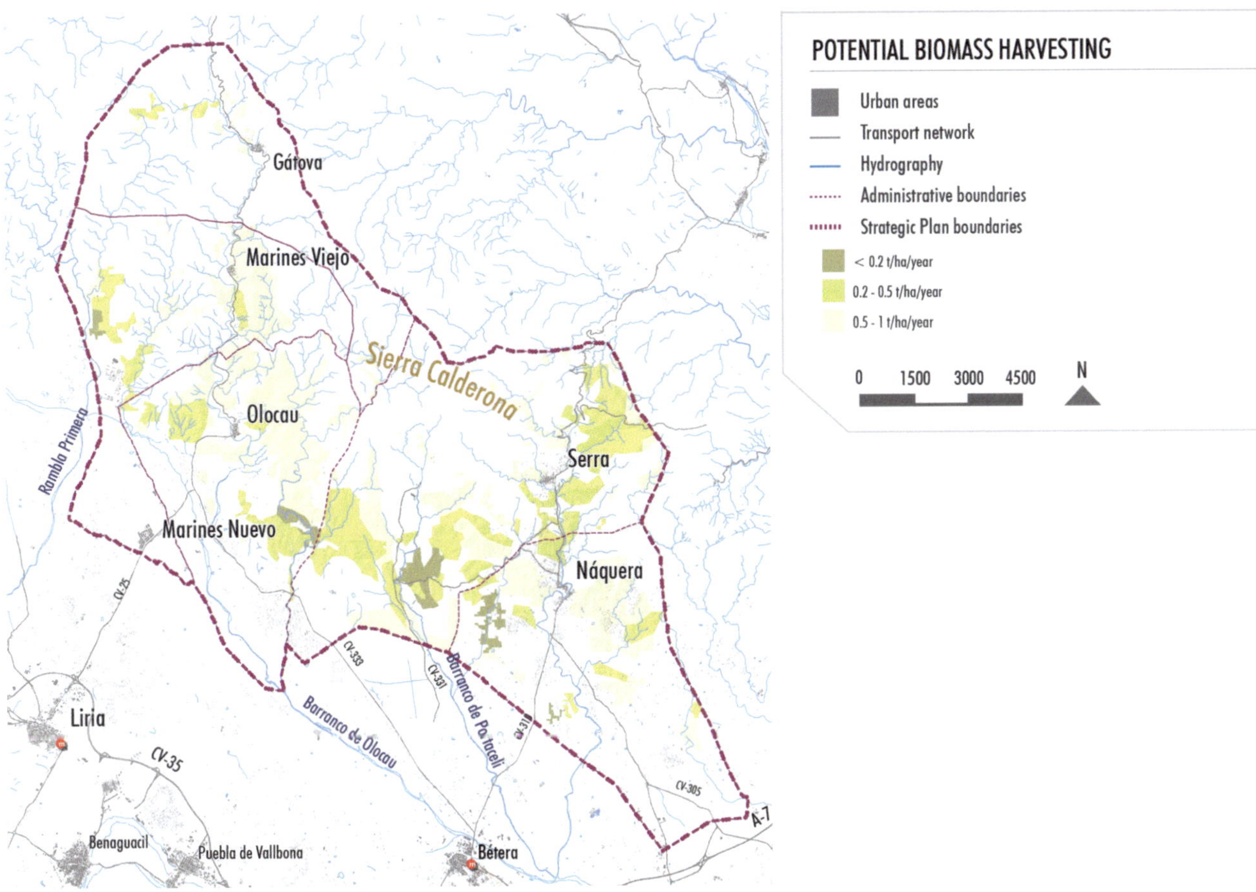

Fig. 3.5 Biomass harvesting. *Source* Own elaboration based on cartography from PATFOR, Vaersa

3.2.2.3 Cork Harvesting

Current and potential situation: The cork oak forests are one of the most valuable plant communities in Sierra Calderona, where they grow due to the siliceous nature of its soils. This tree species is found in the innermost areas and has a very variable degree of development, but generally, most specimens are young and should be allowed to grow larger in order to become a sustainable source of cork.

Harvesting areas: 235.75 ha could be suitable for use in this activity (1% of the total area of the Sierra Calderona Strategic Plan), from which the Special Protection Areas of the Natural Park have to be excluded. The remaining land is mostly privately owned and would have a low potential for cork productivity (Fig. 3.6).

Production: It is estimated that every 10 years, 83 tons of cork could be extracted.

Economic impact: Difficulty of access to the cork oak trees would raise the costs of extraction and decrease the margins of benefit. As will be described in the local economy subchapter, the annual revenue is estimated to be €133,040.00 every 10 years (13,304 €/year).

Necessary facilities: The area lacks cork processing industries, the nearest being located in the municipality of Soneja.

Environmental impact: Cork extraction increases by a factor of five the production of cork by the cork oak trees. Due to its high CO_2 sequestration power, cork extraction has a remarkable positive environmental impact. However, the extraction should not be done in trunks with a diameter of less than 65–80 cm and at least every 9 years. The extraction activity must be developed according to the slope, and it would negatively affect the shrubs growing around.

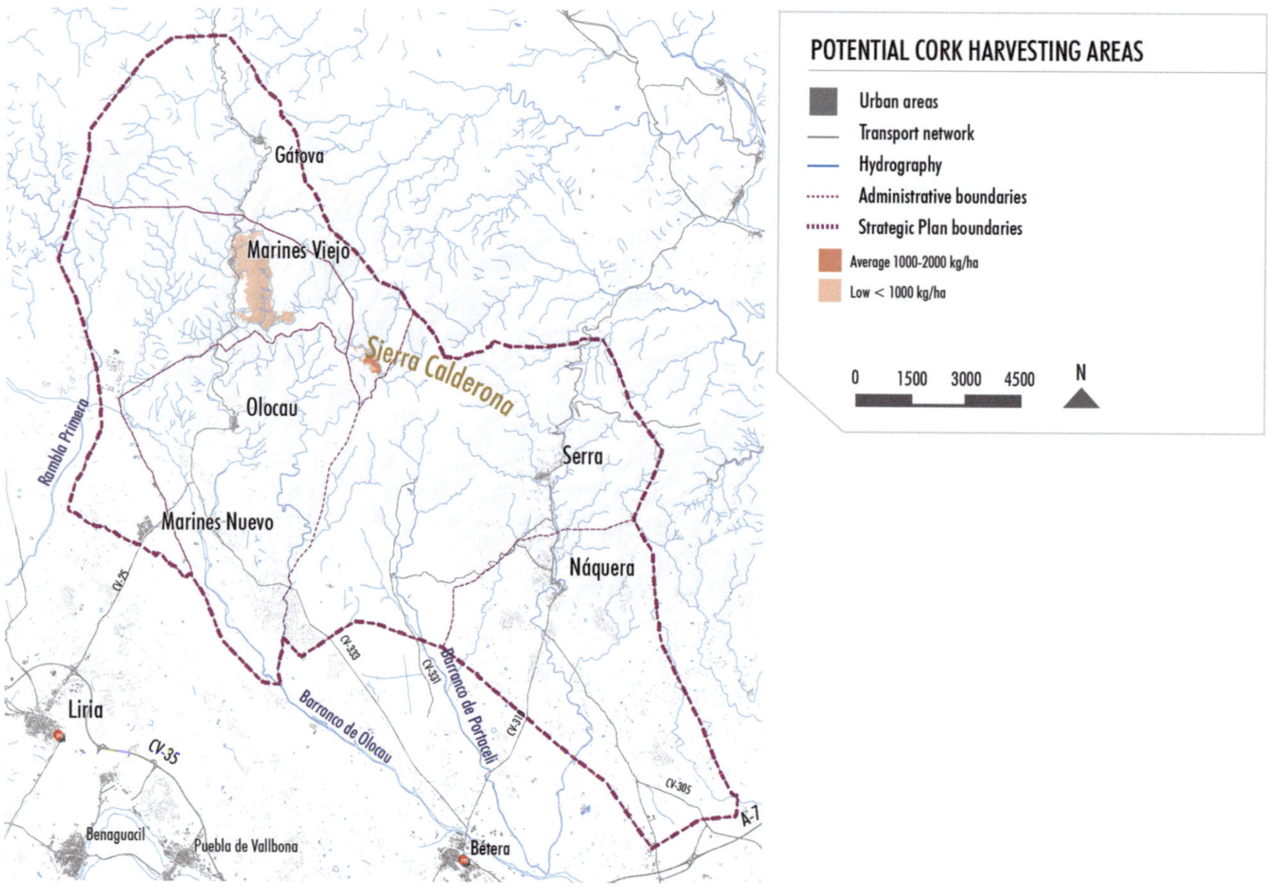

Fig. 3.6 Cork harvesting. *Source* Own elaboration based on cartography from PATFOR, Vaersa

3.2.2.4 Truffle Harvesting

Current and potential situation: The forests in the area generally have a very low capacity for mycological production and particularly for truffle production. However, due to the diversity of soils and existing microclimates, there are areas with an average performance (4% of the area of the Sierra Calderona Strategic Plan) and areas with a low performance (25%). The most suitable species for exploitation in the area would be: holm oak (*Quercus ilex*), Portuguese oak (*Quercus faginea*), oak (*Quercus robur*), kermes oak (*Quercus coccifera*) and hazelnut tree (*Corylus avellana*).

Harvesting areas: The areas with an average and low exploitation potential occupy, respectively, 719.8 ha and 4.462.42 ha (Fig. 3.7).

Economic impact: Given the high degree of uncertainty in the implementation and production of truffles, and in order not to inflate their possible impact on the local economy, only the areas with an average productive potential are considered (720 ha). At the same time, the production performance is estimated as low (0.5 kg/ha/year). These estimates would generate a potential benefit of 179,967.70€/ year.

Environmental impact: The production of truffle requires the implementation of mycorrhizal inoculations and of similar silvicultural treatments as those used for the prevention of fire. This implies a certain loss of biodiversity and vegetable mass, but an improvement of the phytosanitary condition of the forest.

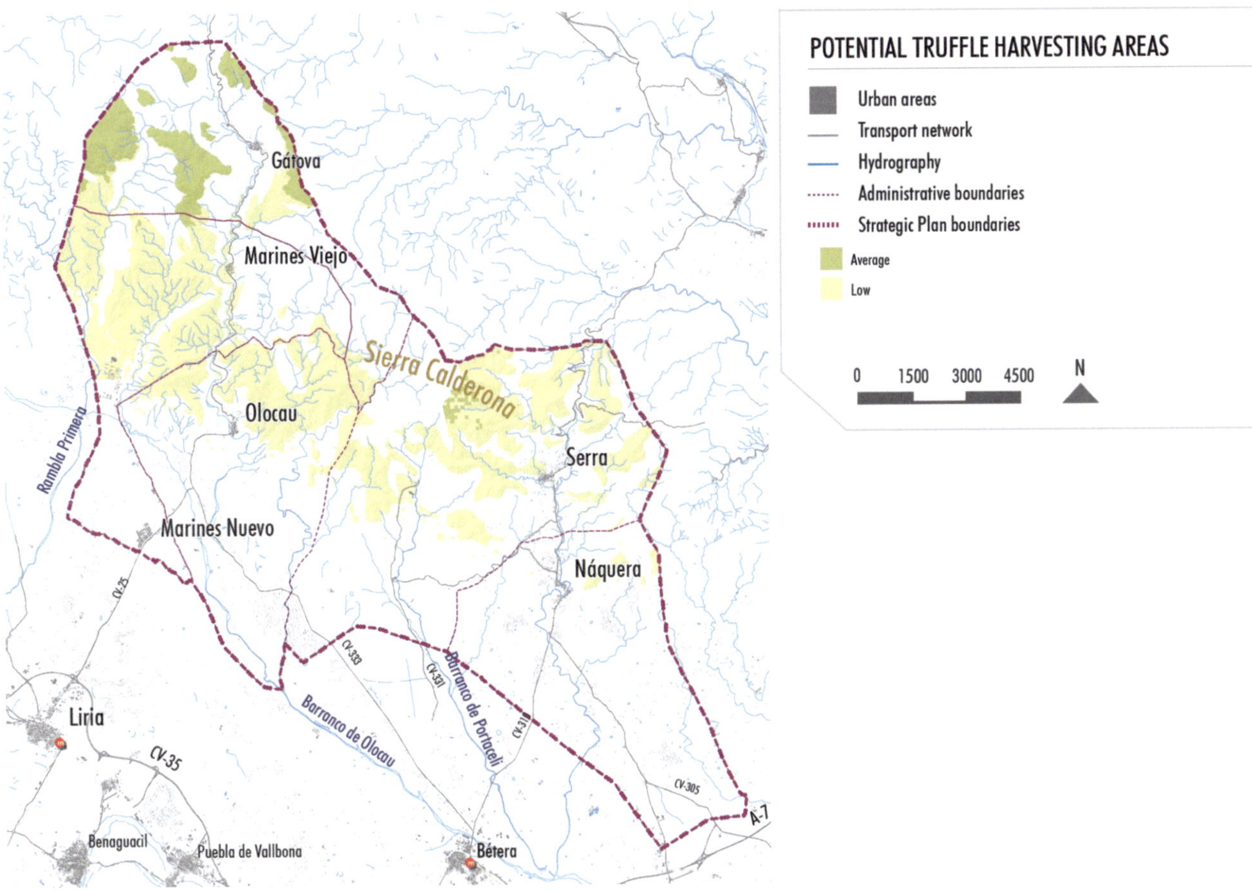

Fig. 3.7 Truffle harvesting. *Source* Own elaboration based on cartography from PATFOR, Vaersa

3.2.2.5 Honey Harvesting

Current and potential situation: At present, there are beehives distributed in the foothills and interior areas of Sierra Calderona. However, this activity is developed by private individuals, with no support from any kind of marketing or distribution network.

Harvesting areas: Honey harvesting would be possible on 2,637 ha (13% of the area covered by the Sierra Calderona Strategic Plan), 76% of which would be suitable for producing monofloral and 24% for multifloral honey (Fig. 3.8).

Economic impact: If six groups of beehives (250 hives/group) were placed with a separation of at least 1.5 km, an income of 31,050€/year could be generated.

Environmental impact: The environmental benefits of beekeeping are considerable when compared to the possible negative impacts that beehive management could have on the environment.

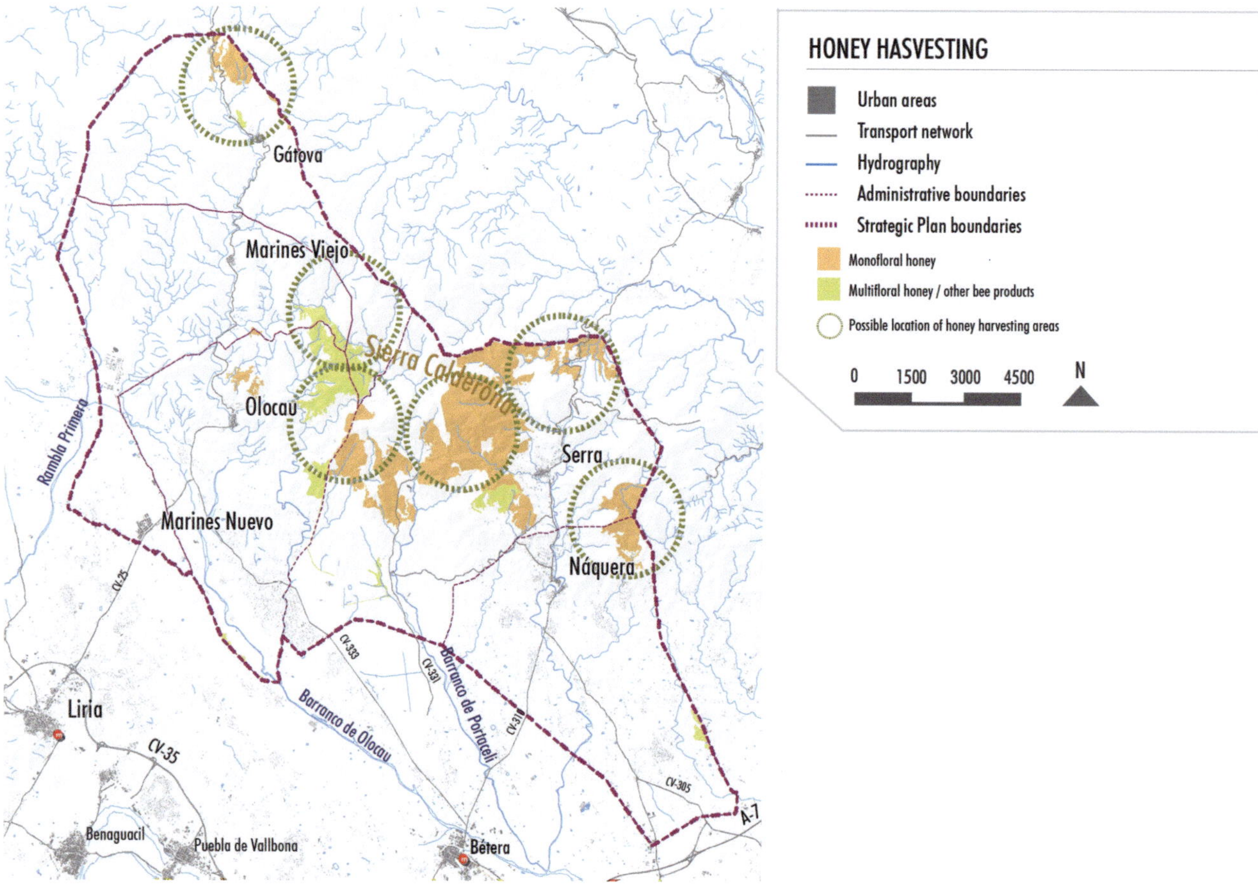

Fig. 3.8 Honey harvesting. *Source* Own elaboration based on cartography from PATFOR, Vaersa

3.2.2.6 Hunting

Current and potential situation: Hunting is an activity that is strongly rooted in the region. There are six hunting grounds and five hunting clubs with a powerful social presence. The hunting clubs help manage the Natural Park by maintaining and improving both their own facilities (drinking troughs, feeders, fences, etc.) and the infrastructure of common use (tracks and forest roads). These are maintained using their own funds, as well as managing subsidies for the improvement of the rural environment granted by the regional government.

Hunting reserves: They cover 16,240 ha (81% of the area included in the Sierra Calderona Strategic Plan), 90% of which is estimated to be suitable for small game hunting and 10% for big game hunting (Fig. 3.9).

Necessary facilities: Hunting, in its various forms, requires the existence and maintenance of drinking troughs, feeders, fences, paths, etc.

Economic impact: As can be seen in the local economy subchapter of this chapter, the annual revenue is estimated to be 147,270.82 €/year (estimating an average of 10 €/ha in small game hunting and of 0.76 €/ha in big game hunting).

Environmental impact: Public opinion on hunting is varied. On one hand, its supporters justify that hunting helps preserve the balance of certain animal populations and the whole management of the area. On the other hand, its critics argue that as a result of hunting, biodiversity has been lost in favour of species targeted by hunters, scrubland now predominates over woodlands, and imported game specimens are periodically released into the area.

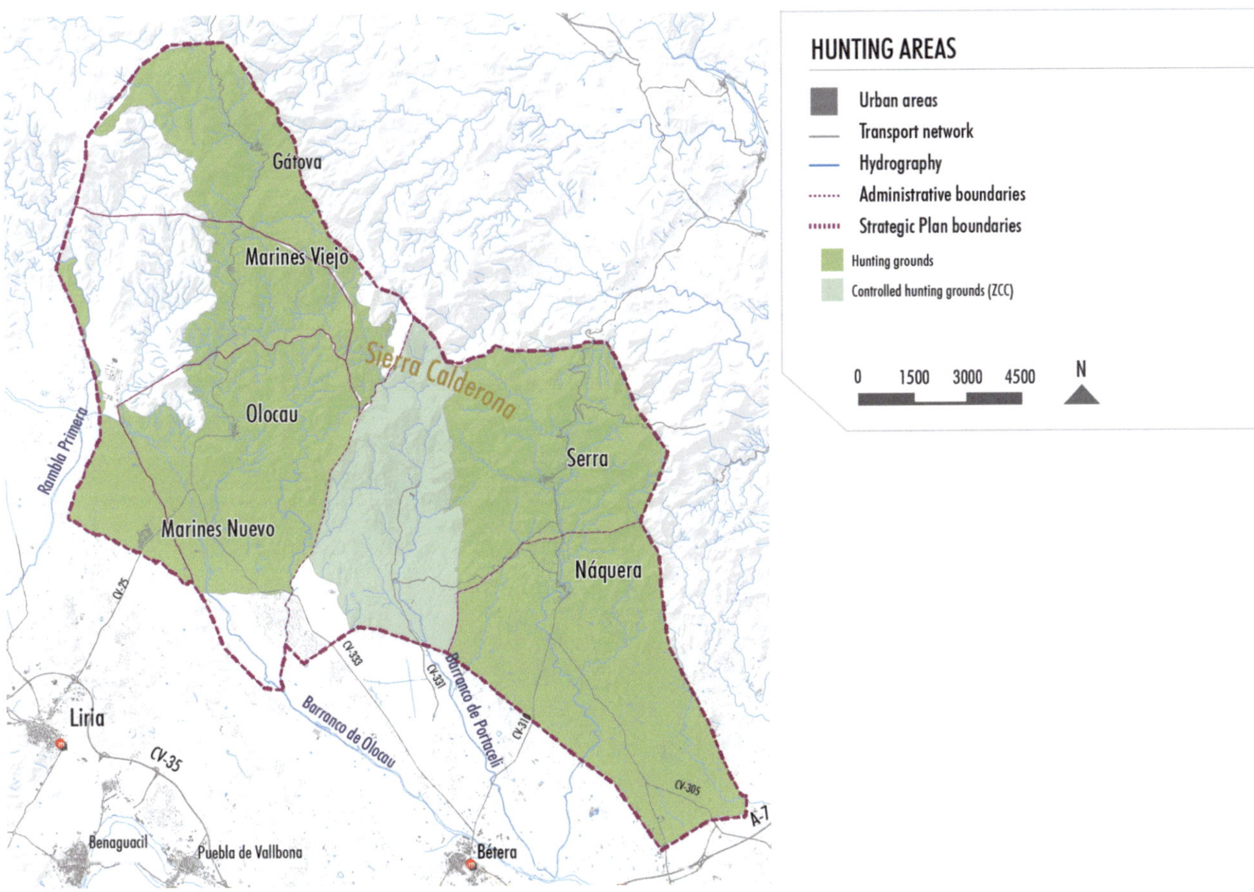

Fig. 3.9 Hunting. *Source* Own elaboration based on cartography from PATFOR, Vaersa

3.2.3 Fire Prevention

Most of the area covered by the Sierra Calderona Strategic Plan includes forest land with strong human presence that, together with the inherent risks of the Mediterranean forest, make it necessary to plan, measure and implement resources and infrastructure to both prevent and extinguish fires.

3.2.3.1 Hazard and Vulnerability

Considering "hazard", the inherent ease of a forest system to propagate a fire, and "vulnerability" the degree of human, material or environmental damage produced by that fire; it can be concluded that almost all the wooded areas included in this Plan can be classified as "highly hazardous and vulnerable" for fire risks, mainly due to the strong presence of dense pine forest mixed with scrubland. In addition, the presence of numerous scattered residential and recreational areas in the southern part of Sierra Calderona (Olocau and Náquera), and the heavily frequented areas around Serra, increase both the level of hazard and vulnerability of the region (Figs. 3.10 and 3.11).

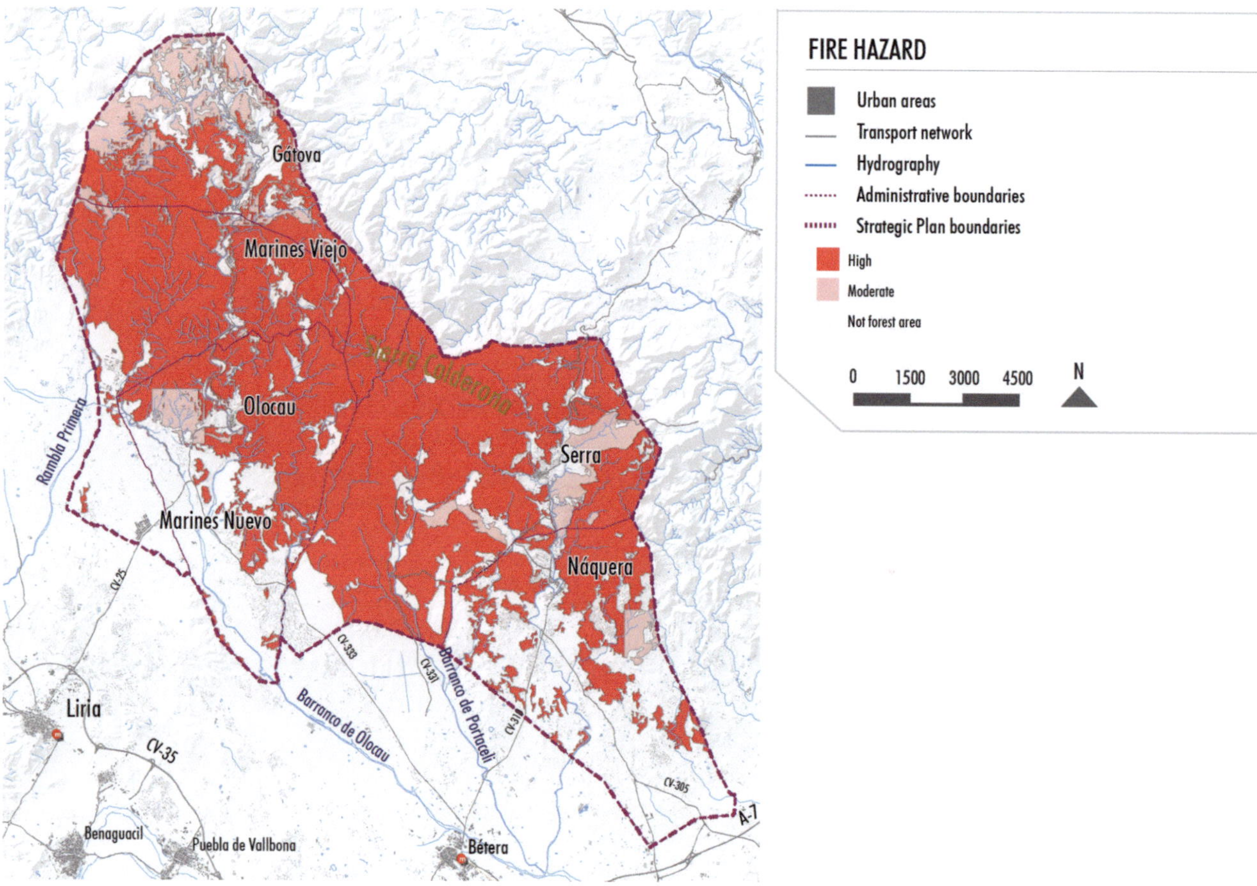

Fig. 3.10 Fire hazard. *Source* CITMA

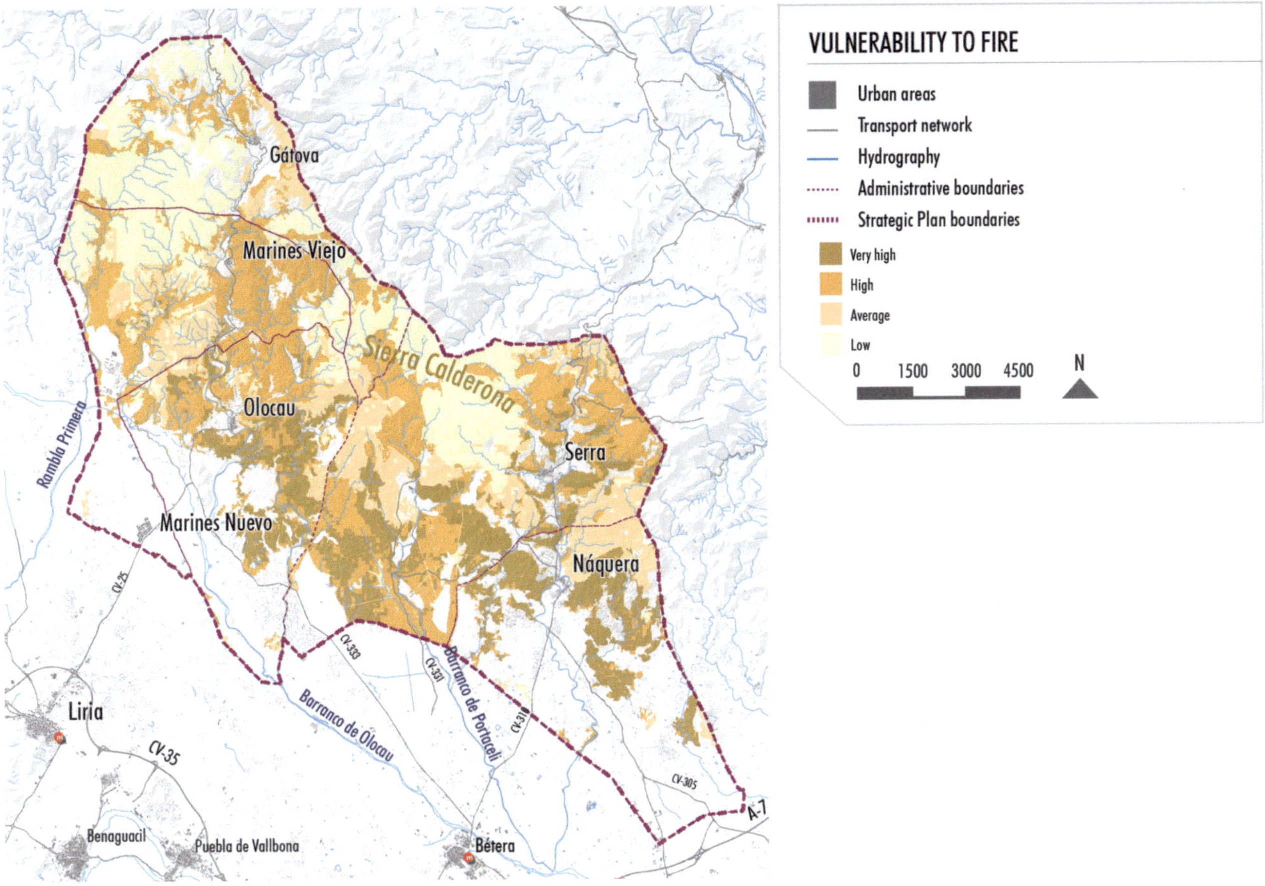

Fig. 3.11 Vulnerability to fire. *Source* CITMA

3.2.3.2 Fire Prevention and Extinction Network

The *Plan de Actuaciones de Prevención de Incendios Forestales* (in English "Forest Fire Prevention Plan") of the Sierra Calderona Natural Park defines the locations and development of forest firebreaks, access roads for emergency vehicles and the location of fire hydrants and water reservoirs. The analysis of that Plan and its associated cartography reveals the following (Fig. 3.12):

- The forest firebreaks network relies, mostly, on the road network and forest tracks. This overlapping of uses is positive, as it reduces their impact on the environment and the landscape. Additionally, it is observed that many of these tracks are used for recreation purposes (hiking, mountain biking, hunting, etc.) and are located over pre-existing rural paths, near to remaining rainfed agriculture plots.
- Some of the tracks and roads have a very broad section (GR10), producing a significant impact on the landscape.
- The hydrants are concentrated in the surroundings of the towns of Náquera, Olocau and Serra, while the water reservoirs are located in the northern area of the Natural Park.
- The role of rainfed agriculture as firebreaks has been seriously undermined due to the progressive disappearance of this kind of agriculture in the inner mountainous areas.
- The existence of housing estates and scattered settlements next to poorly managed forest areas compromises the security of both people and properties.

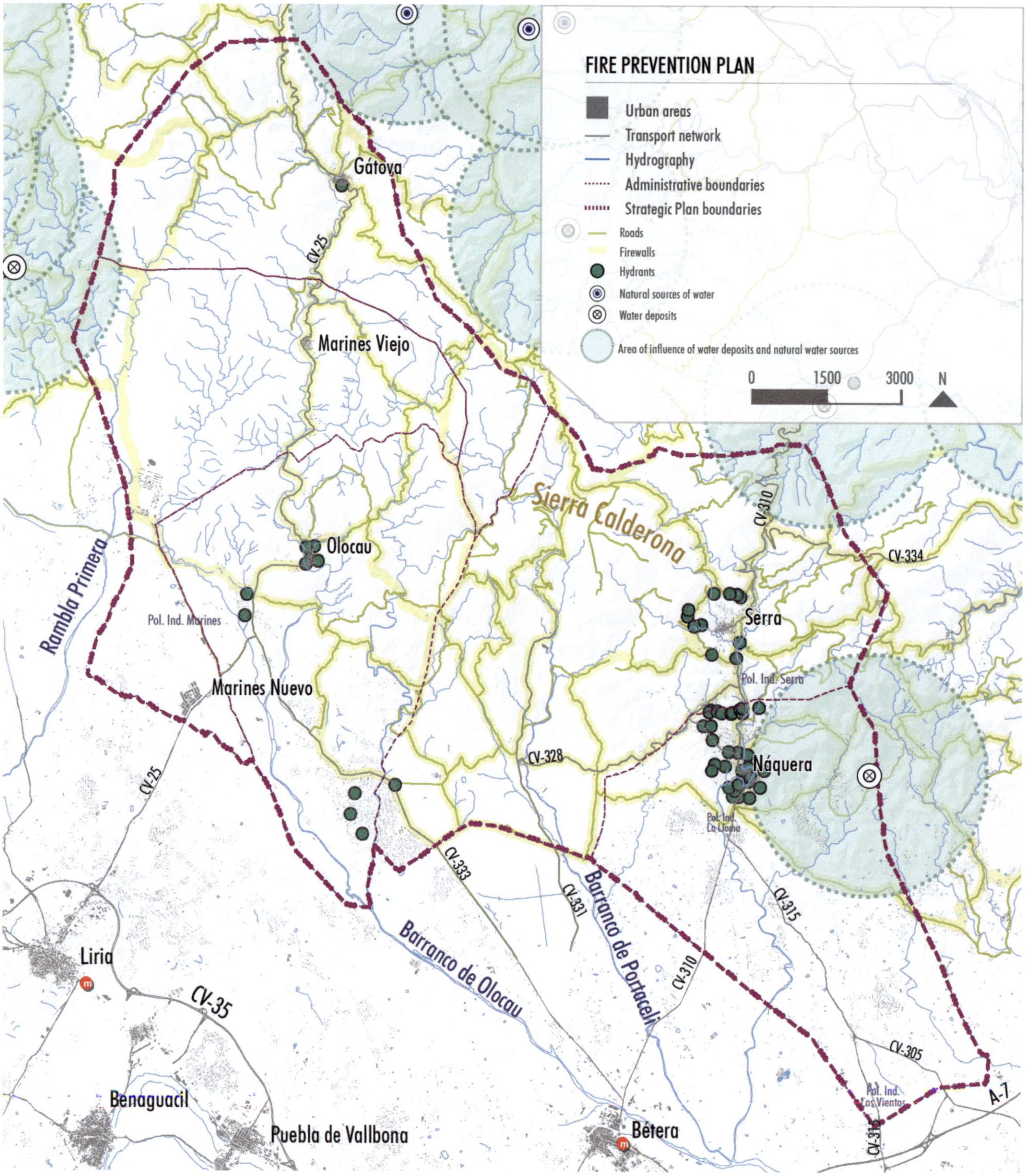

Fig. 3.12 Fire prevention (firewalls, roads and water network). *Source* Own elaboration based on cartography from PATFOR, Vaersa

3.2.4 Public and Recreational Use

Sierra Calderona attracts a large number of visitors throughout the year as a result of its numerous housing estates, its proximity to the metropolitan area of Valencia and its remarkable scenic landscape. As it will be explained in the subchapter "Tourism and public use", these visitors develop different types of activities, generally based on a daily visit pattern.

In order to facilitate its use by the public, a network of paths, recreational areas and picnic areas has been developed (Fig. 3.13).

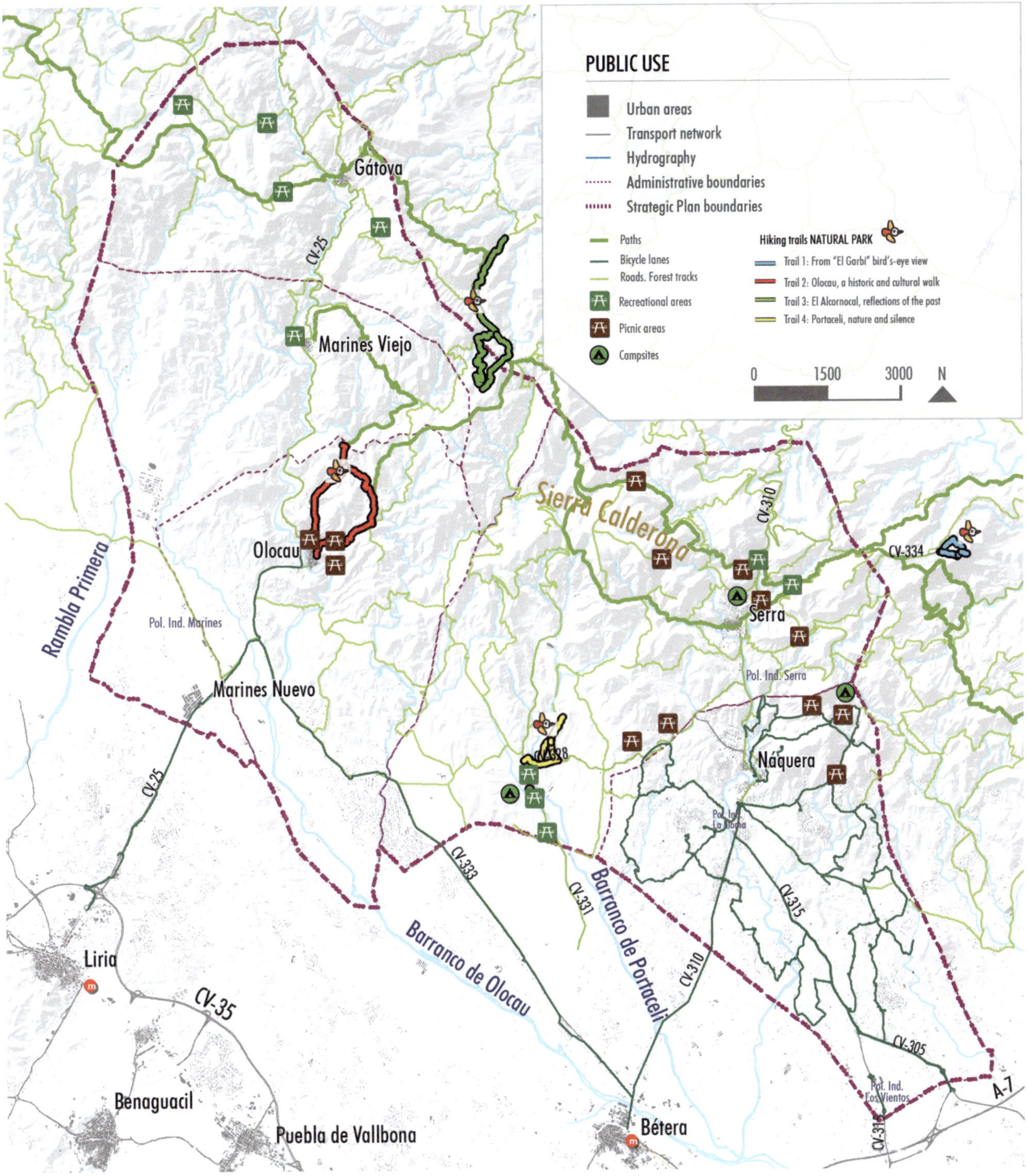

Fig. 3.13 Public use (network of paths, recreational areas, picnic areas, springs, fountains and camping areas). *Source* Own elaboration based on CITMA, FEMECV, VAERSA

3.2.4.1 Roads and Paths Network

The forest areas included in the Sierra Calderona Strategic Plan have a dense network of roads and paths, mostly located over existing rural trails. Some of them have been significantly expanded in order to serve as primary access routes for fire brigades. Additionally, some minor paths have been opened to facilitate access for hikers to enclaves of high scenic interest (Fig. 3.14). The existing path and road network presents several problems, such as lack of continuity, poor connectivity with the rural villages and towns and

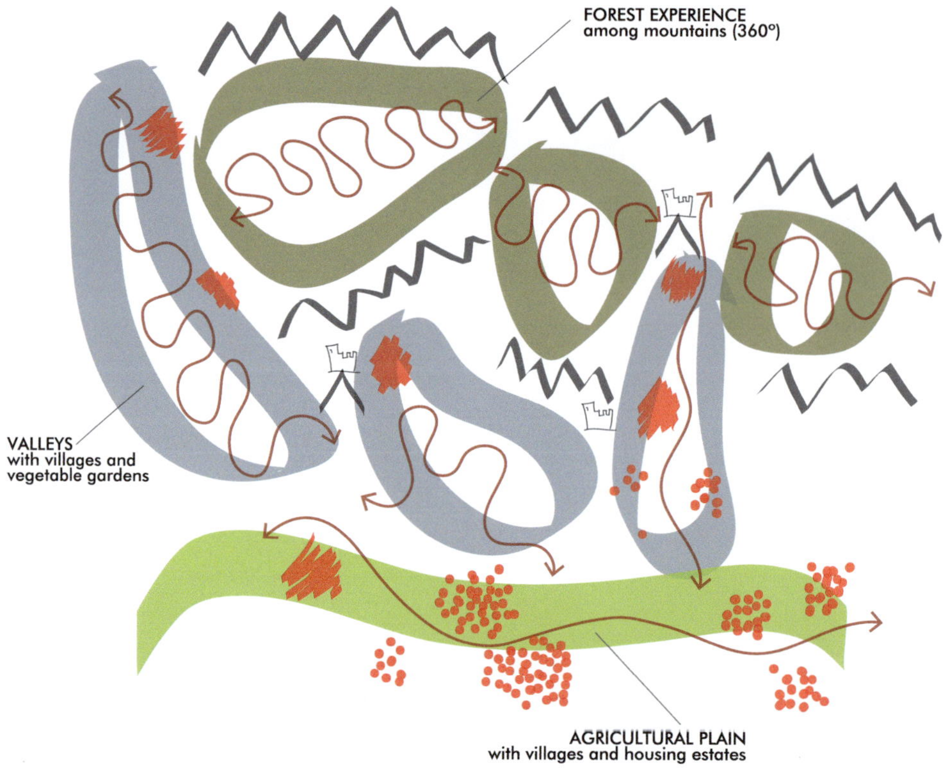

Fig. 3.14 Perception of the rural forest environment. *Source* Own elaboration

precarious signposting in both towns and along roads. The network is structured by three kinds of paths:

- Dirt tracks accessible to all kinds of vehicles: These are wide enough for two vehicles crossing, and their surface is generally in good condition. They are mostly located along the GR10 (*Gran Recorrido* or Major Route), around Portaceli's recreational area, connecting the main roads with residential units or housing estates and giving access to agricultural plots in mountainous areas.
- Dirt tracks accessible only by 4WD: with enough width for two vehicles crossing but an irregular surfacing. They are mainly branches or interconnections of the above-mentioned tracks.
- Paths: not wide enough for a conventional vehicle. They give visitors access to enclaves of high scenic interest (peaks, viewpoints, valleys, etc.).

3.2.4.2 Recreational Areas, Picnic Areas and Other Services

The forest areas included in the Sierra Calderona Strategic Plan are accompanied by a network of recreational areas concentrated mostly at the "Pla del Lluch" (an area of intensive use in the vicinity of Portaceli's Carthusian monastery), by the water springs in the surroundings of Gátova (as a tourist attraction in the most isolated village included in the Plan) and around the town of Serra. Some picnic areas complement the provision of recreational spaces and are located near Olocau's centre and in mountainous areas north of Náquera. The use of water springs as centres of the existing recreational sites is also perceived as a positive aspect. However, the following weaknesses have been detected:

- Absence of recreational sites in central areas of the area covered by the Sierra Calderona Strategic Plan (although there are fountains and great quality settings).
- Lack of campsites and rural accommodation in inner areas.
- Absence of signposting and orientation maps to access the recreational areas from the roads (people get there because they already know where they are located) (Tables 3.2, 3.3 and 3.4).

Table 3.2 Campsites. *Source* Own elaboration

	Campsites	
	Name	No. of tents
NÁQUERA	Náquera	–
SERRA	Portaceli	35–40
	Les Aliaguetes	–

Table 3.3 Recreational areas. *Source* "Plan de Uso Público" of the Sierra Calderona natural park

	Recreational areas				
	Name	No. tables with benches	No. Barbecues	Fountains	Parking
SERRA		78	6	3	2
	Portacoeli	57	4	YES	YES
	Les Aliguetes	14	1	YES	YES
	Font d l'Ombria	7	1	YES	NO
MARINES		39	6	1	1
	Fuente del Rincón	8	1	YES	YES
GÁTOVA		28	4	4	1
	Fontfría	3	1	SI	NO
	Fuente del Rebollo	5	1	YES	NO
	Fuente de la Alameda	20	2	YES	YES
	Fuente del Tormo	3	1	YES	NO
Total plan		145	16	8	4

Table 3.4 Picnic areas. *Source* Own elaboration

	Picnic areas			
	Name	No. of tables with benches	Fountains	Parking
NÁQUERA		10	1	2
	Font de L'Or	4	YES	YES
	Pi de Salt	4	NO	YES
	Fenassar	2	NO	NO
SERRA		21	3	3
	Broseta-Pedrera	8	NO	NO
	Font del Marianet	1	YES	NO
	Merendero del Castell	3	NO	YES
	Font de Sant Josep	7	YES	YES
	Font del Llentiscle	2	YES	YES
OLOCAU		8	3	1
	Font de la Salut	2	YES	NO
	Font de La Carrasca	3	YES	NO
	Font de L'Arquet	3	YES	YES
Total plan		145	16	8

3.2.5 Public Participation Plan

See Table 3.5.

Table 3.5 Survey results. Public Participation Plan_Phase 1: natural environment. *Source* Own elaboration

Forest and livestock farming areas	Total (%)	GÁTOVA (%)	OLOCAU (%)	MARINES (%)	NÁQUERA (%)	SERRA (%)
Conflicts						
Mountains are not clean enough	48.5	50.0	44.0	39.1	59.6	77.8
Areas at risk of fires and burnt down areas	28.1	50.0	24.0	43.5	29.8	30.6
Inefficiency/insufficiency of fire brigades	14.0	35.7	16.0	13.0	12.8	16.7
Forest vegetation: not diverse. lack of trees, uncontrolled vegetation	12.9	0.0	8.0	4.3	27.7	16.7
Political indifference, insufficient resources and economic management	12.9	14.3	28.0	21.7	14.9	2.8
None	12.3	14.3	28.0	13.0	12.8	8.3
Lack of economic activity and profitability	11.1	35.7	24.0	21.7	4.3	2.8
Excessive bans	11.1	28.6	12.0	26.1	2.1	13.9
Lack of maintenance of recreational areas	7.0	7.1	4.0	8.7	4.3	13.9
Does not know/no answer	21.1	0.0	8.0	13.0	6.4	11.1
Opportunities						
To link maintenance jobs to the local population	17.0	28.6	8.0	21.7	14.9	30.6
To allow economic activities in the mountains that will produce yields	15.2	50.0	12.0	30.4	10.6	11.1
To restore the existing management of the area as it was before it was declared a natural park	7.6	14.3	0.0	8.7	4.3	19.4
To develop activities that will place value on the mountains	6.4	21.4	8.0	8.7	6.4	0.0
To create new recreational and picnic areas	4.1	7.1	12.0	8.7	2.1	0.0
To promote extensive livestock farming and the production of local products related to it	4.1	7.1	0.0	8.7	2.1	5.6
To manage properly forest areas and reforest them	3.5	7.1	4.0	4.3	4.3	2.8
Does not know/no answer	60.8	21.4	76.0	39.1	61.7	52.8

3.2.6 Synthesis and Diagnosis

Forest land: This represents 68% of the surface area of the five municipalities included in the Sierra Calderona Plan, of which 56% is strategic and 44% common forest land. However, common forest land plays an important environmental and landscape role, as it is mainly located between settlements on the south of Náquera and Olocau. While forest land is mostly privately owned (71%), in Serra there are significant pockets that are owned by the municipality or the regional government.

Protection: The Sierra Calderona Natural Park is mainly constituted by a matrix of forest land, including a network of rainfed agricultural crops and small urbanised areas. The Natural Park covers 49% of the area included in the Sierra Calderona Strategic Plan and 64.5% of its total forest land.

Economic activity: The study of potential forest economic activities shows that they could contribute significantly to the local economy, and in particular, they could improve the relationship of the population with the surrounding natural environment. The economic activity has been estimated to comprise:

- Timber harvesting: €120,000/year
- Harvesting for Biomass (chips): €34,000/year
- Cork harvesting: €13,000/year
- Truffle harvesting: €179,000/year
- Honey harvesting: €30,000/year
- Hunting €140,000/year

Fire risk: Fire is the greatest hazard in the area, as the housing estates and towns surrounded by forest are the most vulnerable areas. Currently, the prevention and extinction system includes a system of tracks and paths that constitute a network of forest firebreaks. This firebreak network could be reinforced with rainfed agriculture, which is now in serious decline. Hydrants and water reservoirs are distributed irregularly. There is a strong concentration of hydrants around towns and housing estates in the surroundings of Náquera, Serra and Olocau and a concentration of water reservoirs on the northern side of Sierra Calderona.

Public use: The path and road network (forest tracks and trails) is extensive but discontinuous. Some areas of high interest are difficult to locate and/or access, and there is a lack of recreational sites. There are few campsites and rural accommodation options, as well as shelters in the interior area.

3.3 Agricultural-Livestock Farming Environment

3.3.1 Generalities

The study of the agricultural land in the area covered by the Sierra Calderona Strategic Plan reveals the presence of two basic types of agriculture highly responsive to the orography, microclimatic conditions and water availability:

- **Mountain agriculture**: This mainly has a relict character. It includes historically irrigated plots located on river corridors, surroundings of towns and small interior valleys, as well as rainfed plots located along the paths crossing the mountains and in their adjacent slopes.
- **Agriculture in the Sierra Calderona's foothills and southern plains**: This includes those agricultural areas with gentle slopes and a strong sun exposure due to their southern orientation. It integrates a mosaic of rainfed crops (in recession) and irrigated crops (expanding). Within this mosaic, different urban settlements (housing estates, industry and village centres) are located.

3.3.2 Agricultural Mosaic

The systematic study of land uses (SIOSE_2011) and in-situ observations provided the following conclusions (Fig. 3.15 and Table 3.6):

- Currently, agricultural land covers 20% of the area of the Strategic Plan.
- The presence of citrus trees is increasing and is progressively occupying, and bringing irrigation to the remaining rainfed fields in the Sierra Calderona's foothills and the southern plains. Currently, they represent 10% of the Plan's area and are mainly located on plains and foothills (30% in Náquera, 10% in Marines and Olocau, and they are practically non-existent in Gátova and Serra).
- Olive trees are mainly located in the most central and western areas (Marines, Olocau and Gátova), where they occupy between 6 and 10% of the total surface.
- The areas covered by vegetable gardens are located primarily between the ravines and the historic towns (Gátova, Olocau, Marines Viejo, and with a greater degree of pressure and fragmentation, in Náquera and Serra). They can also be found in residual pockets in small inner valleys.
- Carob trees are the most common non-citrus fruit trees in the area and are located in the foothills with coastal influence. Almond trees are also frequent; they can be found in both foothills and mountainous areas. To a lesser extent, cherry trees can be found in the mountains; they usually assume a very low percentage of the total amount of agricultural land, except in Gátova, where they cover up to 7.5% of its surface.

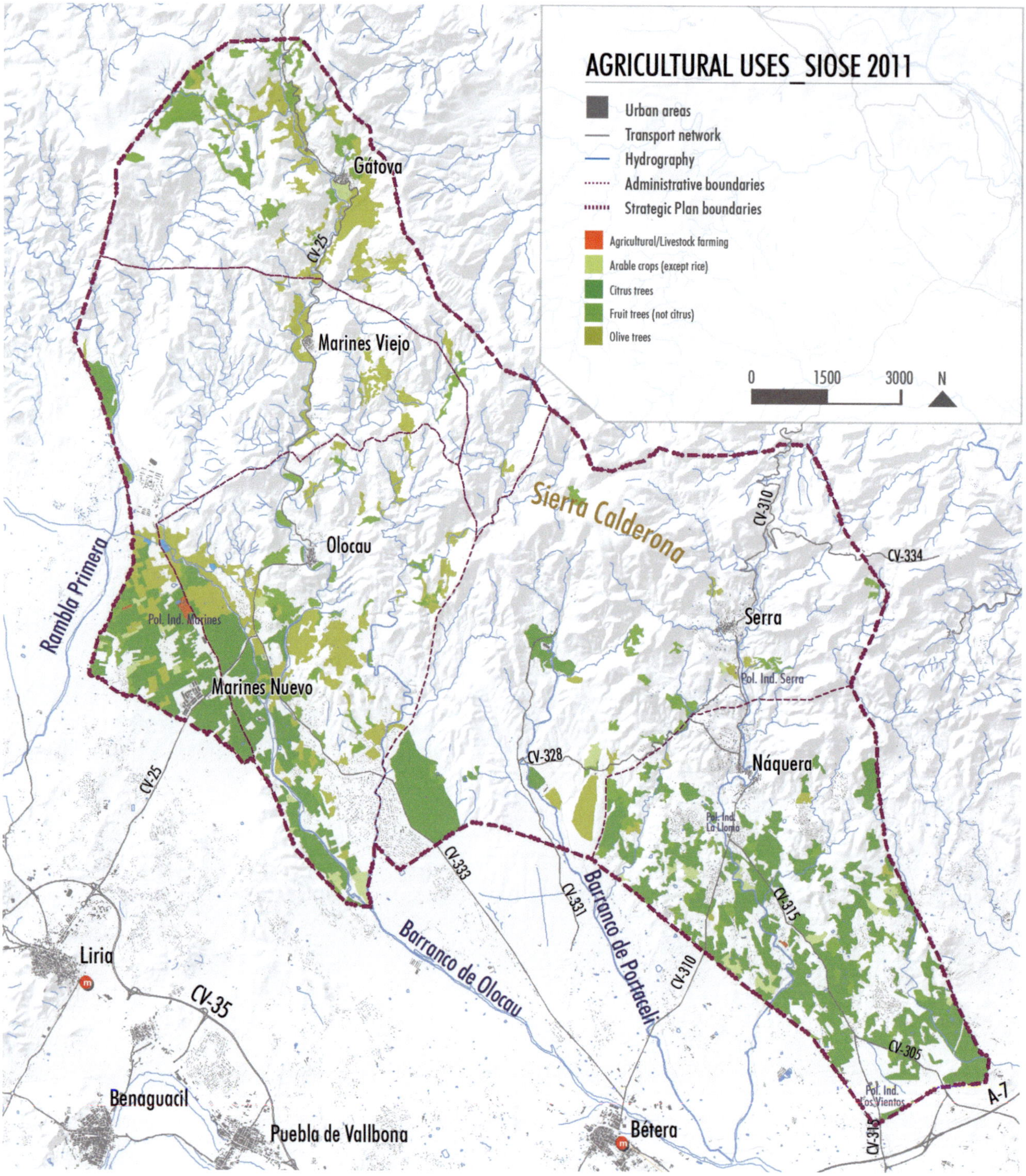

Fig. 3.15 Agricultural uses SIOSE_2011. *Source* CITMA

Table 3.6 Types of crops. *Source* Own elaboration based on SIOSE 2011

Crops	GÁTOVA		MARINES		OLOCAU		NÁQUERA		SERRA		Total	
	Area (ha)	(%)	Area (ha)	(%)	Area (ha)	(%)	Area (ha)	(%)	Area (ha)	(%)	Area (ha)	(%)
Arable crops (different than rice)	10	1.96	22	3.24	34	3.35	59	4.65	15	3.62	140	3.60
Citrus trees	–	–	386	56.93	370	36.42	1061	83.61	235	56.76	2052	52.79
Fruit trees	229	44.90	51	7.52	177	17.42	127	10.01	85	20.53	669	17.21
Olive trees	271	53.14	219	32.30	435	42.81	22	1.73	79	19.08	1026	26.40
Total	510	100.00	678	100.00	1.016	100.00	1.269	100.00	414	100.00	3.887	100.00

3.3.3 Evolution of the Agricultural Mosaic

As displayed diagrammatically in Fig. 3.16, since the second half of the twentieth century, the amount of agricultural land and the crops has changed drastically (Figs. 3.17, 3.18 and Table 3.7).

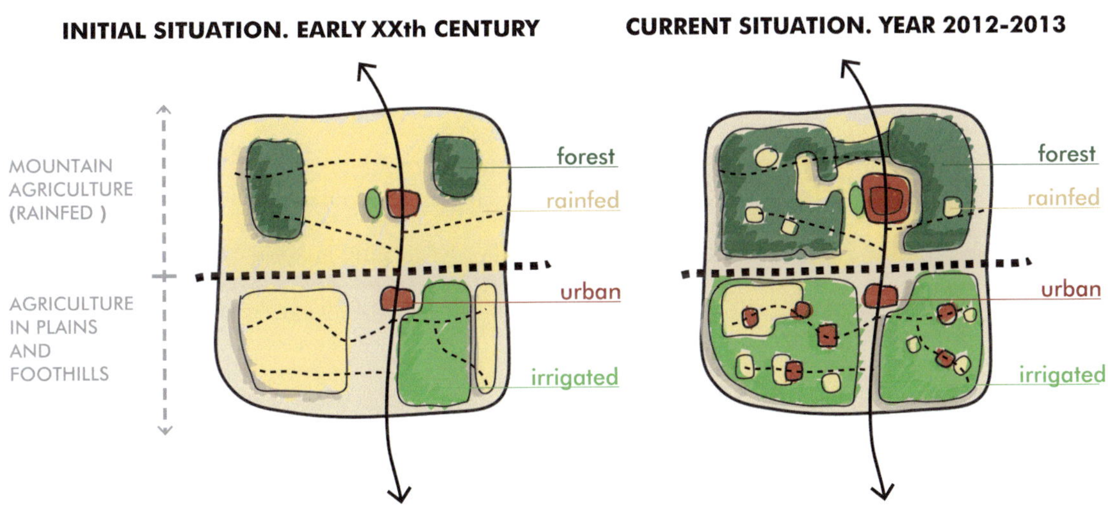

Fig. 3.16 Evolution of the Sierra Calderona agricultural mosaic. *Source* Own elaboration

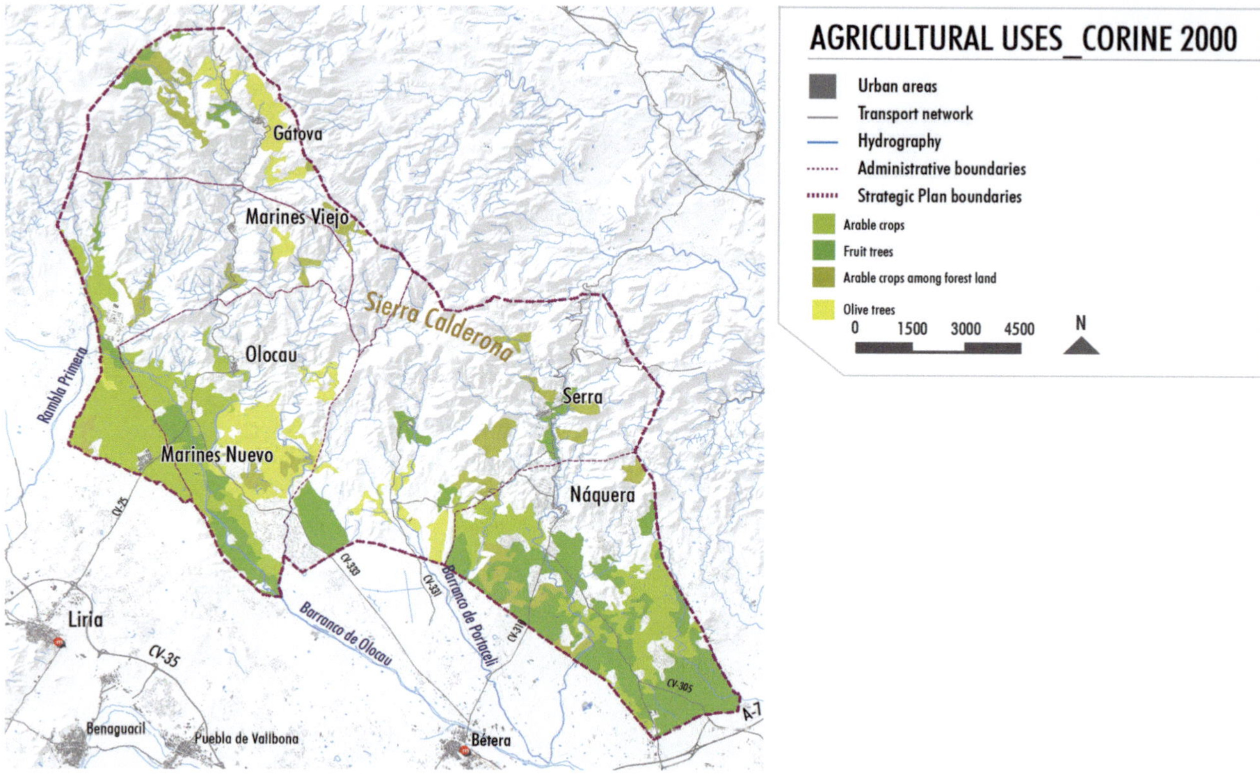

Fig. 3.17 Agricultural uses (CORINE 2001). Source CITMA

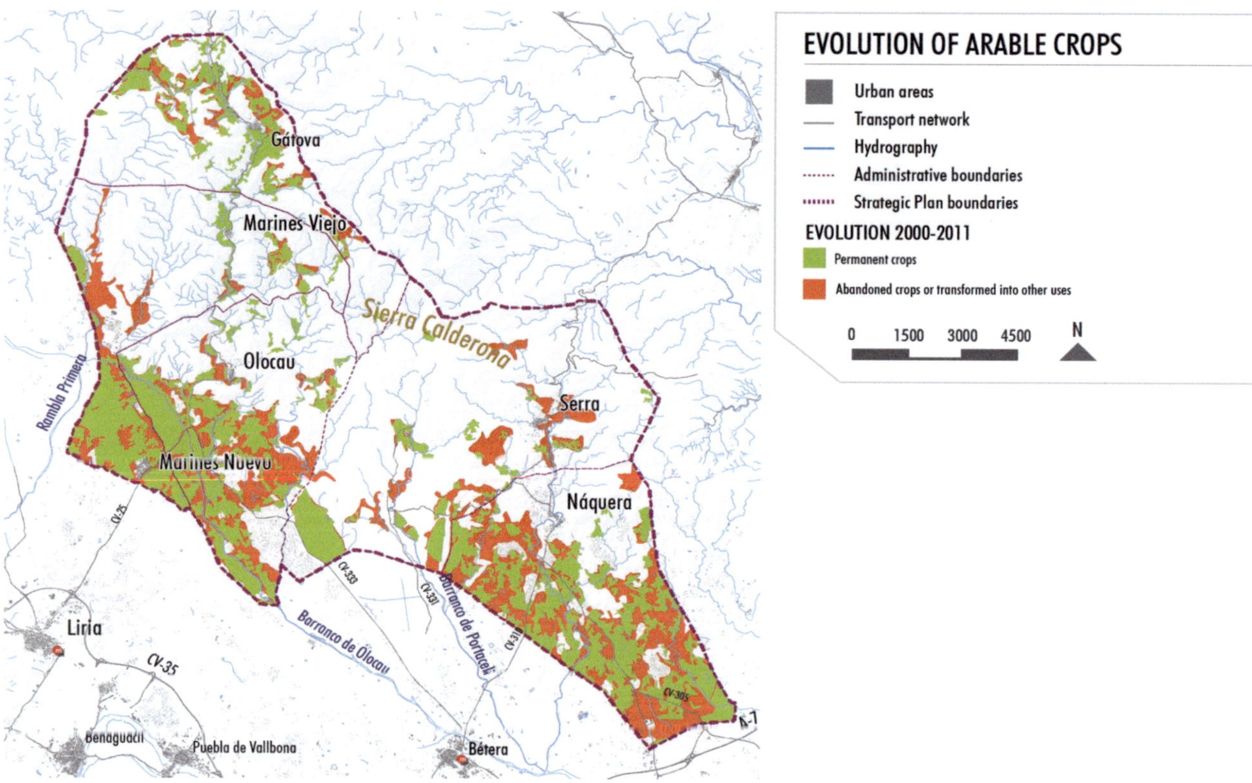

Fig. 3.18 Evolution of arable crops 2000–2011. *Source* Own elaboration

Table 3.7 Agricultural land. Abandonment or land-use change of agricultural areas in plains, foothills and mountainous areas. *Source* Own elaboration

	% abandonment before 2000	% abandonment or change of use between 2000 and 2011
Foothill agriculture		
Zone I_Náquera	19.00	40.00
Zone II_Marines-Olocau	9.00	33.00
Mountain agriculture		
Zone I_Náquera, Serra-Portaceli	83.22	50.00
Zone II_Marines-Olocau-Gátova	84.10	20.00

3.3.3.1 Mountain Agriculture

Mountain agriculture is disappearing due to its low profitability and difficulties to modify the traditional systems of production:

- 80–85% of the plots have been abandoned, though traces of the terraces are still visible.
- A substantial percentage (around 40%) has been recently abandoned (between 2000 and 2011).
- Abandoned plots have been occupied by wild vegetation, increasing the extension and connectivity of forest areas.

3.3.3.2 Foothills and Southern Plains Agriculture

The foothills and southern plains have experienced a rapid transformation from rainfed agriculture to irrigated agriculture (mainly citrus trees).

- Agricultural land (19% east of Náquera and 9% in Olocau–Marines) is increasingly occupied by infrastructure and urban uses (housing, industry). This transformation has happened recently in most cases, as can be seen when comparing the land cover from 2000 and 2011.
- Fragmentation as a result of the proliferation of housing areas.

3.3.3.3 Areas with Soils of High Fertility

Urban sprawl is mainly taking over land with medium to high agrological capacity; while in mountainous areas, the forest is taking over lands with low soil fertility (Fig. 3.19).

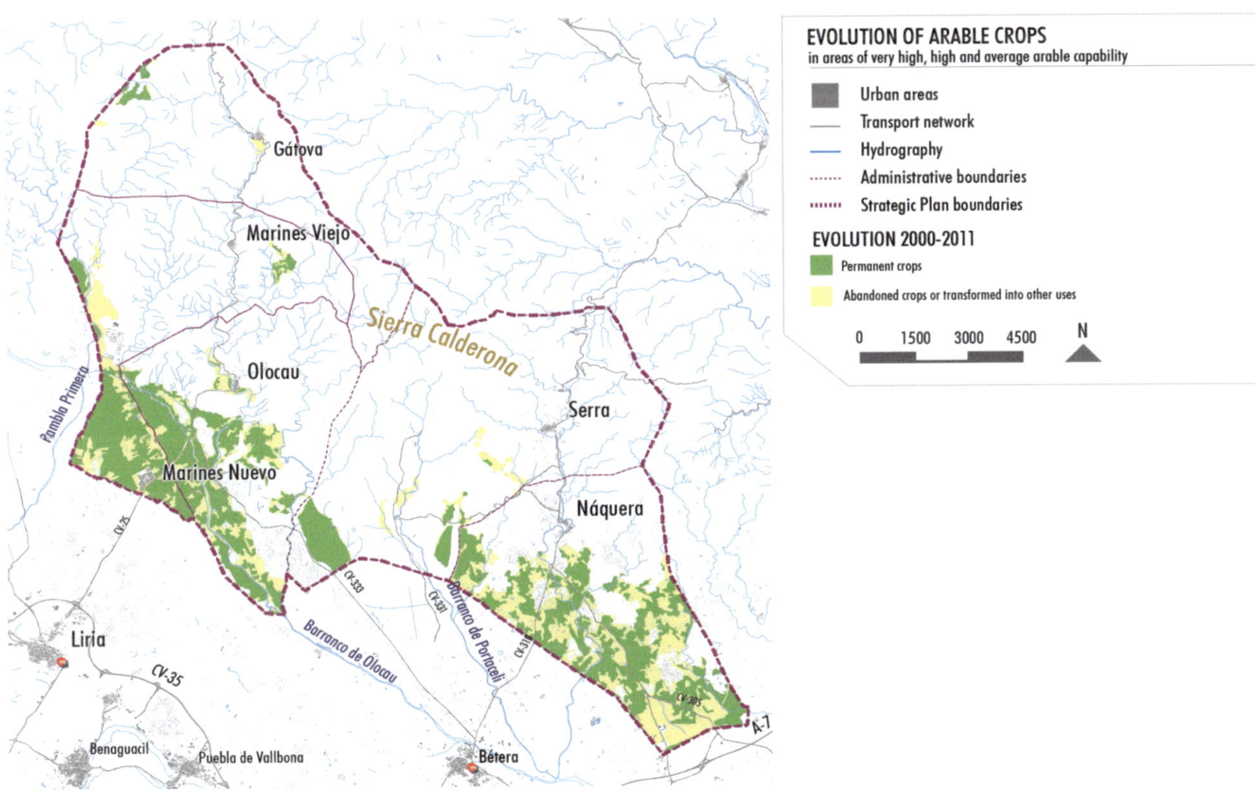

Fig. 3.19 Abandonment and permanence of crops in soils with very high, high or medium agrological capability. *Source* Own elaboration

3.3.4 Productivity and Profitability of the Agricultural Mosaic

In order to determine the viability of different kinds of crops in the area, an estimation of their profitability was made with the following conclusions (Fig. 3.20 and Tables 3.8, 3.9):

- The profitability of irrigated and well-managed citrus trees is estimated to be around €2,000/ha. This explains their increasing introduction on previously rainfed and other arable plots, as well as the development of costly landform transformations and infrastructural investments, where future urbanisation could be envisioned in the medium or long run.
- Rainfed crops produce lower yields and profits, which become especially critical in the case of carob trees.
- Irrigated crops produce average yields around the historic towns that are mainly intended for self-consumption.
- Certain fruit trees, like cherry trees (traditionally growing in Serra), could have high yields and revenues.

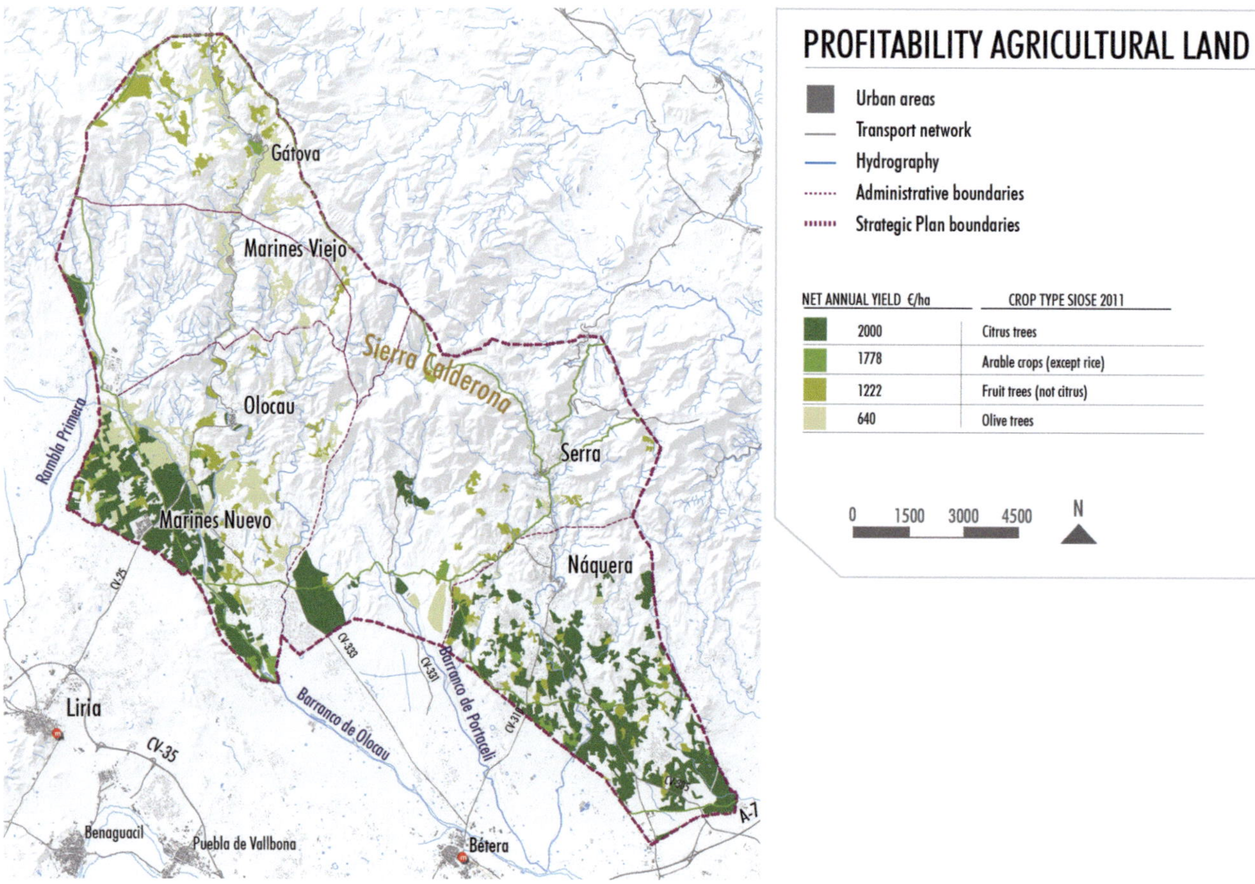

Fig. 3.20 Profitability in agricultural land. *Source* Own elaboration

Table 3.8 Agricultural land. Abandonment or transformation by municipality. *Source* Own elaboration based on SIOSE 2011 and CORINE 2000, CITMA

Crop	SIOSE 2011		CORINE 2000		% abandonment or change of use since between 2000 and 2011	
	Area (ha)	(%)	Area (ha)	(%)	Area (ha)	(%)
Total plan	3,887.00	19.53	6,408.00	32.10	−2,521.00	39.34
NÁQUERA	1,269.00	32.80	2,422.00	62.55	−1,153.00	47.61
SERRA	414.00	7.22	775.00	13.52	−361.00	46.58
MARINES	678.00	19.25	1,027.00	28.74	−349.00	33.98
OLOCAU	1,016.00	27.18	1,611.00	43.10	−595.00	36.93
GÁTOVA	510.00	16.76	573.00	18.83	−63.00	10.99

Table 3.9 Estimation of economic yields on agricultural land. *Source* Own elaboration

Crops	Econɔmic yields (€/ha)	GÁTOVA		MARINES		OLOCAU		NÁQUERA		SERRA		Total	
		Area (ha)	Yield (€)	Area (ha)	Yield (€)	Area (ha)	Yield (€)	Area (ha)	Yield (€)	Area (ha)	Yield (€)	Area (ha)	Yield (€)
Arable crops different than rice	1778	10	17,780	22	39,116	34	60,452	59	104,902	15	26,670	140	248,920
Citrus trees	2000	–	–	386	772,000	370	740,000	1061	2,122,000	235	470,000	2052	4,104.000
Fruit trees	1222	229	279,838	51	62,322	177	216,294	127	155,194	85	103,870	669	817,518
Olive trees	64C	271	173,440	219	140,160	435	278,400	22	14,080	79	50,560	1026	656,640
	Total	510	471,058	678	1,013.598	1.016	1,295,146	1.269	2,396,176	414	651,100	3.887	5,827,078

3.3.5 Livestock Farming

Extensive livestock farming takes place in forest land and the associated system of rainfed agriculture. Intensive livestock farming is located in dedicated premises.

3.3.5.1 Extensive Livestock Farming

Current and potential situation: The area covered by the Sierra Calderona Strategic Plan has virtually no extensive livestock farming, although grazing could play an important role in the control of vegetation in forest firebreaks. This is despite the subsidies offered by the regional government "Consejería de Infraestructuras, Territorio y Medio Ambiente" to encourage grazing in forest areas. The lack of extensive livestock farming can be explained by its low profitability when compared to intensive agriculture, the lack or dispersion of pastures and the lack of infrastructure.

Areas where the activity could be developed: Theoretically, 55% of the area (11,025 ha) could be use for extensive sheep or goat livestock farming. However, a large proportion of this area corresponds to pine forest with low grazing yields, while some of the areas with better suitability for livestock farming correspond to spaces of high environmental value, where the uncontrolled presence of livestock, especially goats, could be harmful (Fig. 3.21).

Necessary facilities: Extensive livestock farming requires the availability of pastures, watering holes or drinking troughs, sheepfolds and cattle trails.

Economic impact: The profitability of livestock farming in the Valencian Region is either zero or negative. Its continuity would only be feasible if the selling prices increase in recognition of the higher quality and ecological value of the products (meat) and if the production costs fall as a consequence of the improvement of the production processes and the generation of economies of scale.

Environmental impact: Grazing can contribute to the management and cleaning of the mountains but, if uncontrolled, it could damage the vegetation cover, especially if the livestock includes goats.

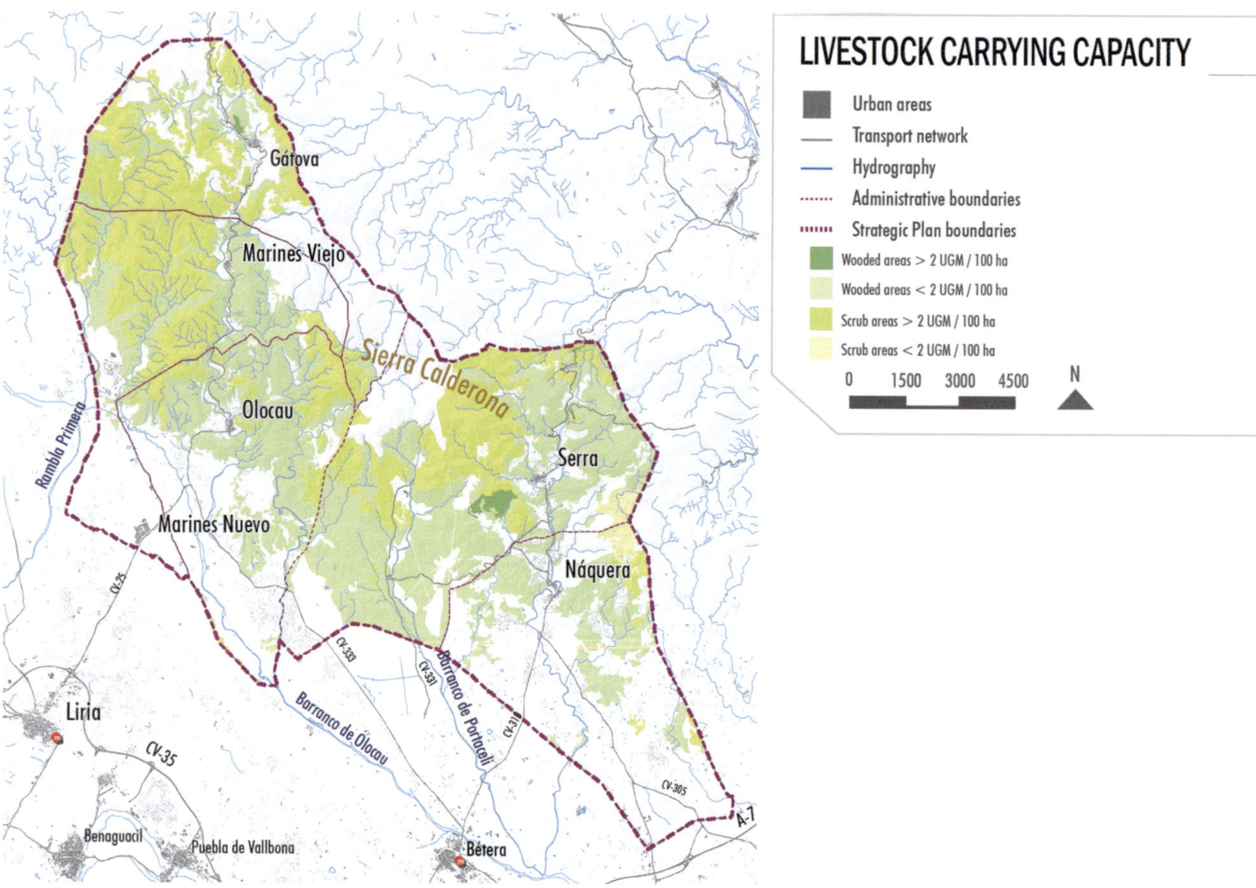

Fig. 3.21 Livestock carrying capacity (1 UGM = 1 livestock unit (10–20 sheep or goat heads)). *Source* Own elaboration based on PATFOR, Vaersa

3.3.5.2 Intensive Livestock Farming

Its presence in the area of the Sierra Calderona Strategic Plan is irregular. In Náquera, there are six farms dedicated to beef cattle, two to sheep, two to goats and one to horses; in Marines, there is a pig farm, and in Gátova one of rabbits.

3.3.6 Public Participation Plan

See Table 3.10.

Table 3.10 Survey results. Public Participation Plan_Phase 1: agriculture and livestock farming. *Source* Own elaboration

Rainfed agriculture	Total (%)	GÁTOVA (%)	OLOCAU (%)	MARINES (%)	NÁQUERA (%)	SERRA (%)
Conflicts						
Abandonment of crops	37.4	64.3	48.0	65.2	29.8	36.1
Not enough yields	22.2	42.9	32.0	34.8	19.1	19.4
Prohibitions	9.9	14.3	4.0	8.7	12.8	16.7
Loss of valuable crops such as cherry trees, almond trees and vines	8.8	14.3	16.0	8.7	4.3	13.9
Poor access infrastructure	6.4	28.6	0.0	13.0	0.0	8.3
Lack of subsidies	6.4	7.1	0.0	4.3	10.6	11.1
None	5.8	14.3	8.0	4.3	8.5	2.8
Personal consumption > not enough production to commercialised	5.3	21.4	4.0	8.7	2.1	5.6
Lack of food processing industries in the area	5.3	0.0	16.0	13.0	2.1	2.8
Do not know/no answer	42.1	14.3	32.0	13.0	40.4	47.2
Opportunities						
To promote a productive landscape	11.1	7.1	16.0	0.0	12.8	19.4
To increase subsidies	11.1	14.3	0.0	4.3	14.9	25.0
To subsidy those aiming to start to cultivate	6.4	0.0	4.0	4.3	6.4	16.7
To encourage not to abandon the crops	6.4	28.6	0.0	8.7	4.3	8.3
To create oil mills. wineries and agriculture cooperatives in the area of the plan	4.1	0.0	0.0	8.7	6.4	5.6
Do not know/no answer	76.6	78.6	80.0	69.6	72.3	69.4
Irrigated agriculture	**Total** (%)	**GÁTOVA** (%)	**OLOCAU** (%)	**MARINES** (%)	**NÁQUERA** (%)	**SERRA** (%)
Conflicts						
Low yields: low prices. market competition	15.2	7.1	16.0	43.5	19.1	5.6
Personal consumption > not enough production to commercialised	12.9	35.7	12.0	17.4	6.4	19.4
Abandonment of crops	12.3	7.1	16.0	21.7	8.5	13.9
None	8.8	28.6	0.0	13.0	14.9	2.8
Water restrictions (irrigation). water is expensive	7.6	0.0	4.0	8.7	10.6	13.9
Low incomes	6.4	7.1	0.0	21.7	8.5	2.8
Monoculture citrus trees	5.3	0.0	8.0	4.3	12.8	0.0
Natural Park restrictions	4.1	0.0	0.0	0.0	12.8	2.8
Lack of cooperativism	2.9	0.0	0.0	0.0	6.4	5.6
Do not know/no answer	42.7	21.4	44.0	13.0	34.0	47.2
Opportunities						
None	6.4	50.0	4.0	13.0	0.0	0.0
To manage from "Comunidades de regantes" and "Consejo Agrario"	4.1	0.0	0.0	0.0	6.4	11.1

(continued)

Table 3.10 (continued)

Rainfed agriculture	Total (%)	GÁTOVA (%)	OLOCAU (%)	MARINES (%)	NÁQUERA (%)	SERRA (%)
To shield and promote self-consumption	2.3	0.0	0.0	8.7	2.1	2.8
To produce following ecological standards	2.3	0.0	0.0	0.0	6.4	2.8
To professionalise agriculture. To merge plots	1.8	0.0	4.0	0.0	4.3	0.0
To export to new markets	1.2	0.0	4.0	0.0	0.0	2.8
To introduce new kinds of crops	1.2	0.0	0.0	0.0	2.1	0.0
To create a system to hand over land	0.6	0.0	0.0	0.0	2.1	0.0
Do not know/no answer	81.9	50.0	88.0	78.3	80.9	83.3

3.3.7 Synthesis and Diagnosis

Agricultural land: This accounts for 20% of the area of the Sierra Calderona Strategic Plan and could be divided into two main categories: mountain agriculture, mainly represented by rainfed crops (olive, almond, carob trees) in a very strong recession; and lowland agriculture, mainly irrigation crops (mostly citrus trees, as they represent 52% of the total agricultural land).

Abandonment of soils with high agricultural fertility: Urbanisation, especially during the last decades, has mainly developed on soils with a medium, high or very high agricultural fertility.

Economic possibilities of agriculture: The study of the profitability of different crops in the area, set between €2,000/ha for citrus trees and €600/ha for olive trees, allows us to conclude about the feasibility of professional agriculture. However, agriculture should be opened to more profitable crops, and production, processing and marketing activities should be optimised. These would provide the products with an added value and would make them recognisable as high-quality products. It has also been observed that currently there are neither facilities dedicated to the production of processed agricultural products, nor proper initiatives for the reintroduction of profitable crops, such as cherry trees.

Livestock farming: Extensive livestock farming has virtually disappeared from Sieera Calderona. However, its importance in the management of the mountains supports its reintroduction in duly confined areas. Such reintroduction would only be feasible if combined with a strong marketing strategy and proper processing of the products (meat, sausages, cheese, etc.).

3.4 Urban Environment and Well-Being

3.4.1 Generalities

The study of urban areas, residential uses, public facilities, services and industry in the area covered by the Sierra Calderona Strategic Plan suggests that:

There are three kinds of residential urbanised areas (Figs. 3.22, 3.23 and Tables 3.11, 3.12):

- **Compact and multifunctional villages and towns (residential, services, commerce, public facilities)**: the villages of Gátova, Olocau, Marines Viejo, Marines Nuevo, Serra and Náquera.
- **Housing estates of low/medium density and mono-functional character (residential)**: either annexed to towns (especially vast in Náquera) or distributed among the agricultural-forest matrix (especially frequent in the southern plains and light slopes south of Náquera and Olocau and in the south-western part of Serra).

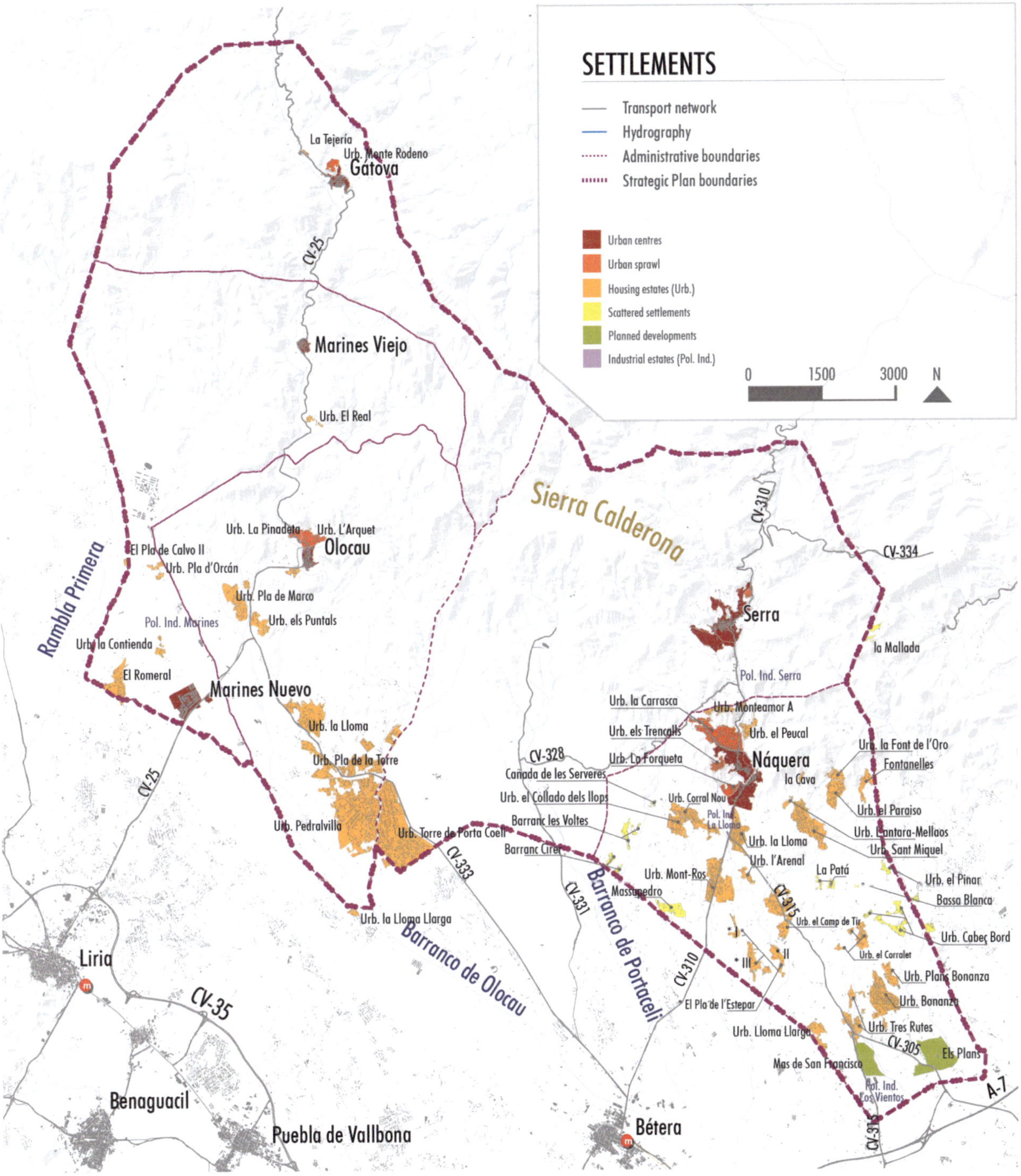

Fig. 3.22 Villages and towns, housing estates and scattered settlements. *Source* Own elaboration

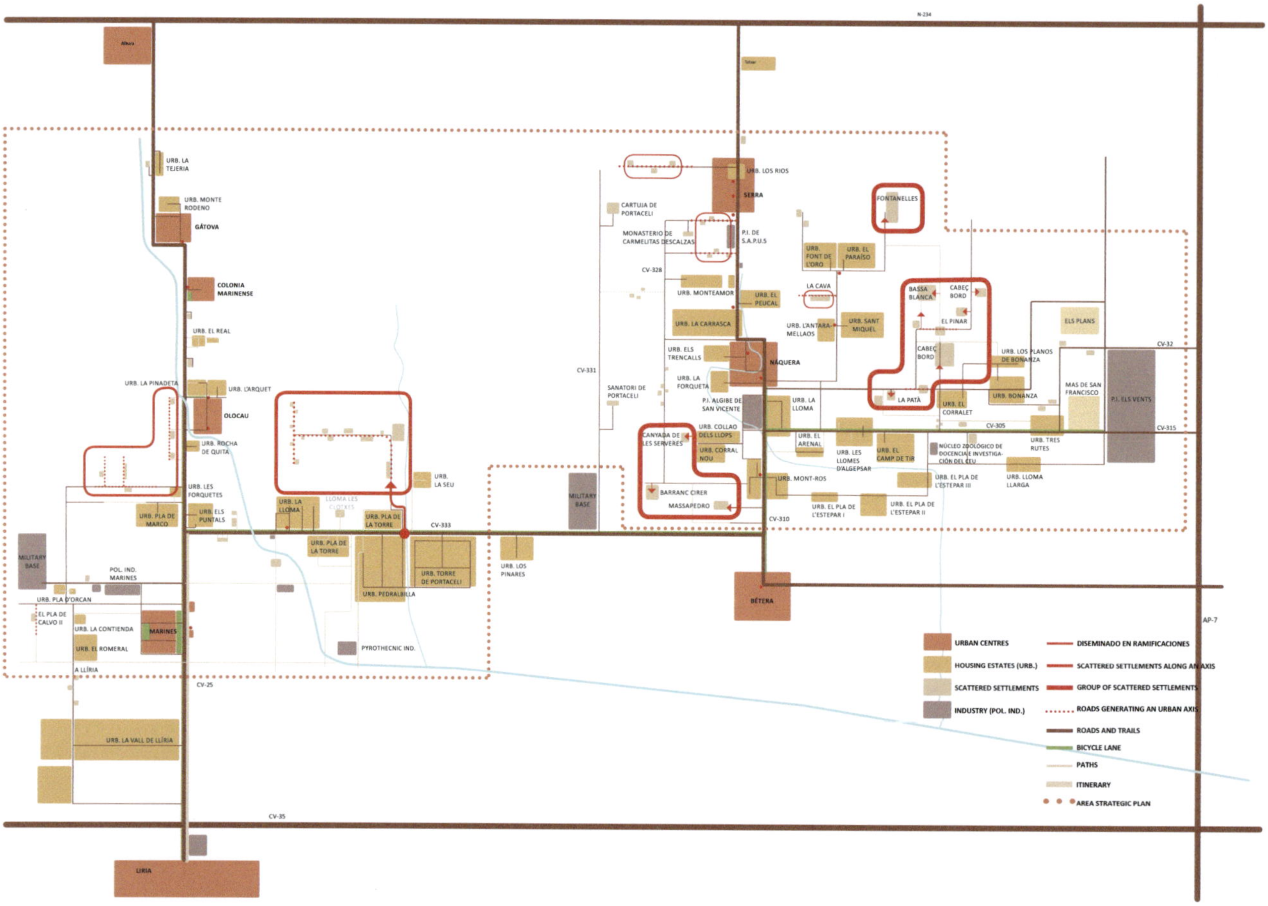

Fig. 3.23 Territorial model (without scale). *Source* Own elaboration

Table 3.11 Percentage of land in towns (urban centres), housing estates, scattered settlements and industrial areas in each of the municipalities. *Source* Own elaboration

	Villages and towns		Housing estates		Scattered settlements		Industrial	
	Area (ha)	(%)	Area (ha)	(%)	Area (ha)	(%)	Area (ha)	(%)
TOTAL	237.1	1.19	799.48	4.01	42.60	0.21	107.20	0.54
NÁQUERA	58.5	1.51	423.02	10.93	42.60	1.10	92.20	2.38
SERRA	66.7	1.16	13.36	0.23	0.00	0.00	5.40	0.09
MARINES	90.5	2.53	57.6	1.61	0.00	0.00	9.60	0.27
OLOCAU	10.2	0.27	299.1	8.00	0.00	0.00	0.00	0.00
GÁTOVA	11.2	0.37	6.4	0.21	0.00	0.00	0.00	0.00

Table 3.12 Data from towns (urban centres) and housing estates (numbers of dwellings, permanent and temporary inhabitants, educational/sport/health/cultural facilities). *Source* Own elaboration

	Name	No. dwellings	Permanent population (no. INHAB)	Seasonal population (no. INHAB)	Population ratio (permanent/seasonal)	No. Educational centres	No. sports centres	No. health centres	No. cultural centres
TOTAL		11,675	13,198.3	40,047	32.96	14	90	8	17
NÁQUERA		5,540							
	Urban centre		2,990	8,970	33.33	3	36	2	2
	Housing estates		2,910	11,640	25.00	0	10	0	0
SERRA		2,337							
	Urban centre		2,351	7,053	33.33	4	13	1	2
	Housing estates		998	3,992	25.00	0	1	1	0
MARINES		875							
	Urban centre		1,581	2,055	76.92	4	12	1	6
	Housing estates		251	628	40.00	0	0	0	0
OLOCAU		2,180							
	Urban centre		599	1,557	38.46	2	13	1	4
	Housing estates		918	2,295	40.00	0	2	1	0
GÁTOVA		743							
	Urban centre		458	1,145	40.00	1	1	1	3
	Housing estates		142.3	712	20.00	0	2	0	0

- **Scattered settlements**: groups of dwellings, often illegal or unregulated, with non-existent or very precarious infrastructures (roads, sewage, lighting, etc.). Some of the settlements included in this category have become part of the second category by adding new growth and by unevenly improving the existing infrastructures.

There are two kinds of industrial areas:

- **Industrial estates**: of small size, usually located in urban fringes. The industrial estate "Los Vientos" is the main exception due to its large size and remote location, but it is still connected to the AP7 motorway.
- **Scattered industry**: isolated, mainly located along the main or secondary roads. The pyrotechnic industry, located in the south of the municipality of Olocau, stands out because of its importance.

Dwellings used as secondary residences are mainly located in housing estates or in scattered low-/medium-density settlements. They cover a very significant portion of the urbanised land, conferring to the territory a marked seasonality, since they are mainly used in summer or during weekends.

The location of housing estates and scattered settlements is highly determined by the following factors:

- Proximity to motorways, rural roads, villages or towns
- Views
- Prior existence of scattered settlements.

Regarding public services and considering the number of permanent residents in each one of the five municipalities included in this Plan, it can be concluded that the existing public facilities are appropriate and adequate. However, the following points should be noted:

- The sharp increase in the population during the summer results in temporary insufficiencies in some services and facilities.
- The location of public facilities in villages and towns, or in their periphery, provides services to the permanent population, though they remain remote from the housing

estates and the temporary residents. The lack of public facilities in those areas has been partially compensated through the creation of private facilities that, from the perspective of the temporary residents, should be jointly managed and financed by the municipal administrations).

The vertiginous urban development in recent decades has generated an inflated forecast for new housing that does not follow any sustainable pattern, since it consists mainly of monofunctional settlements with single-family houses.

3.4.2 Villages and Towns, Housing Estates and Scattered Settlements

For analysis and diagnostic purposes, villages or towns and housing estates/scattered settlements are studied separately, though this division causes a certain overlap when discussing housing estates attached to towns.

3.4.2.1 Villages and Towns

The historical towns of Gátova, Olocau and Marines Viejo are configured as a fabric of traditional compact blocks, while in Marines Nuevo, a late example of a village of colonisation, the fabric is defined by rows of single-family patio-houses.

In Náquera and Serra, from an a priori historical centre with the same characteristics as the previous ones (in terms of surface, typologies, etc.), there has been an expansive growth of monofunctional residential typologies of detached and semi-detached houses that has ended up enclosing their historical cores. Their disposition has mainly been determined by landform and the proximity of major roads (Fig. 3.24 and Tables 3.13, 3.14).

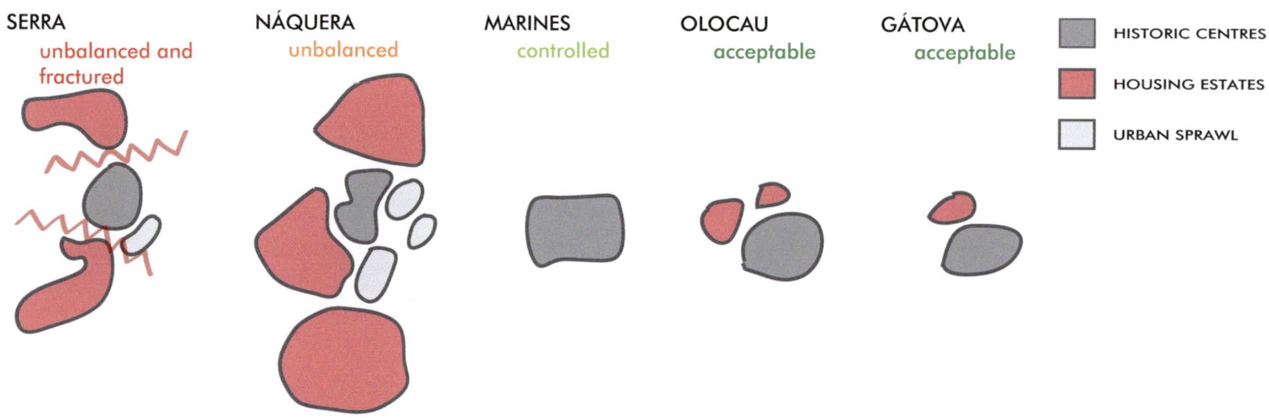

Fig. 3.24 Schematic structure of Sierra Calderona's towns. *Source* Own elaboration

Table 3.13 Land distribution in each town (total, built, free public space, free private space). *Source* Own elaboration

	Urbanised land	Built		Public open space		Private open space	
	Area (m²)	Area (m²)	% total	Area (m²)	% total	Area (m²)	% total
Total	1,833,051.70	521,135.09	28.43	636,463.82	34.72	675,452.79	36.85
NÁQUERA	585,084.00	144,600.00	24.71	177,640.00	30.36	262,844.00	44.92
SERRA	650,107.06	138,249.43	21.27	229,958.66	35.37	281,898.97	43.36
MARINES VIEJO	58,684.41	25,472.45	43.41	19,557.26	33.33	13,654.70	23.27
MARINES NUEVO	315,582.72	110,883.94	35.14	135,569.28	42.96	69,129.50	21.91
OLOCAU	101,558.51	52,823.68	52.01	29,627.60	29.17	19,107.23	18.81
GÁTOVA	122,035.00	49,105.59	40.24	44,111.02	36.15	28,818.39	23.61

Table 3.14 Public facilities in each municipality. *Source* Own elaboration

	NÁQUERA	SERRA	MARINES	OLOCAU	GÁTOVA
Health—social services					
Heath centre	YES	YES	YES	YES	YES
Residence for the elderly	YES				
Hospital		YES			
Administrative-institutional					
Town hall	YES	YES	YES	YES	YES
Magistrate's court	YES				
Sports					
Court wall	YES				YES
Public swimming pool	YES	YES	YES	YES	YES
Public football pitch	YES				YES
Public sports centre	YES		YES	YES	
Public sports outdoor area	YES	YES			
Public sports area	YES	YES	YES	YES	
Bullfighting arena				YES	
Firing range	YES				
"Tentadero" (informal bull fighting)	YES				
Equestrian centre	YES				
Security					
Local police	YES	YES	YES		
Military base			YES		
Civil protection	YES				
Socio-cultural					
Civic centre					YES
Multifunctional					YES
Municipal agency for reading		YES	YES	YES	
Day-care centre for the elderly	YES			YES	
Tourist information	YES	YES		YES	
Environmental education office		YES			
Telecentre				YES	
Cultural space	YES			YES	
Information centre for the youth	YES		YES		
Women's centre			YES		
Public theatre			YES		
Public library	YES				
Educational					
Public school	YES	YES	YES	YES	YES
Farm school				YES	
Education centre for the youth				YES	
Public nursery	YES	YES	YES		
Adult school		YES	YES		
Music school	YES	YES	YES		
University centre (teaching/research)	YES				

(continued)

Table 3.14 (continued)

	NÁQUERA	SERRA	MARINES	OLOCAU	GÁTOVA
Green areas–parks–open spaces					
Park			YES		YES
Infrastructure					
Wastewater treatment plant	YES		YES	YES	YES
Waste collection point	YES				
Religious					
Cemetery	YES	YES	YES	YES	YES

Building Types:

In Gátova, Olocau and Marines Viejo, more than 80% of buildings are traditional terraced or attached single-family dwellings which define continuous lines and compact closed blocks. In Gátova and Olocau, some of these traditional buildings have been replaced by multi-family residential buildings with a greater number of storeys, increasing the density of population and changing the visual character of the affected locations, as well as generating poorly finished dividing walls and breaking the street width-building height ratio.

In those three villages, there is a strong mix of functions and land uses (residential, tertiary and public facilities).

In Marines Nuevo, and following the general pattern used in many rural Spanish villages of colonisation from the 1950s and 1960s, terraced housing with backyards is the predominant building type. According to the strict building regulations of the village, the same type has been used in recent buildings. Densification has occurred through the subdivision of the initial patio-houses and the construction of building extensions, as the backyards lost their agricultural function. In relation to land uses, there is a balanced mixture of functions as envisioned in this prototypical example of completely planned villages.

The evolution of Serra's historical centre is similar to that of Gátova and Olocau, but differs in that in Serra, the historical and compact core represents only 20% of its total urban area, since most of its recent development has taken place in low-density areas with single-family houses (semi-detached and detached) located expansively and defining a diffuse urban edge. The centre and, partially, its surrounding developments have a multifunctional character, while the larger area of single-family houses is monofunctional residential. Public facilities (sports, education, etc.) could eventually operate in this case as connectors between the two kinds of fabrics, but, so far, this possibility has not yet been materialised, due to the absence of an intermediate multifunctional fabric.

In Náquera's historical centre, the replacement of traditional houses with multi-storey residential buildings has been especially common, producing the aforementioned effects on density and street section. At the same time, the historical core has been surrounded by a monofunctional residential fabric, mostly of detached and semi-detached houses, but also of multi-family residential blocks with commercial functions on the ground floor. These developments continue now along the CV310 road, generating a continuous and linear built-up fabric that eventually might reach other urbanised areas.

This linear growth along the CV310 road could be attenuated by generating denser developments in some other strategic locations, by using the system of articulated free spaces located along Náquera's ravine more purposefully and by taking advantage of some scattered public services and facilities (educational, cultural and sport). In addition, the CV310 road also has potential to be transformed into a boulevard and there is a pocket of strategic and developable land available that was delimited after the construction of a bypass road (See spatial structure of towns and villages and spatial distribution of building types and public facilities in Figs. 3.25, 3.26, 3.27, 3.28, 3.29 and 3.30) (See percentual distribution of land in towns and villages, and list of public facilities in Tables 3.13 and 3.14).

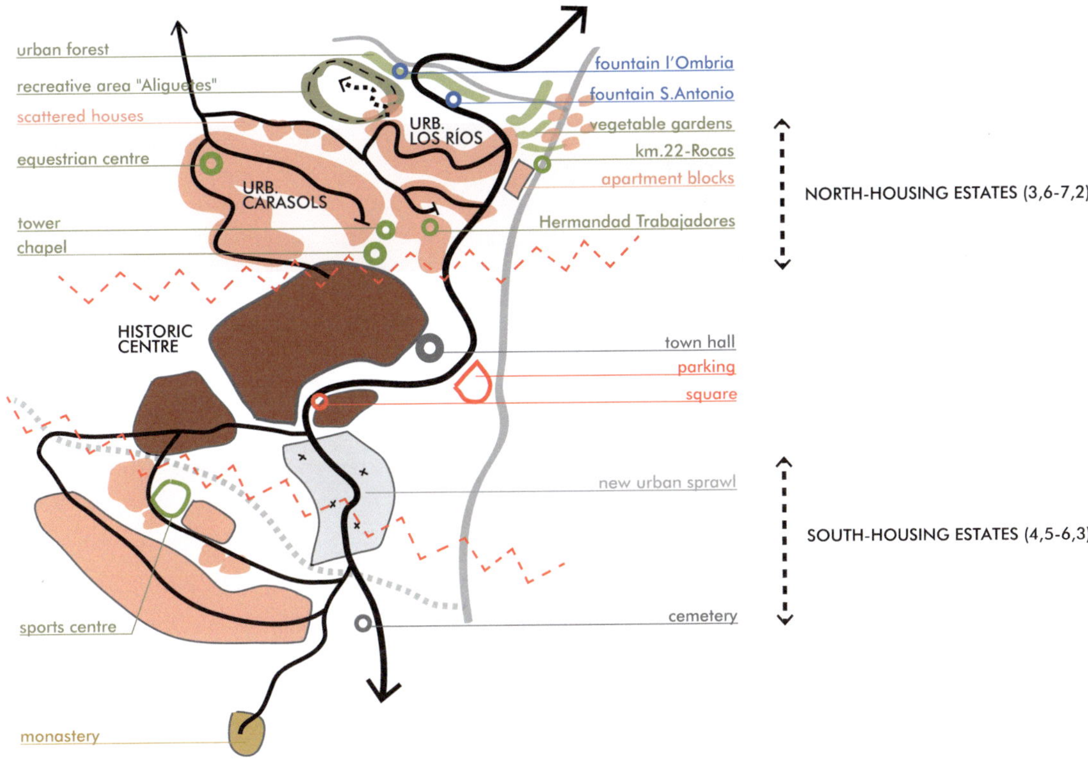

Fig. 3.25 Schematic structure of Serra. Source Own elaboration

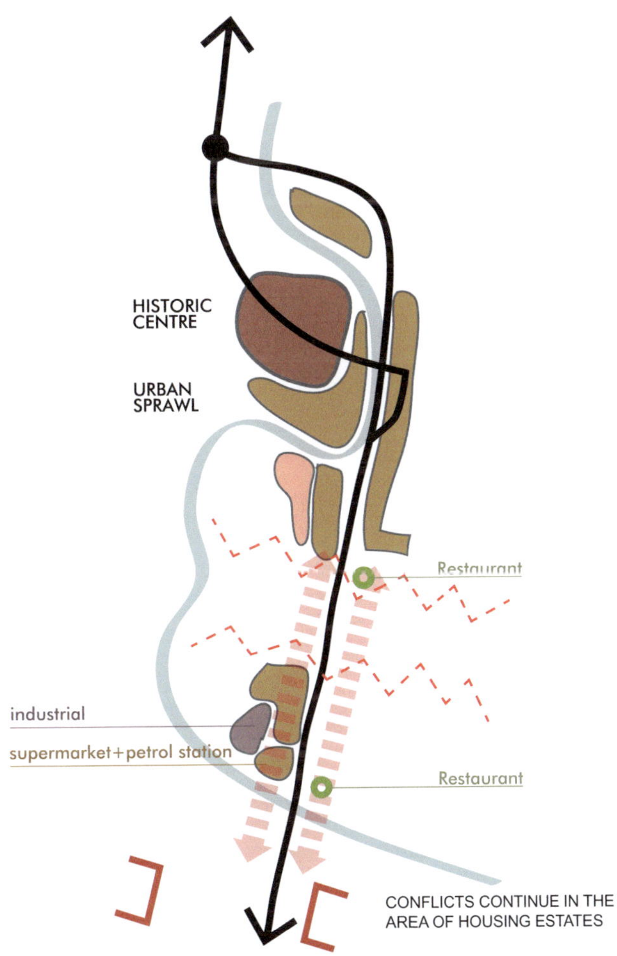

Fig. 3.26 Schematic structure of Náquera. *Source* Own elaboration

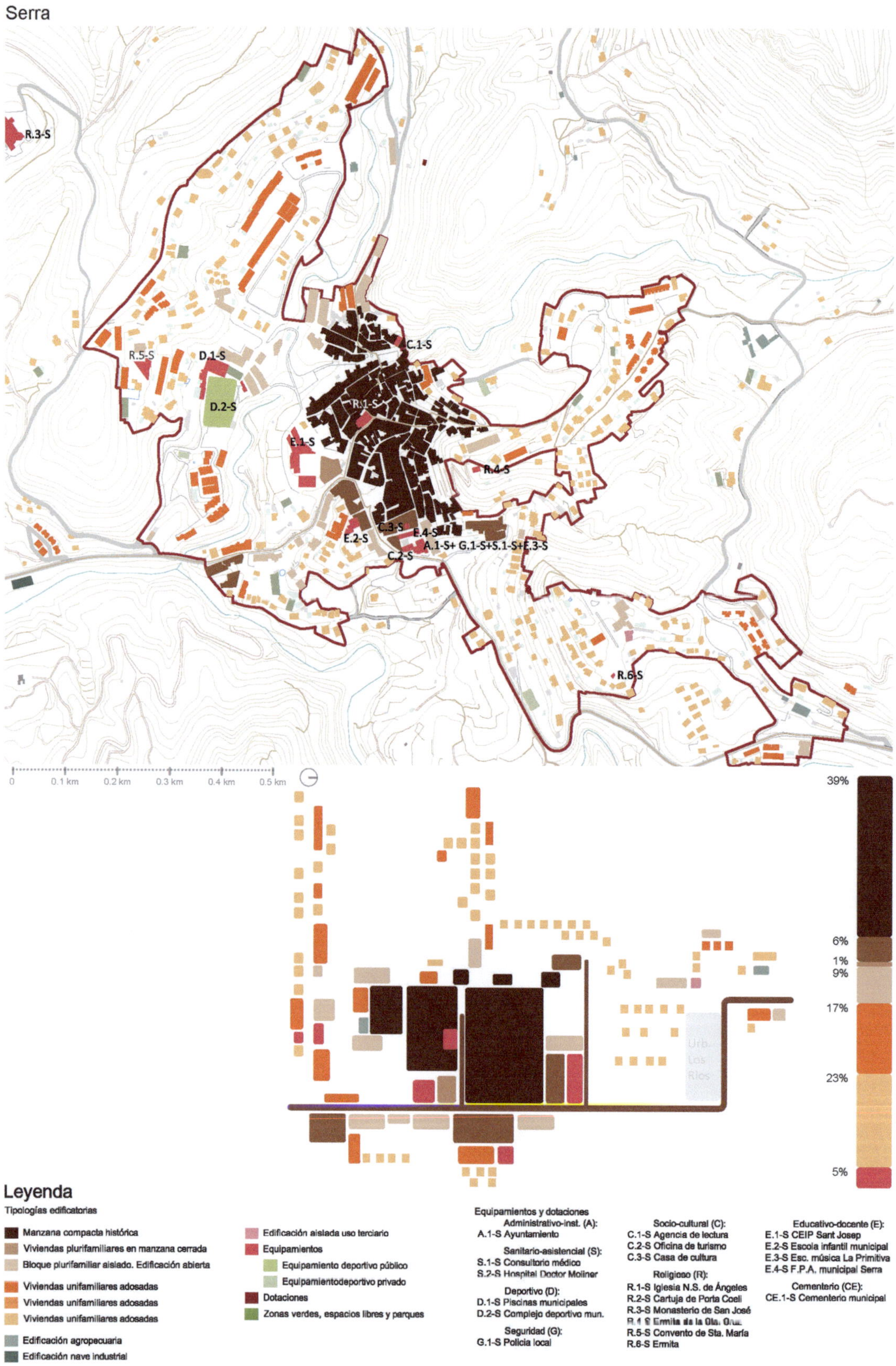

Fig. 3.27 Building types and public facilities in the town of Serra. *Source* Own elaboration

Fig. 3.28 Building types and public facilities in the town of Náquera. *Source* Own elaboration

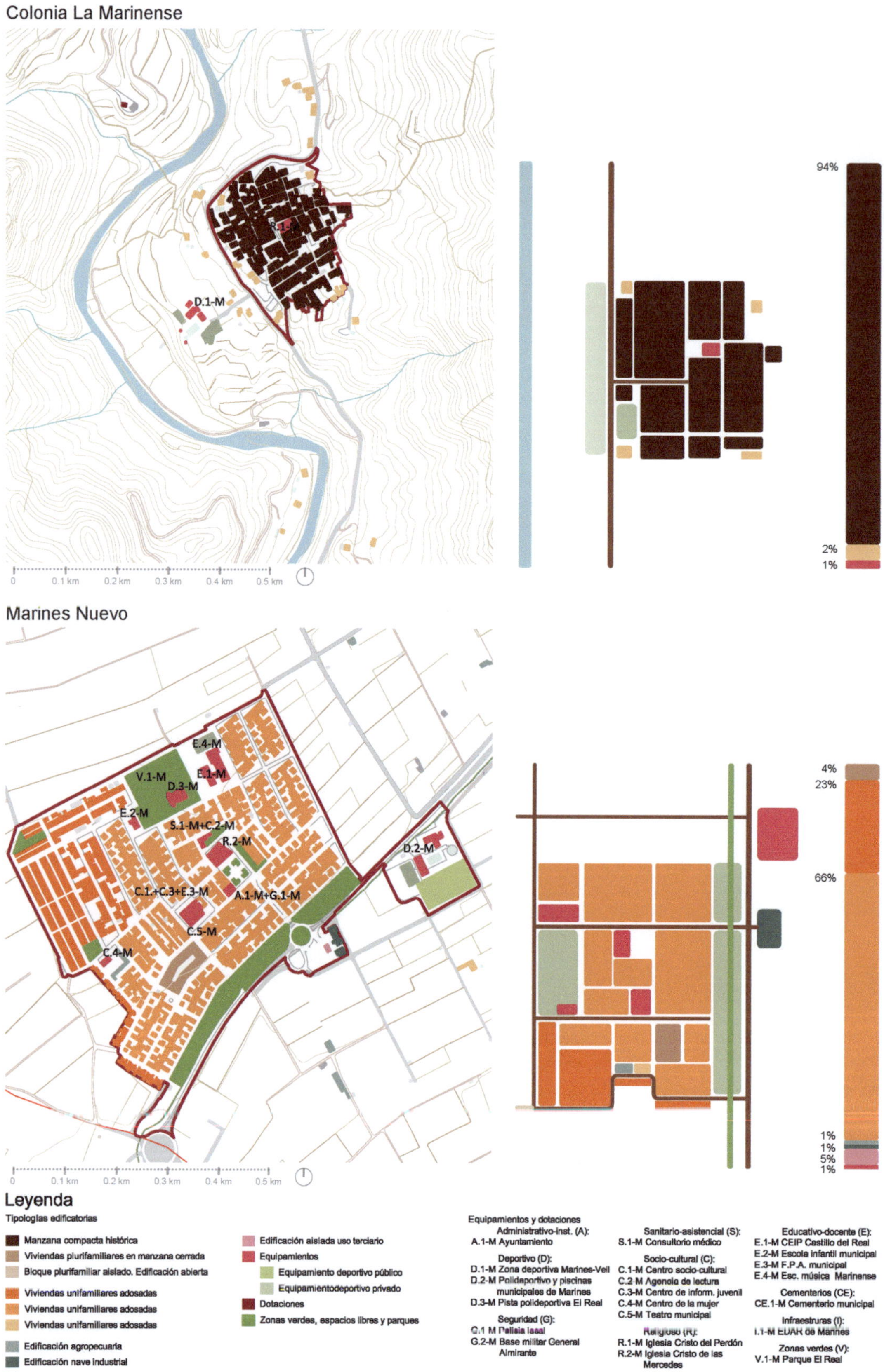

Fig. 3.29 Building types and public facilities in the towns of Marines Viejo (Colonia La Marinense) and Marines Nuevo. *Source* Own elaboration

Fig. 3.30 Building types and public facilities in the towns of Gátova and Olocau. *Source* Own elaboration

3.4.2.2 Urban Public Space

Gátova and Olocau have a very similar network of public spaces, with most of their public activity concentrated along the old road and around the town hall square. In both cases, the road has been treated correctly. The largest facilities are usually located at the urban fringe (e.g. sports fields) and are not well connected with the main network of public spaces.

In Marines Viejo, the public space has been poorly maintained. The village has two main poles of attraction: "Plaza de España" and the sports complex, which are not well connected with each other.

Marines Nuevo has a well-planned and maintained network of public spaces. The most important are the main square and the linear park that separates the town from the CV25 road. The sports complex is the only facility not well connected, as it lies on the opposite side of the CV25 road.

In Serra, there are two kinds of situations matching the morphology already described. In the historical centre, public space is scarce but intensively used due to its compact character and to the invasive presence of parking spaces. Outside the historic centre, the situation is quite the opposite as a result of its low density. Serra's core, developed along its main street and the CV25 road, is strongly disassociated from the housing estates around it. Neither the public facilities nor the absence of an intermediate urban fabric have helped to resolve that problem. In spite of the potential of Serra's ravines to articulate a continuous public space network, the green areas remain disconnected in the town.

The densification of Náquera's historical centre has caused the visual quality of the streets to deteriorate and has given a dominant role to cars. The village has two main nodes, one to the north, where education and sports facilities are located, and one more institutional to the south. These nodes are poorly connected to each other. The urban fabric is discontinuous, being especially fragmented in the south, where it has developed along different secondary roads. Despite the diversity of building types, there is a lack of connecting urban tissue, and the green areas and public facilities are also poorly connected. Nonetheless, Náquera's ravine still has the potential to act as an articulator of the public space network (Figs. 3.31, 3.32, 3.33 and 3.34).

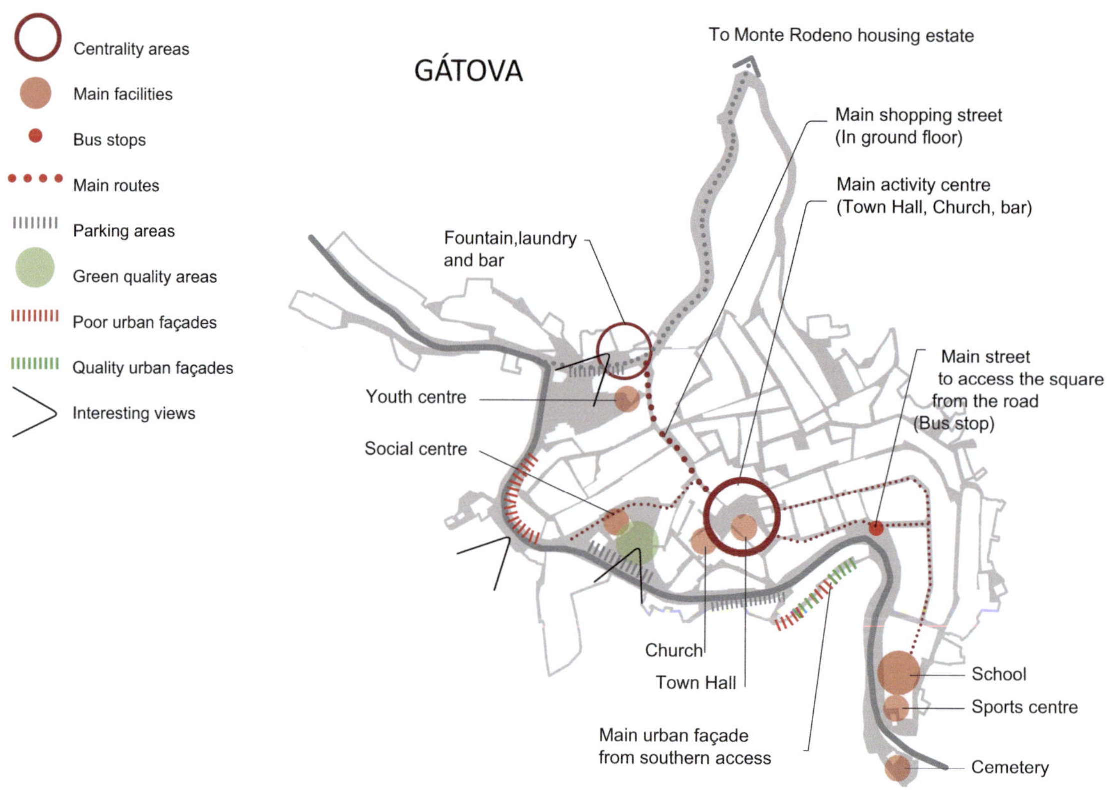

Fig. 3.31 Urban public space in Gátova. *Source* Own elaboration

Fig. 3.32 Urban public space in Olocau, Marines Viejo and Marines Nuevo. Source Own elaboration

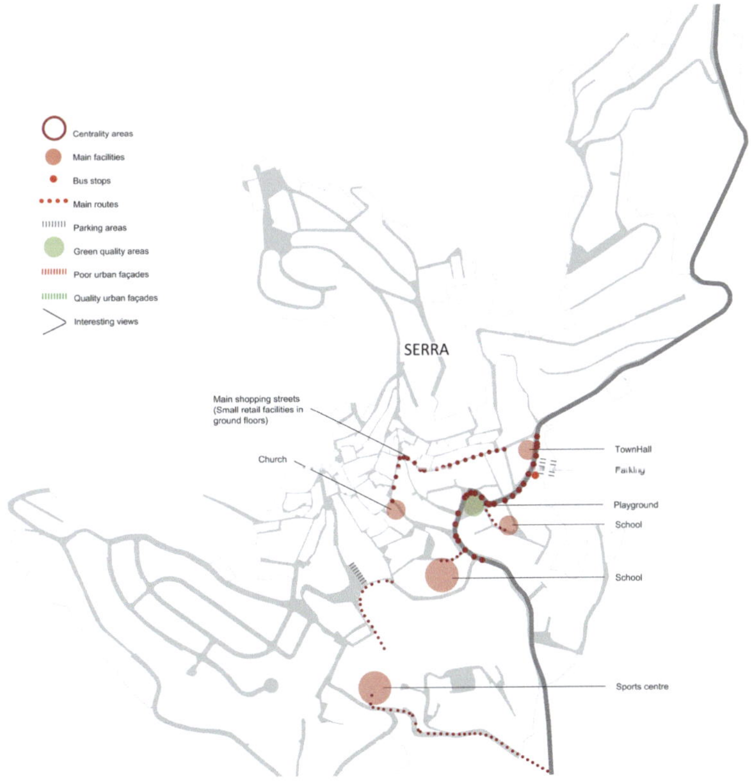

Fig. 3.33 Urban public space in Serra. *Source* Own elaboration

Fig. 3.34 Urban public space in Náquera. *Source* Own elaboration

3.4.2.3 Housing Estates and Scattered Settlements

These tissues represent a very important part of the urbanised land within the scope of the Plan, and during the high season and weekends, their population exceeds the population living in villages or towns (Figs. 3.35 and 3.36).

Building Types:

In general, every housing estate is structured around a network of streets, giving access to fenced private plots on both sides. The most common housing type is detached housing with private gardens, although in the most recent developments, semi-detached houses, and exceptionally multi-storey residential blocks, can also be found (see examples in Figs. 3.37, 3.38 and 3.39).

Many recently built housing estates stem from pre-existing scattered settlements, as can be appreciated when looking at the street layouts partially based on old rural tracks, and at the precarious to non-existent level of urbanisation (street surfacing, lighting, sewage, etc.). Although nowadays the construction of illegal housing in land designated as non-developable is strictly controlled, still some illegally built scattered settlements remain unregulated. They can be categorised into three main types:

- Scattered building units: dispersed single buildings in an agricultural or forest-based matrix.

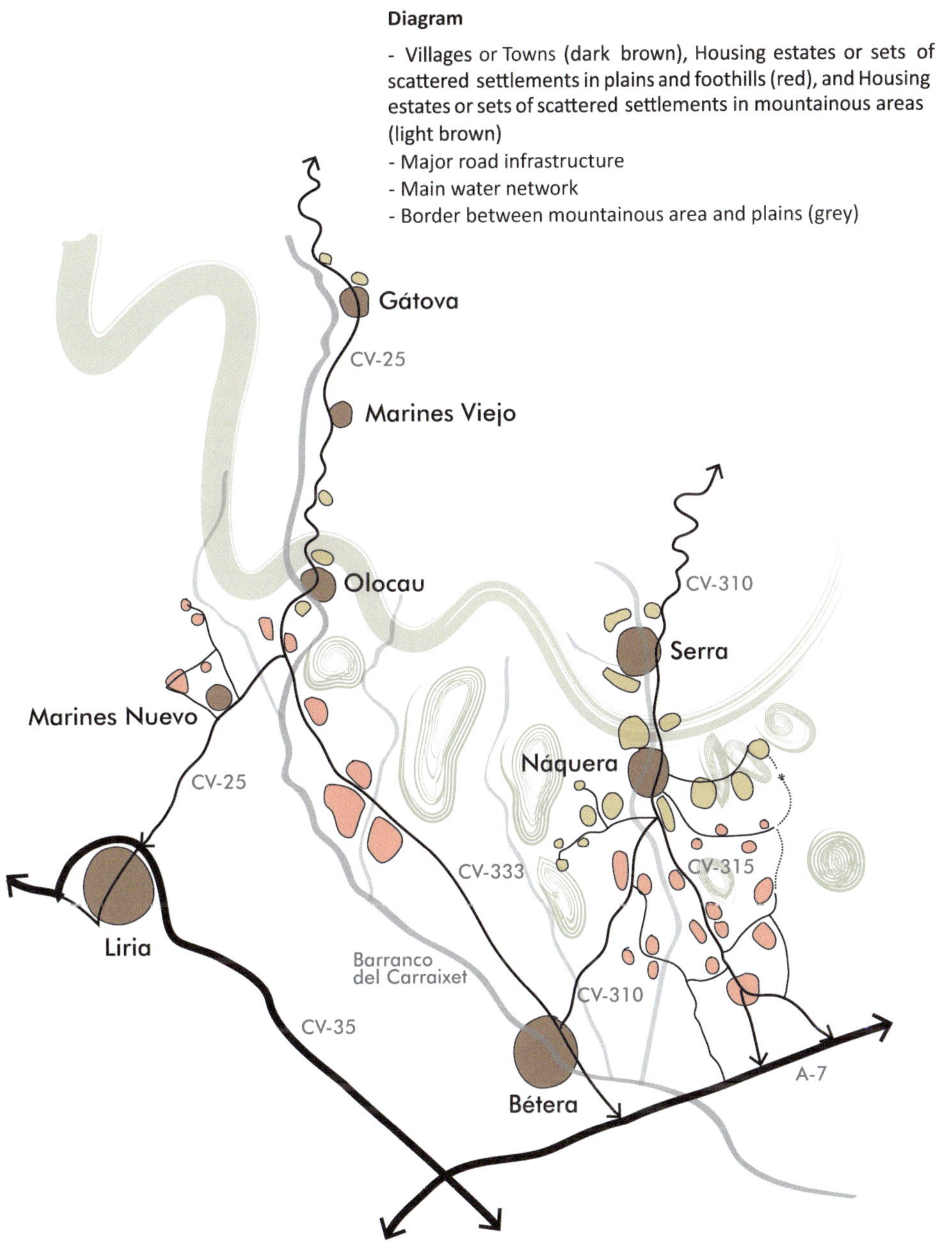

Fig. 3.35 Diagrammatic distribution of towns, housing estates and scattered settlements in the overall scope of the plan. *Source* Own elaboration

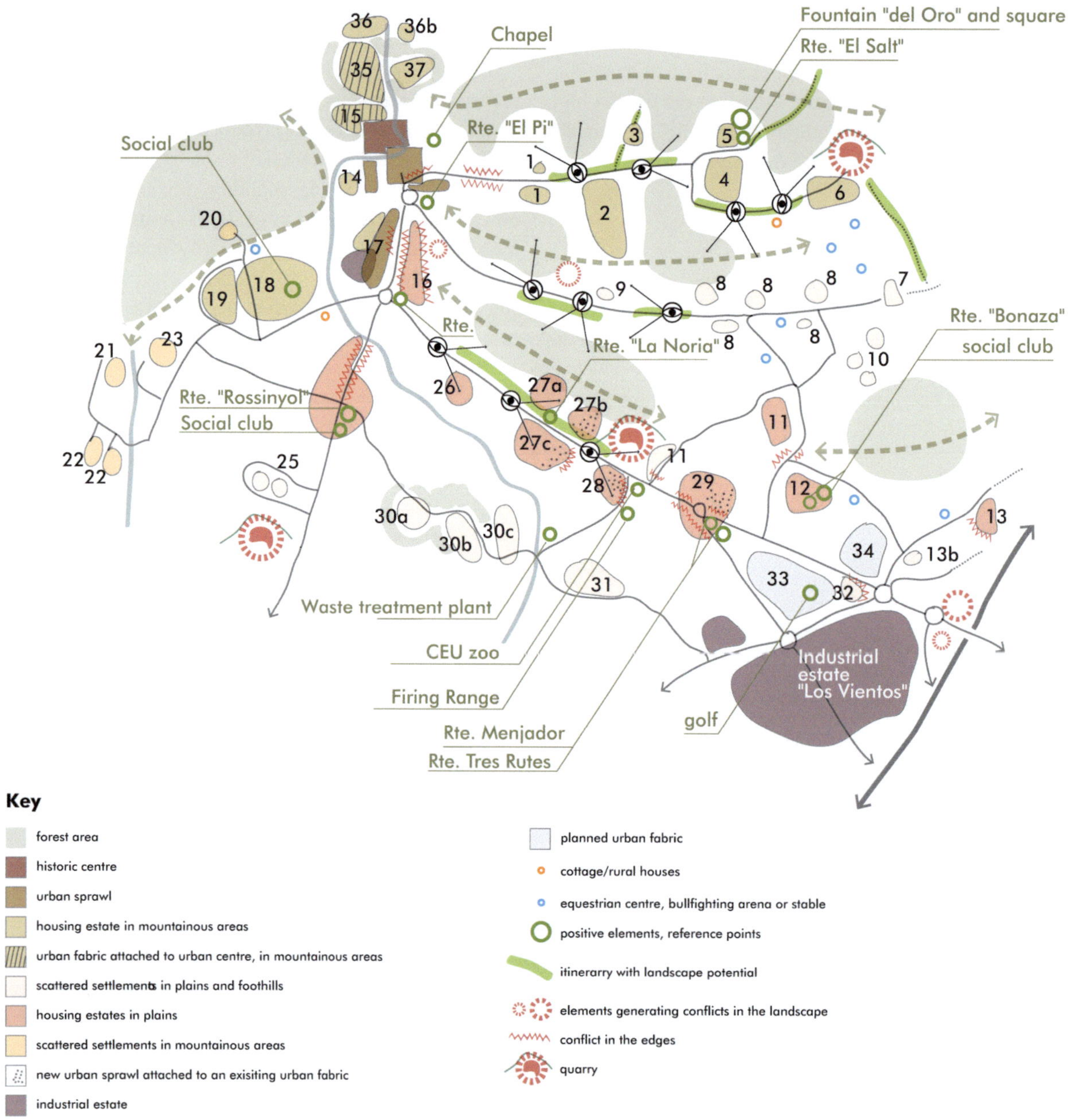

List of housing estates

1 L'Antara
2 San Miguel
3 La Cava
4 Paraíso
5 Fuente del Oro
6 Fontanelles
7 El Pinar
8 Bassa Blanca
9 La Patá
10 Cabeç Bord
11 Corralet
12 Bonanza

13 no name
14 Forqueta
15 Trencalls
16 Lloma
17 blocks/supermarket/ industrial
18 Corral Nou
19 Collado dels Llops
20 Cañada de les Serveres 21 Barranco de Les Voltes 22 Barraco Cirer
23 no name
24 Mont Ros

25 Massa Pedro
26 L'Arenal
27 La Lloma
28 Campo de Tiro. Nuevo, Viejo
29 Tres Rutas
30 Pla de L'Estepar I, II y III
31 La Lloma
32 Els Plans
33/34 Future projects: Mas de S. Francisco and Els Plans
35 La Carrasca
36 Monteamor
37 El Peucal

Fig. 3.36 Diagrammatic distribution of towns, housing estates and scattered settlements in the municipality of Náquera. *Source* Own elaboration

Fig. 3.37 Building types and public facilities in the housing estate of Torre Portaceli (Serra). *Source* Own elaboration

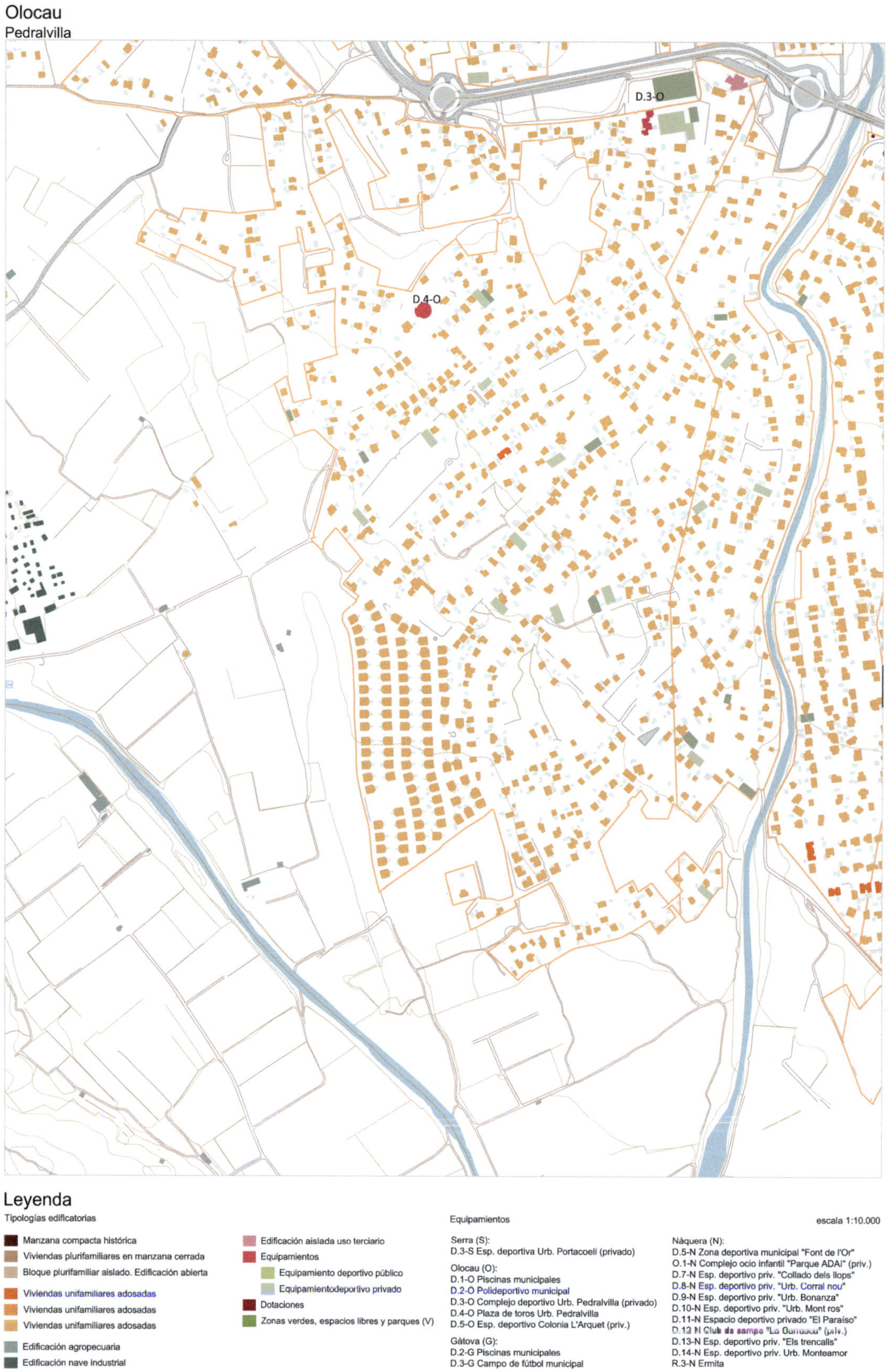

Fig. 3.38 Building types and public facilities in the housing estate of Pedralbilla. *Source* Own elaboration

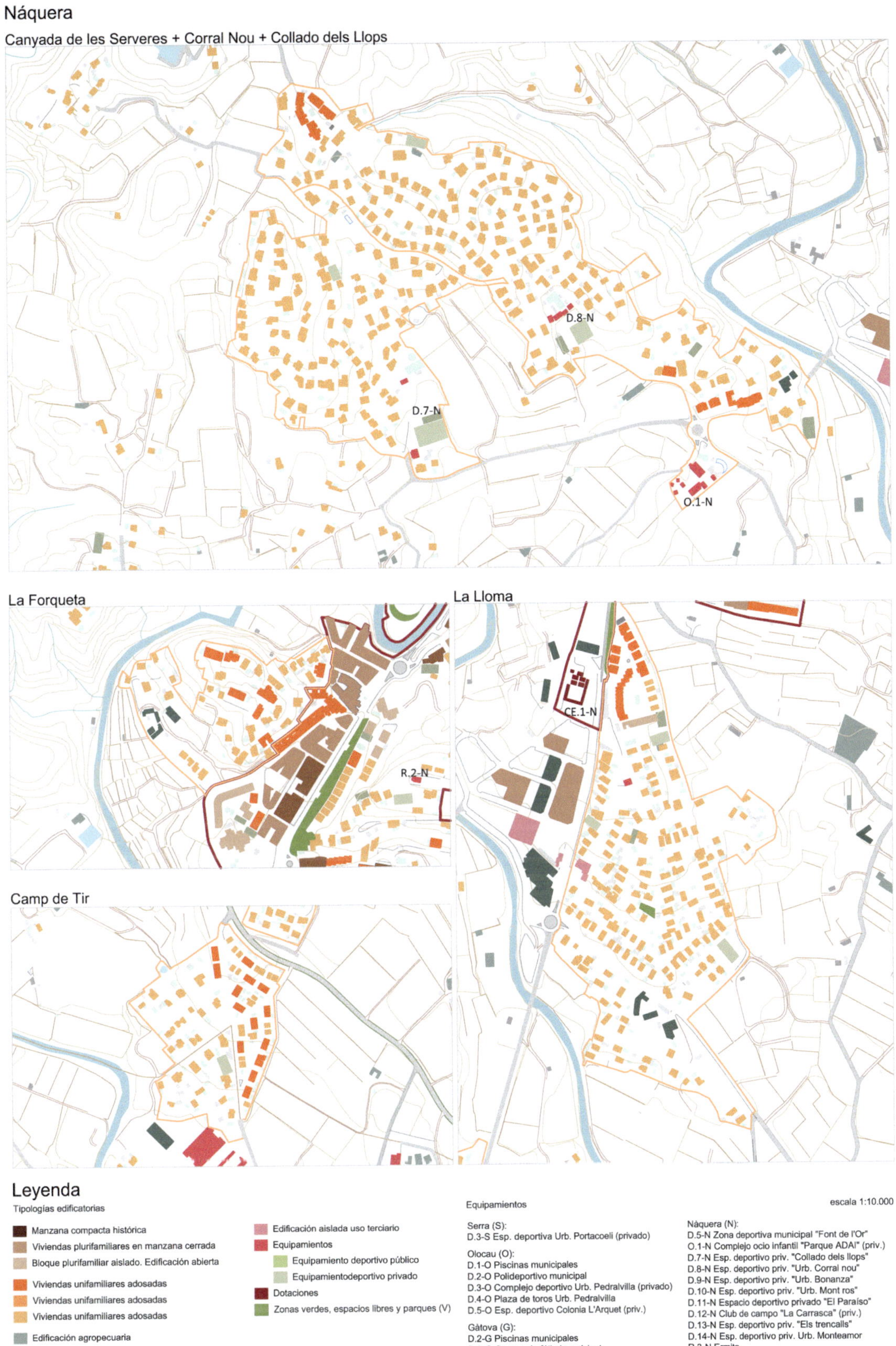

Fig. 3.39 Building types and public facilities in the housing estates of Canyada de les Serveres, Corral Nou, Collado dels Llops, La Forqueta, Camp de Tir and La Lloma (Náquera). *Source* Own elaboration

- Scattered buildings along rural roads: an intermediate stage between the previous and next.
- Scattered conglomeration: type generated by the ramification and aggregation of the above-mentioned types. This type usually relegates the agricultural and forest land use to an interstitial role.

The regularisation of some illegal and scattered settlements has been achieved in many cases by adding new built fabrics with the appropriate urbanisation conditions. This strategy, however, has not substantially improved the existing tissues, whose accesses, facilities and services often remain in poor condition.

In relation to land uses, housing estates are mostly monofunctional residential areas, with few specific spaces for tertiary use, such as gas stations and shops that only open during high season. This, together with the lack of public transport connections, makes private vehicles necessary for basic domestic supply. Public facilities are also scarce both due to the precarious origin of these developments and their seasonal use as secondary residences. In some of the larger and more consolidated estates, privately owned social clubs have been created.

Moreover, many of the estates lack proper sewage networks, and waste is collected with a varying frequency, depending on the season.

As the permanent population increases on housing estates, so does their aspiration to participate and be integrated in the general functioning of the municipalities. Nevertheless, the provision of new public facilities and services should consider their low density (as it increases the per capita costs for public services), strong seasonality and their distance to the towns or villages.

In scattered settlements, this situation is more critical, since the roads are precarious and facilities, sewage and public lighting often non-existent. The residential use is, in this case, usually combined with agricultural uses.

Public Space

As explained in the previous point, public space was not generally considered in the progressive growth of many of the existing housing estates, resulting in the absence of adequate sidewalks, public lighting, greenery or open spaces associated with services or public facilities.

The lack of regulation when fencing residential plots and the poor treatment of their edges has only aggravated the precarious situation of public spaces. These aspects will be discussed later on, in the subchapter "Landscape and territorial models" of the current chapter.

3.4.3 Industry

3.4.3.1 Location and Connections

In Gátova and Olocau, there are no industrial estates and industrial activity is developed in isolated facilities. The most important, due to its dimensions and singularity, is the fireworks factory located at the south of Olocau's municipal area. In Gátova, there is a small industrial estate within the area of influence of the Sierra Calderona Natural Park. In Serra, the industrial activity plays also a minor role and it is concentrated in the industrial estate SAPU5, located next to the village centre and occupying 0.09% of the municipal area. Marines has a larger industrial estate that covers 0.27% of the municipal land and where an important paint factory is located. The industrial activity is more significant in Náquera with two industrial estates, "Aljibe de San Vicente", located next to the town, and "Los Vientos", much larger, next to the AP7. Scattered industrial uses are scarce; the existing ones are located along the main roads and usually associated with the storage of building materials.

All the industrial estates in the area are well connected, given their proximity to the transport infrastructures, but only "Aljibe de San Vicente" (Náquera) is accessible by public transport (bus) (Fig. 3.40 and Table 3.15).

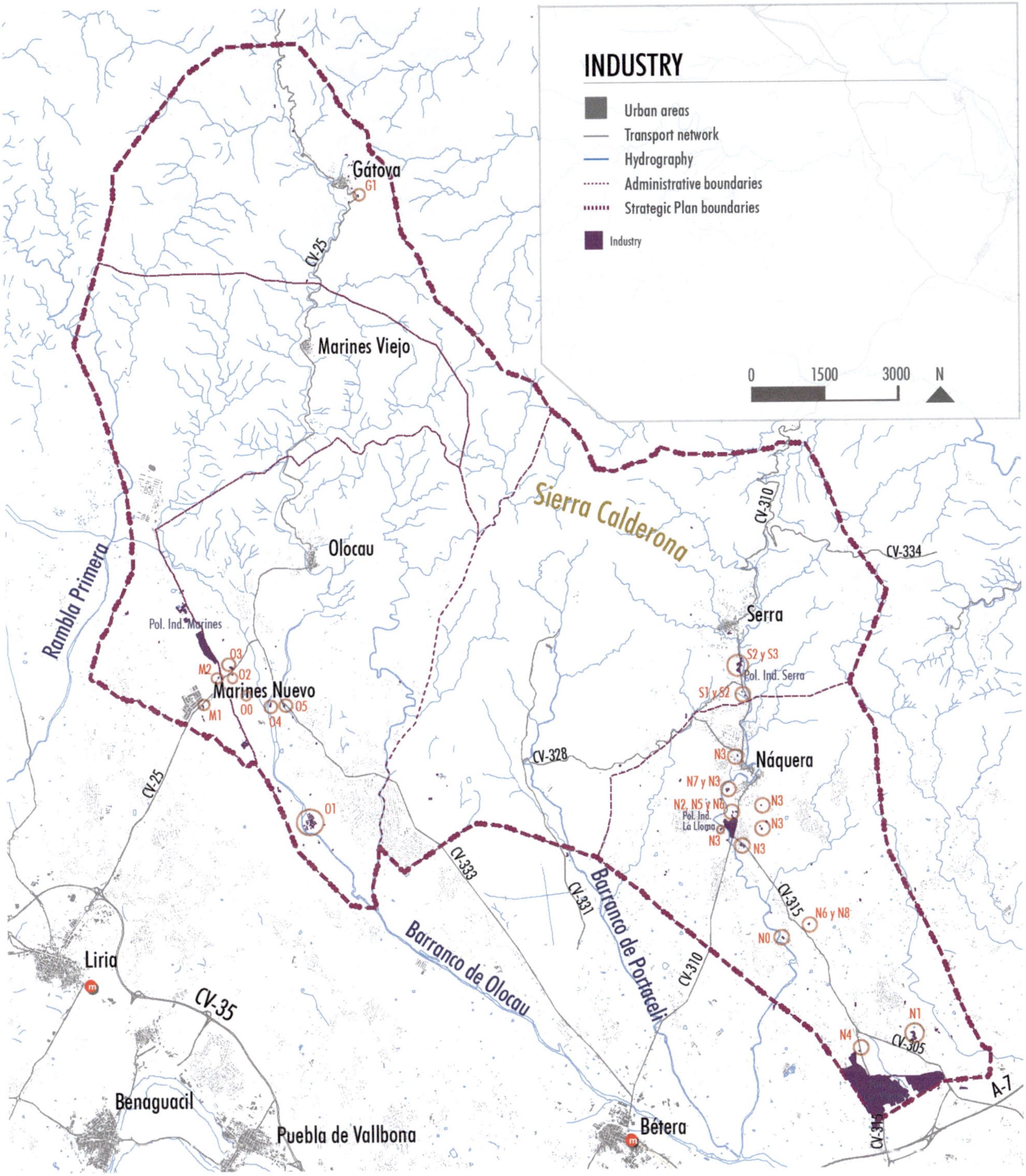

Fig. 3.40 Industrial areas. See codes in Table 3.15. *Source* Own elaboration

Table 3.15 Scattered industrial uses. *Source* Own elaboration

Municipality	Code plan	Code	Name	Area (m^2)	Total area (m^2)
Total strategic plan					180,653.00
NÁQUERA	N1	1	Pyrotechnic	48,822.00	83,591.00
	N2	2	Garage	720.00	
	N3	3	Storage construction materials	13,078.00	
	N4	7	Transport and excavations	3,166.00	
	N5	8	Treatment plant of inert waste	11,767.00	
	N6	9	Garden centre	3,259.00	
	N7	10	Abandoned (old warehouse)	2,769.00	
	N8	11	Storage	10.00	
SERRA	S1	0	Unknown	1,090.00	3,852.00
	S2	3	Storage construction materials	2,445.00	
	S3	7	Transport and excavations	317.00	
MARINES VIEJO					
MARINES NUEVO	M1	2	Garage	769.00	5,799.00
	M2	3	Storage construction materials	5,030.00	
OLOCAU	O1	1	Pyrotechnic	69,343.00	83,790.00
	O2	3	Storage construction materials	6,213.00	
	O3	4	Foundry, glass, aluminium	3,687.00	
	O4	5	Storage and parking	2,650.00	
	O5	6	Fabrication construction materials	1,897.00	
GÁTOVA	G1	3	Storage construction materials	3,621.00	3,621.00

Code	
0	Unknown
1	Pyrotechnic
2	Garage
3	Storage construction materials
4	Foundry, glass, aluminium
5	Storage and parking
6	Fabrication and storage construction material
7	Transport and excavations
8	Treatment plant of inert waste
9	Garden centre
10	Abandoned old storage
11	General storage

3.4.3.2 Industrial Activity and Evolution

A large proportion of the industrial activity in the area is linked to the construction sector or activities related to it (metalwork, prefabricated materials, etc.). In the industrial estate "Los Vientos", these activities become more diversified and include several warehouses and a scrapyard. The paint factory in Marines and the fireworks factory in Olocau deserve a special mention due to their singularity and international character.

3.4.4 Expected Urban Development

3.4.4.1 Occupancy Level in Villages and Towns

Two different situations can be identified

- Marines Viejo, Marines Nuevo and Olocau: Despite their different morphology, the centres of these three villages are almost completely constructed (only 2–6% of the lots are empty). This implies that land designated as not for building might need to be transformed into land suitable

Table 3.16 Vacant lots in towns. *Source* Own elaboration

Municipality	Urbanised land (m^2)	Land designated for building (m^2)	Vacant lots (m^2)	Relation areas (%) vacant lots/land designated for building	Relation areas (%) vacant lots/urbanised land
SERRA	650,107.06	420,148.40	78,662.26	18.72	12.10
OLOCAU	101,558.51	71,930.91	4,008.46	5.57	3.95
NÁQUERA	585,084.00	407,444.23	49,118.58	12.06	8.40
MARINES NUEVO	315,582.72	180,013.44	4,807.3	2.67	1.52
MARINES VIEJO	58,684.41	39,127.15	1,447.78	3.70	2.47
GÁTOVA	122,035.00	77,923.98	18,730.39	24.04	15.35
Total plan	1,833,051.70	1,196,588.11	156,774.77	13.10	8.55

for building or developable land, especially after studying the amount of empty dwellings, the estimated demographic growth, the current planning bylaws and the determinations that may arise from the inter-municipal planning ordinances.

- Náquera, Serra and Gátova: In these villages, 12, 18 and 24% of their developable or developed land corresponds to vacant lots. Therefore, there seems to be no reason to define new developable land (Table 3.16).

3.4.4.2 Occupancy Level in Housing Estates

Obviously, the developments with a greater percentage of vacant lots are those still in construction: "Mas de San Francisco" and "Els Plans". Among the already consolidated estates, the level of occupation varies depending on the location, accessibility and others aspects. Globally, and by municipality, it can be concluded that 7% of lots are vacant in Marines and Náquera, 11% in Serra and 15% in Gátova and Olocau (Table 3.17).

Table 3.17 Vacant lots in housing estates. *Source* Own elaboration

By municipality	Urbanised land (m^2)	Land designated for building (m^2)	No. vacant lots	Area vacant lots (m^2)	Relation areas (%):vacant lots/land designated for building	Relation areas (%) vacant lots/urbanised land
NÁQUERA	4,605,898.39	3,646,765.57	149	254,086.72	6.97	5.52
Urb. Monteamor A	80,943.05	52,772.39	5	9,456.20	17.92	11.68
Urb. El Peucal	87,141.11	78,830.67	1	3,684.21	4.67	4.23
Urb. La Carrasca	399,605.64	355,569.52	8	10,244.22	2.88	2.56
Urb. Els Trencalls	57,903.35	50,080.22	2	1,470.11	2.94	2.54
Urb. La Forqueta	66,551.18	60,001.41	5	16,314.32	27.19	24.51
Urb. La Cava	13,337.69	12,183.18	1	1,348.41	11.07	10.11
Urb. Font de l'Oro	70,437.10	60,790.86	4	4,991.95	8.21	7.09
Urb. Fontanelles	89,686.70	84,230.76	3	8,533.90	10.13	9.52
Urb. El Collado dels Llops	226,475.80	196,573.35	15	20,651.63	10.51	9.12
Urb. L'Antara-Mellaos	26,575.17	20,579.33	1	954.11	4.64	3.59
Urb. El Paraíso	162,600.80	127,411.39	9	11,198.86	8.79	6.89
Urb. Corral Nou	147,434,07	137,056.80	5	8,049.31	5.87	5.46
Urb. La Lloma	207,192.34	178,285.23	11	11,751.51	6.59	5.67
Urb. Sant Miguel	359,836.26	313,327.72	12	16,027.52	5.12	4.45
Urb. Mont-Ros	402,323.53	337,025.67	12	22,566.96	6.70	5.61
Urb. L'Arenal	52,019.97	44,545.85	2	3,531.32	7.93	6.79
Urb. Ampliación Las Lomas/Urb. El Camp de Tir	70,416.83	58,377.45	0	0.00	0.00	0.00
Urb. El Corralet	136,545.90	121,295.22	5	6,747.57	5.56	4.94
Urb. Les Llomes de l'Algepsar	232,315.91	195,227.59	15	19,278.17	9.87	8.30
Urb. El Pla de l'Estepar I	282,308.57	249,771.27	9	43,472.25	17.40	15.40
Urb. Bonanza	363,315.74	296,219.64	12	15,795.33	5.33	4.35
Urb. Plans Bonanza	95,290.22	206,340.24	5	8,659.24	4.20	9.09
Urb. Els Plans/Urb. Tres Rutes	185,293.48	156,059.05	3	2,661.52	1.71	1.44
Urb. Lloma Llarga	108,006.54	254,210.76	4	6,698.10	2.63	6.20
Urb. Mas de San Francisco	238,198.64	–	–	–	–	100.00
Urb. El Plans	444,142.80	–	–	–	–	100.00
SERRA	1,129,108.35	954,762.26	46	110,139.25	11.54	9.75
Urb. Los Ríos	22,773.67	20,955.80	1.00	3,330.22	15.89	14.62
Urb. Torre de Portacoeli	1,106,334.68	933,806.46	45.00	10,809.03	11.44	9.65
MARINES	326,203.96	274,710.49	12	19,311.88	7.03	5.92
Urb. El Real	26,322.97	23,837.77	0	0.00	0.00	0.00
Urb. Pla d'Orcán	47,852.66	44,023.89	1	1,355.13	3.08	2.83
Urb. La Contienda	42,757.47	39,190.00	1	1,093.00	2.79	2.56
Urb. El Romeral	209,270.86	167,658.83	10	16,863.75	10.06	8.06
OLOCAU	2,674,696.24	2,334,615.34	124	360,379.07	15.44	13.47

(continued)

Table 3.17 (continued)

By municipality	Urbanised land (m^2)	Land designated for building (m^2)	No. vacant lots	Area vacant lots (m^2)	Relation areas (%):vacant lots/land designated for building	Relation areas (%) vacant lots/urbanised land
Urb. La Pinadeta	54,702.39	49,540.08	2	3,976.37	8.03	7.27
Urb. L'Arquet	128,942.00	110,089.96	7	28,631.47	26.01	22.20
Urb. Rocha de Quita y el Colmenar	35,859.14	30,997.77	2	2,735.15	8.82	7.63
Urb. Pla de Marco—Les Forquetes	243,434.00	211,206.69	15	34,092.67	16.14	14.00
Urb. Els Puntals	113,158.54	100,599.82	6	30,835.63	30.65	27.25
Urb. La Lloma—Los Picapiedra	348,821.75	295,383.51	21	73,106.72	24.75	20.96
Urb. Pla de la Torre—La Seu	541,187.49	479,420.03	22	47,149.24	9.83	8.71
Urb. Pla de Pedralbilla	1,208,590.93	1,057,377.48	49	139,851.82	13.23	11.57
GÁTOVA	63,960.15	51,981.30	6	7,966.99	15.33	12.46
Urb. Monte Rodeno	50,868.15	43,272.20	3	6,744.38	15.59	13.26
Urb. La Tejería	13,092.00	8,709.10	3	1,222.61	14.04	9.34
Total plan	8,799,867.09	7,262,834.96	337.00	751,883.91	10.35	8.54

3.4.4.3 Occupancy Level in Industrial Estates

The industrial estates of Marines and Serra are practically full, while in Náquera both estates have many vacant lots: in "Aljibe de San Vicente", 40% of the lots are vacant, and in "Los Vientos", 17% of the area is still undeveloped (Fig. 3.41 and Table 3.18).

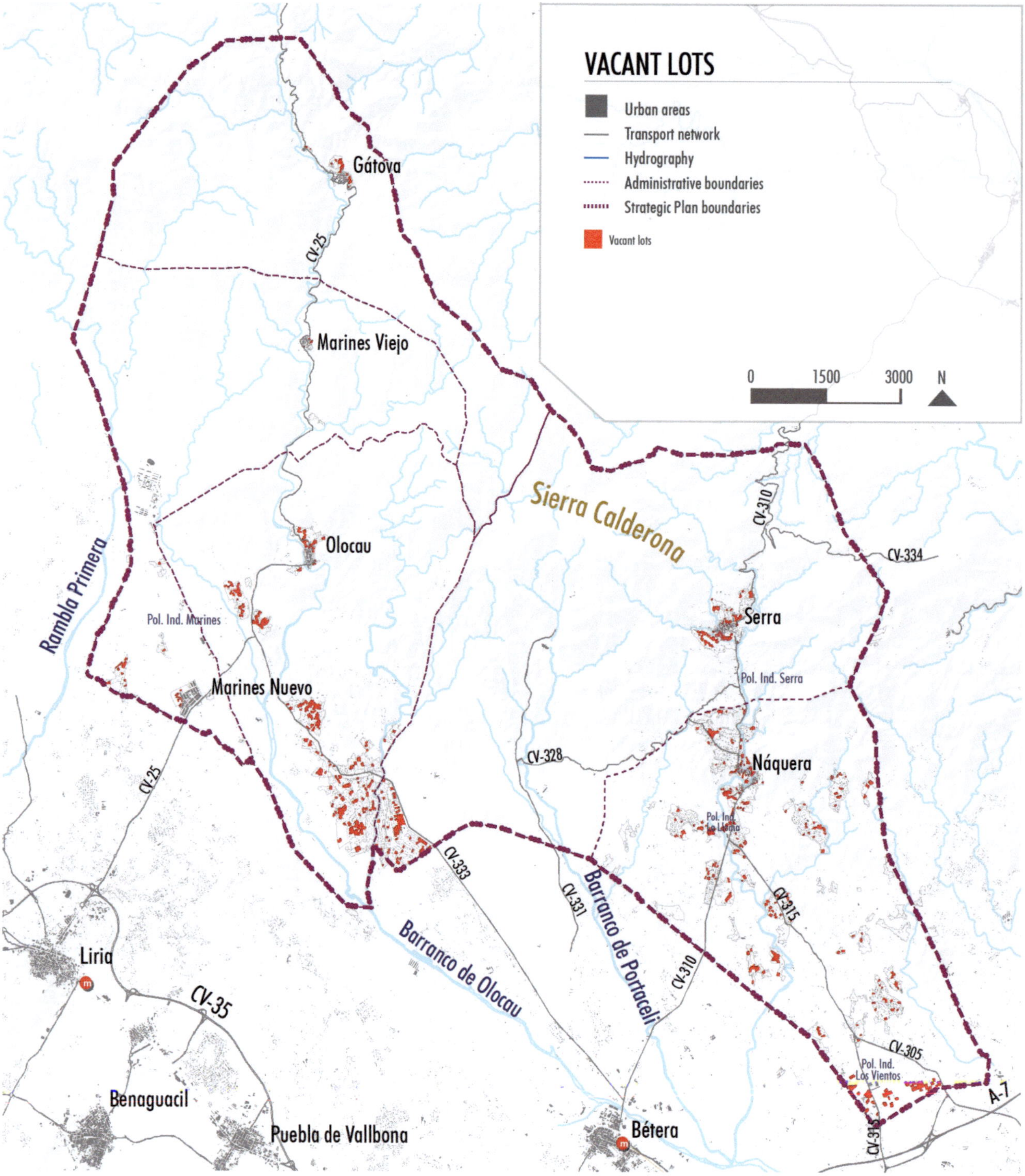

Fig. 3.41 Vacant lots. *Source* Own elaboration

Table 3.18 Vacant lots in industrial estates. *Source* Own elaboration

By municipality	Urbanised land (m²)	Land designated for building (m²)	No. of vacant lots	Area vacant lots (m²)	Relation areas (%) vacant lots/land designated for building	Relation areas (%) vacant lots/urbanised land
NÁQUERA	971,774.97	744,075.93	33	134,284.09	18.05	13.82
P.I. Aljibe de San Vicente	63,951.15	38,310.89	6	15,243.75	39.79	23.84
Els Vents	907,823.82	705,765.04	27	119,040.34	16.87	13.11
SERRA	8,067.69	8,067.69	0	0.00	0.00	0.00
PI de SAPU 5	8,067.69	8,067.69	0	0.00	0.00	0.00
MARINES	78,264.42	78,264.42	1	6,405.98	8.19	8.19
PI Marines	78,264.42	78,264.42	1	6,405.98	8.19	8.19
Total plan	1,058,107.08	830,408.04	34.00	140,690.07	16.94	13.30

3.4.4.4 Vacant Dwellings in Villages and Towns

The large number of dwellings used as secondary residences in both towns and housing estates makes it difficult to estimate the actual number of vacant dwellings. According to the 2001 census, in Gátova there are 472 vacant dwellings, in Marines Viejo 332, in Olocau 314, in Serra 44, in Náquera 115 and none in Marines Nuevo. Nevertheless, the municipalities estimate a much lower number of vacant dwellings, and further surveys may be necessary.

3.4.4.5 New Housing

Local Plans provided the basic information to estimate the amount of new housing that might be constructed in the areas classified as developable land. In those areas where the number of dwellings was not precisely defined, approved housing densities and an average size of 100 m^2 per dwelling have been used. According to these criteria, it is estimated that 7,267 dwellings could be built in the area covered by this Strategic Plan, 42% of which (3,084 units) would be located in the municipality of Serra, 38% (2,742) in Náquera and 20% (1,441) in Olocau. In Marines and Gátova, new dwellings will not be built, since their current Local Plans do not include new developable land. As shown in Fig. 3.42 and Table 3.19, new dwellings would be mainly located in low- or medium-density housing estates that do not follow the sustainability criteria and targets set by the Territorial Strategy of the Valencian Region. If the estimated number of dwellings was built, the housing stock would increase by 62%, rising from 11,675 to 18,942 units (Tables 3.19, 3.20 and 3.21).

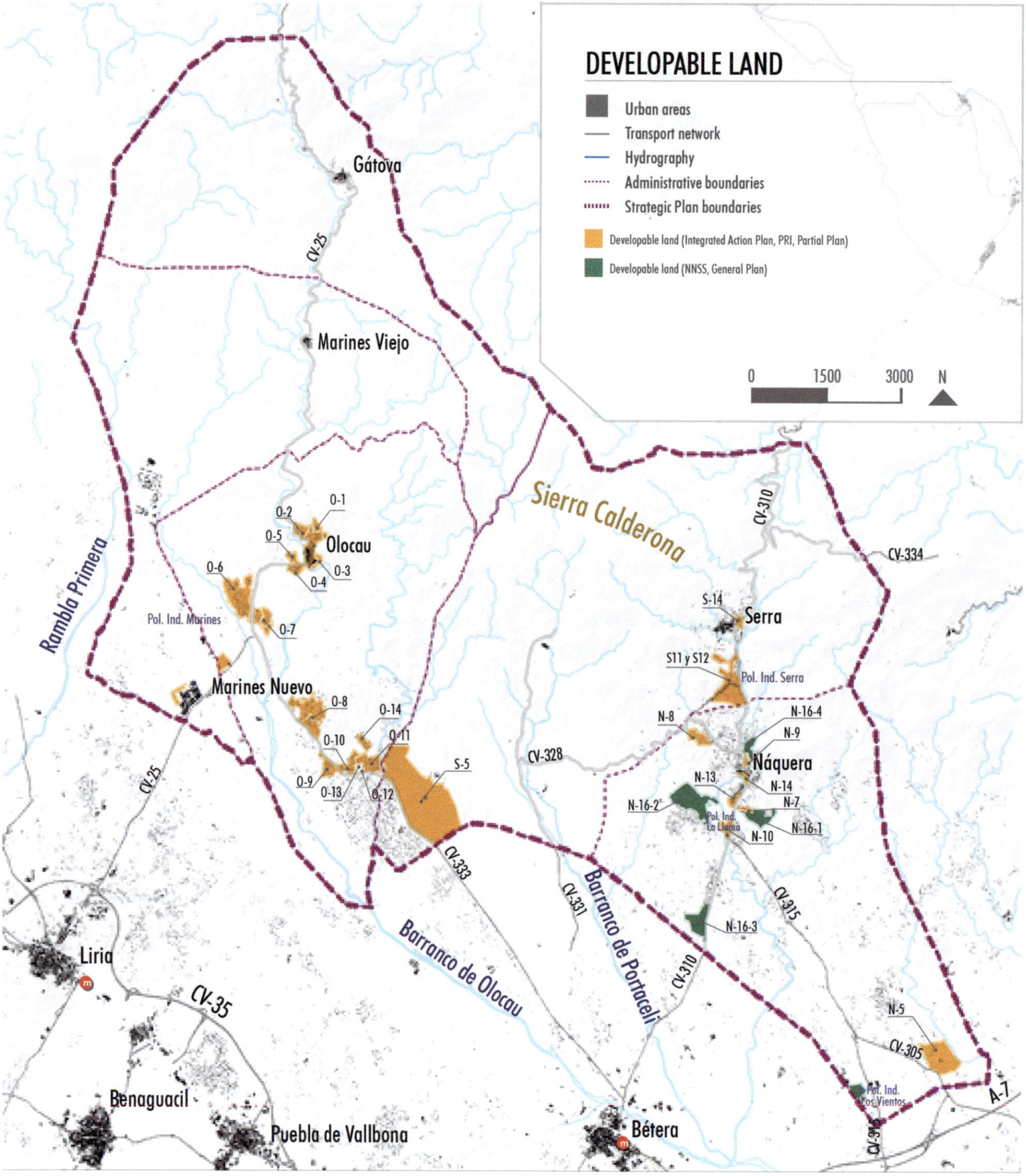

Fig. 3.42 Developable Land. Use codes to see at the Table 3.19 the estimated amount of new dwellings per sector. *Source* Own elaboration

Table 3.19 Developable land. Forecast. *Source* Own elaboration

Municipality	% area municipality/total plan	% area urbanised land	Existing dwellings	Land designated as for building: ID	Land designated as for building (m^2)	% land designated for building	New dwellings	Max. density (dwelling/Ha)	% dwellings new/existing
Total plan	100.00	4.37	11,675.00		4,648,421.41	10.38	7,266.87		
NÁQUERA	19.40	1.51	5,540.00			3.52	2,742.22		49.50
				N8	99,424.15	0.26	195	UNKNOWN	
				N9	20,188.46	0.05	86	UNKNOWN	
				N5	369,957.78	0.96	739	UNKNOWN	
				N16-1	196,421.88	0.51	390	UNKNOWN	
				N16-2	393,745.16	1.02	778	UNKNOWN	
				N16-3	159,493.00	0.41	310	UNKNOWN	
				N16-4	121,879.85	0.31	244	UNKNOWN	
SERRA	28.72	1.13	2,337.00			3.61	3,083.94		131.96
				S5 (Sierras de Serra)	1,655,240.35	2.89	2092	UNKNOWN	
				S11-12	380,470.75	0.66	761	UNKNOWN	
				S14	33,087.61	0.06	231	UNKNOWN	
OLOCAU	18.73	0.27	2,180.00			3.26	1,440.72		66.09
				O6-1 (noreste)	59,122.99	0.16	143	25	
				O6-2	64,851.09	0.17	130	UNKNOWN	
				O6-3 (Hota Baix)	43,489.74	0.12	273	70	
				O6-4 (Rocha Quito)	55,458.92	0.15	25	10	
				O6-5 (Colmenar)	43,361.79	0.12	36	10	
				O6-6 (Pla de Marco y Les Forquetes)	348,977.34	0.93	452	20	
				O6-7	80,638.27	0.22	113	UNKNOWN	
				O6-8 (alrededors dels Puntals)	99,632.65	0.27	45	10	
				O6-9 (Cachirulo del Turo)	99,109.11	0.27	58	10	
				O6-10 (Pla de la Torre)	97,437.45	0.26	39	10	
				O6-11 (Pla Torre II)	138,466.70	0.37	38	10	
				O6-12	27,743.77	0.07	55	UNKNOWN	
				O6-13	7,624.94	0.02	15	UNKNOWN	
				O6-14 (Coveta del Tabaco)	52,597.66	0.14	18	10	
MARINES	17.91	1.05	875.00	–	–	–	0	20	0.00
GATOVA	15.25	0.40	743.00	–	–	–	0		0.00

Table 3.20 Average dwelling price in towns. *Source* Own elaboration

Town	House in village (€/m²)	Flat (€/m²)	Detached house (€/m²)
NÁQUERA	800–1,200	800–1,000	1,500.00
SERRA	800–1,200	800–1,200	1,000–1,400
OLOCAU	550–650	1200–1,600	
MARINES		500–1,200	1,000–1,500
GÁTOVA		500–1,200	500–700

Table 3.21 Average dwelling price in housing estates. Source Own elaboration

Municipality	Housing estate	Area (m²)	Price (€)	Price (€/m²)
NÁQUERA	Urbanización Corral Nou	214.00	300,000.00	1,401.87
		177.00	385,000.00	2,175.14
		333.00	396,000.00	1,189.19
	Urbanización El Collado dels Llops	200.00	215,000.00	1,075.00
		160.00	300,000.00	1,875.00
		220.00	370,000.00	1,681.82

3.4.5 Public Participation Plan

3.4.5.1 Concerning Villages and Towns

See Table 3.22.

3.4.5.2 Concerning Housing Estates and Scattered Settlements

See Table 3.23.

Table 3.22 Results of the Public Participation Plan. Phase 1: villages and towns. *Source* Own elaboration

Villages and towns	Total (%)	GÁTOVA (%)	OLOCAU (%)	Marines (%)	NÁQUERA (%)	SERRA (%)
Conflicts						
Neglected public space	29.2	14.3	36.0	30.4	25.5	50.0
None	29.2	57.1	28.0	60.9	25.5	25.0
Lack of social services (education, healthcare, etc.)	15.2	0.0	4.0	13.0	34.0	13.9
Vacant dwellings/neighbourhood abandonment	14.6	21.4	44.0	17.4	4.3	11.1
Poor accessibility: pedestrians/bicycles/cars	11.7	57.1	8.0	13.0	6.4	11.1
Lack of parking spaces or bad location	7.0	0.0	8.0	0.0	4.3	22.2
Do not know/no answer	22.8	7.1	8.0	8.7	19.1	5.6
Opportunities						
To improve public space	24.0	14.3	40.0	26.1	17.0	41.7
To improve shopping facilities	18.1	0.0	24.0	8.7	14.9	41.7
To improve offer and location of parking spaces	7.0	0.0	4.0	0.0	4.3	25.0
None	5.8	42.9	0.0	4.3	2.1	5.6
To promote rural tourism and the occupancy of vacant houses	4.1	28.6	4.0	4.3	0.0	0.0
Do not know/no answer	56.1	14.3	48.0	65.2	68.1	27.8

Table 3.23 Results of the Public Participation Plan. Phase 1: housing estates and scattered settlements. *Source* Own elaboration

Housing estates and scattered settlements	Total (%)	GÁTOVA (%)	OLOCAU (%)	MARINES (%)	NÁQUERA (%)	SERRA (%)
Conflicts						
Population "socially disconnected" and "politically forgotten"	23.4	21.4	40.0	17.4	31.9	22.2
Lack of services (garbage collection, public transport)	17.5	7.1	24.0	17.4	31.9	11.1
None	17.5	35.7	8.0	34.8	10.6	27.8
Poor accessibility	11.7	7.1	12.0	21.7	21.3	2.8
Too high taxes in comparison with existing services	11.7	0.0	16.0	13.0	14.9	13.9
Public spaces of poor quality	9.9	7.1	28.0	13.0	8.5	2.8
Do not know/no answer	43.9	35.7	36.0	39.1	25.5	44.4
Opportunities						
To improve services (garbage collection, public transport)	15.2	14.3	20.0	21.7	21.3	11.1
To improve planning	7.6	7.1	12.0	8.7	6.4	11.1
To integrate retail facilities in the village	6.4	0.0	24.0	0.0	6.4	5.6
To improve accessibility	5.8	7.1	12.0	8.7	8.5	0.0
To reduce taxes	5.8	0.0	0.0	0.0	14.9	8.3
Do not know/no answer	71.9	42.9	52.0	78.3	66.0	80.6

3.4.5.3 Concerning Well-Being

See Table 3.24.

3.4.5.4 List of Associations and NGOs in Náquera, Serra, Olocau, Marines and Gátova

See Table 3.25.

Table 3.24 Results of the Public Participation Plan. Phase 1: well-being. *Source* Own elaboration

Social welfare	Total %	GÁTOVA %	OLOCAU %	MARINES %	NÁQUERA %	SERRA %
Conflicts						
Displaced pupils	22.2	35.7	4.0	4.3	46.8	25.0
Limited healthcare services	21.1	7.1	8.0	17.4	27.7	41.7
Limited cultural offer	21.1	14.3	32.0	13.0	27.7	27.8
None	17.5	35.7	36.0	34.8	4.3	16.7
Limited offer of training courses (apprenticeships, vocational training)	17.5	21.4	24.0	13.0	19.1	25.0
Lack of opportunities for the youth	17.0	28.6	24.0	13.0	19.1	19.4
Limited sport facilities (located mostly in schools)	14.6	7.1	12.0	8.7	25.5	19.4
Do not know/no answer	30.4	14.3	24.0	26.1	21.3	8.3
Opportunities						
To increase the cultural offer	18.1	14.3	40.0	8.7	19.1	22.2
To increase the offer of training courses	15.8	14.3	28.0	13.0	8.5	30.6
To create a supramunicipal secondary school, to avoid displacements	9.9	7.1	0.0	4.3	12.8	25.0
To improve and increase healthcare services	9.9	0.0	12.0	4.3	10.6	22.2
To improve and increase the offer of sport facilities	9.9	7.1	24.0	4.3	12.8	8.3
Do not know/no answer	55.6	35.7	60.0	65.2	51.1	27.8

Table 3.25 Public Participation Plan. Phase 1: list of associations and NGOs in Náquera, Serra, Olocau, Marines and Gátova. *Source* Own elaboration

Municipality	Neighbours	Cultural, political, religious	Sports	Agricultural
NÁQUERA	Falla Amics de Náquera	Peña Taurina	Unión de Fútbol Náquera	Cooperativa Consum
	Comité del hermanamiento	Peña del Cuerno	Club de Frontón	
	Asociación de amigos, jubilados y pensionistas “San Joaquín y Santa Ana”	Falla l’Encarnació	Club de Pelota Valenciana	
	Asociación Jubilados y Pensionistas “Sant Francesc”	Grup de Rondalla i Danses “La Calderona de Náquera”	Sport Ciclista Náquera	
		Sociedad Musical Santa Cecilia	Sociedad Cazadores El Gavilán	
		Colectivo de mujeres “El Salt”		
		A.C.E.N.A.		
		Grupo Coral Manuel Pérez Cabo		
		AMPA alumnos C.P. Emilio Lluch		
		Nuestra Señora de La Encarnación		
		Parque Natural Sierra Calderona		
		Caritas		
SERRA	Veïns Serra de La Calderona	Amics de Tabal i Dolçaina	Sociedad de Cazadores Alto Pino	Coop. Los Montes de Serra
	Democrática de Jubilados y Pensionistas	Dones en Acció	Correcaminos de Los Parajes de Serra	C.R. Ría-Toixima
	Tot Axerra	Sociedad Musical La Primitiva	Club Deportivo Serra Calderona	Agrupació Pou de L’Horteta de Sant Antoni
	Voluntaris de Protecció Civil	Unió Musical	Club Esports de Muntanya Castell de Serra	
	Junta Local Asociación Lucha contra el Cáncer	AMPA CEIP Sant Josep	Amics del Cavall de Serra	
		AMPA Escola Infantil Municipal		
		Penya Valencianista de Serra		
		Benéfico-Cultural Sant Roc		
		Moros i Cristians de Serra		
OLOCAU	Pla de la Torre i adyacents d’Olocau	Jubilats i Pensionistes d’Olocau	La Alegría Club de Caza y Tiro	
	L’Arenal d’Olocau	AMPA del C.R.A. “Alt Carraixet” aulari d’Olocau	La Melodia, Agrupación Musical d’Olocau	
	Veïns d’Olocau	Ames de Casa d’Olocau		
	Veïns de L’Escopar	Recreativo Cultural “La Calderona”		
	Veïns de L’Escopar	Recreativo Cultural d'Olocau		
	Veïns “Els Puntals” Secc.1ª	Mas del Capellà		
	Vecinos Los Picapiedra	Alberg Muntanyes d'Olocau		
	Vecinos La Lloma de Olocau	Formación Humanitaria Integral		
	Propietarios y Vecinos de Pedralvilla La Nova	Pozo Perla del Pla		
	Vecinos de Pla de Marco-Forquetes	Horta de Dalt		
	Propietarios Pla de Marco-Forquetes	Horta de Baix		

(continued)

Table 3.25 (continued)

Municipality	Neighbours	Cultural, political, religious	Sports	Agricultural
MARINES	Amas de Casa Tyrius	Grupo Scout Cotpaxi	Club de Futbol Marines	C.R. Pla de les Avenes
	Democrática de Jubilados y Pensionistas	Sociedad Musical La Marinense	Sociedad de Cazadores "La Maymona"	
	Jubilados y Pensionistas	Club de Ajedrez Castillo de Marines	Sociedad de Colombicultura	
	Amigos de la Fuente del Morteral	Asociación Moto Antigua de Marines		
	Vecinos Marines Viejo	Madres y padres de alumnos Castillo del Real		
	Vecinos del Romeral	Partido Popular de Marines		
	Lucha Contra el Cáncer	Coalició-Compromís per Marines		
		PSPV de Marines		
GÁTOVA	Jubilados y Pensionistas	Cultural Peña Taurina "La Cepa"	Club de Cazadores "Pico del Águila"	
		Sociedad Musical Santa Cecilia		
		Gátova Avanza		
		Peña Valencianista		
		Cultural Mujeres de Gátova		

3.4.6 Synthesis and Diagnosis

Residential areas: The study of the urban fabric with a predominant residential use reveals the following:

- **Villages and Towns**: The existing historic urban centres in the area included in the Sierra Calderona Strategic Plan are of great quality and have a high potential for tourism. However, some of them are difficult to access and include some poorly integrated new buildings. It should be noted that in spite of being an exceptional example of the rational planning principles that brought the creation of rural colonisation villages, Marines Nuevo is still not well-known and that Marines Viejo is declining very fast. In Náquera and Serra, compact urban tissues of poor quality have been adjoined to their historical centres.
- **Housing estates**: They cover an important part of the build-up land and have experienced a very rapid growth over the past decades. They are characterised by a low–medium-density fabric of residential use with the subsequent negative implications in many sustainability factors. In the area covered by the Strategic Plan, there are two main typologies of housing estates: Firstly, those attached to the villages or towns and, secondly, those located among agricultural plots or forest land. The former kind could be improved and managed together with the villages or towns, while the latter, concentrated predominantly in the south of Náquera and Olocau, pose a more complex problem that might require actions aimed at densification, improvement of public services and facilities, and better connections to towns, especially in those estates with a large permanent population.
- **Scattered settlements**: These often originate from houses that were illegally built in non-developable land, some of which have been later legalised and integrated into housing states by adding new built fabric to them. Despite of the fact that this kind of illegal housing is now strictly controlled, there are still some scattered developments pending regularisation. It must be noted that the addition of new dwellings is a questionable solution, since it reinforces this undesirable pattern of scattered settlements.

Areas of economic activity: This category includes industrial and commercial land. The main industrial estate is located in Náquera, and because of its size and location, by the AP7, it could potentially become an industrial node of regional significance. Marines' industrial estate is also important, while the other industrial areas are smaller and have less activity. In general, their activity is linked to the construction sector or to storage, not to any existing companies related to technology or services.

The commercial activity is mainly located in villages and towns, in the form of small shops and businesses, with the exception of a small shopping mall in the periphery of Náquera town. The lack of shops at the housing estates forces their inhabitants to rely on their own vehicles for obtaining supplies. They tend to visit the shopping malls, located mostly outside the municipalities, and to a lesser extent, the small shops at the local villages or towns.

Housing stock and vacant lots: The construction of dwellings in the areas designated for building (developable land) would increase the amount of dwellings in Náquera by 50%, in Olocau by 67% and in Serra by 110%. According to the current Local Plans, those new dwellings would be primarily located in low density housing estates, producing a negative effect on regional and local sustainability indexes. Since numerous vacant lots exist in compact villages or towns and housing estates, it would be desirable to prioritise the construction of new dwellings in existing urban fabrics before generating new built-up areas.

Seasonality: The population of the area covered by this Strategic Plan varies from 13,000 inhabitants in low season to 40,000 in peak season, generating large fluctuations in the demand for services and resources, as well as inefficiencies in the use of the housing stock. This situation occurs in compact villages and towns, and especially in housing estates. If the current tendency continues, and dwellings used as secondary residency in some housing estates become converted into main residences, it would be necessary to improve the services and facilities offered in those estates.

Public services and facilities: The five municipalities included in the Strategic Plan have a basic and sufficient provision of services and public facilities for the requirements of their permanent populations. Nevertheless, as these facilities are mostly located in the compact villages and towns, they are often physically, functionally and socially disconnected from the housing estates. An inter-municipal strategy could help strengthen some public services by generating economies of scale and by reducing management and maintenance costs. According to the Territorial Strategy of the Valencian Region, the supramunicipal facilities should be located in the town of Liria, as capital of the Campo de Turia county and secondary metropolitan hub. However, given the proximity to the cities of Valencia and Sagunto and the absence of certain facilities in Liria, both Valencia and Sagunto function as specialised services areas.

3.5 Infrastructure and Mobility

3.5.1 Generalities

The aim of this subchapter is to analyse the transport infrastructure and current mobility in the area covered by the Sierra Calderona Strategic Plan in order to support sustainable mobility and promote better use and enjoyment of the road and path systems by motor vehicles, bicycles and/or pedestrians.

3.5.2 Roads

The following variables have been studied:

- **Existing roads**: administration, code, width, surface condition, existence of hard shoulder and presence of bicycle lanes. The roads CV25 and CV333, in the municipalities of Marines, Olocau and Gátova, are part of the basic network of the Valencian Region; while the roads CV305, CV315, CV310, CV328, CV331 and CV333, located in the municipalities of Serra and Náquera, belong to Diputación de Valencia. There two main infrastructure corridors running south–north: the CV25 connecting Liria—Marines Nuevo—Marines Viejo—Olocau—Gátova—Altura, and the CV310 connecting Bétera—Náquera—Serra—Algimia de Alfara. These two corridors, when crossing Sierra Calderona, are located in the valleys and ravines of Náquera and Olocau and are connected with each other through the CV333 (Fig. 3.43 and Table 3.26).
- **Average daily intensities**: The stretches with more intensive use are located in the southern part of the area covered by the Strategic Plan, in the sections connecting urban centres and housing estates with the roads CV35, A7 and V21. The intensity of use decreases towards the north (access to Gátova, section of CV310 to the north of Serra), as well as in interior roads (roads to Portaceli or Garbí) (Fig. 3.44).
- **Conflict zones**: There are no black spots in the area covered by the Strategic Plan, the most problematic sections being those where the coexistence of motor vehicles and bicycle traffic is difficult. The road to Gátova has a great scenic value, its numerous curves obliging drivers to drive slowly. The non-paved accesses to housing estates have also been identified and listed (Fig. 3.45 and Table 3.27).

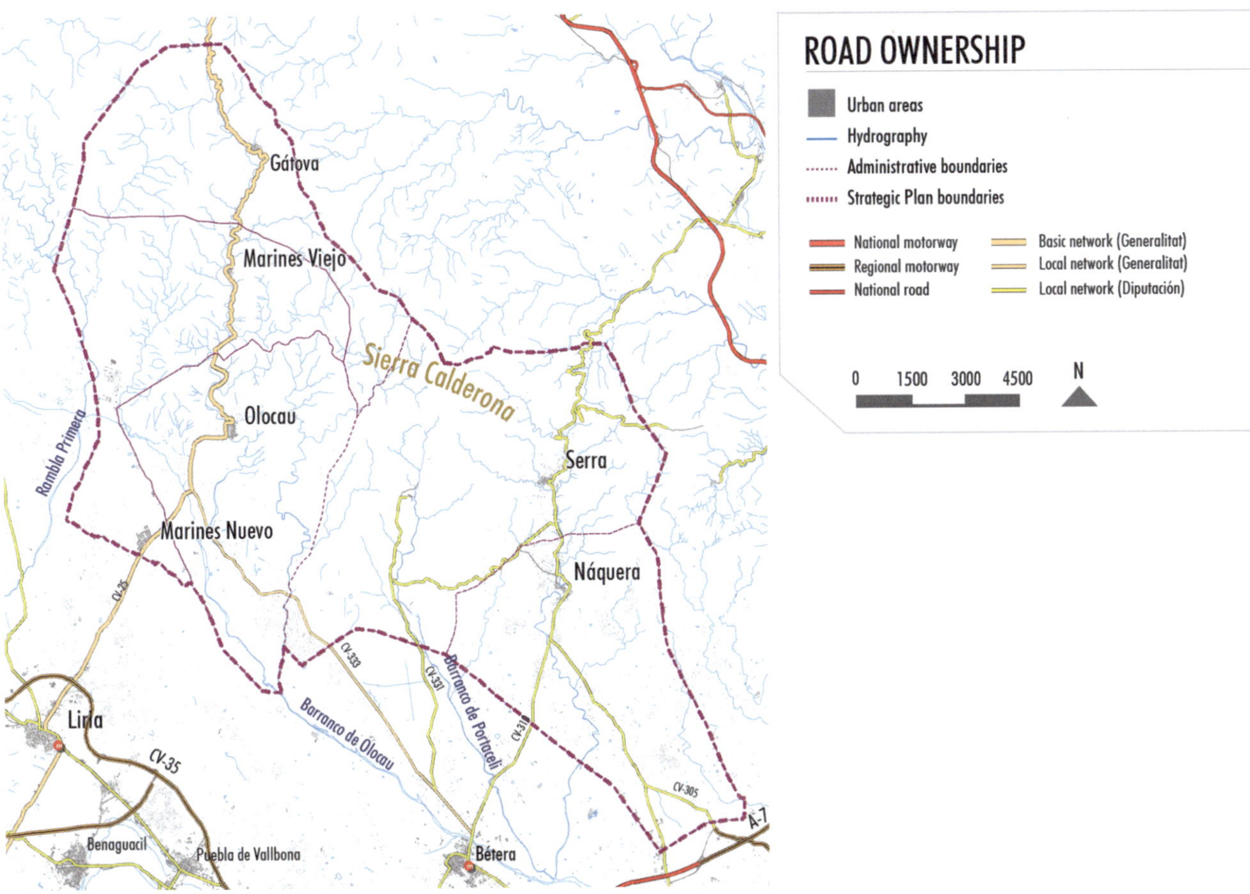

Fig. 3.43 Road ownership and categories. *Source* Own elaboration based on CNIG and BCN200

Table 3.26 Road descriptors. *Source* Own elaboration

Name	Ownership	From–to	Width (m)	Surface condition	Hard shoulder	Bicycle lane
CV-25	G.V.	CV33-Olocau	11	Good	Yes	Yes
CV-25	G.V.	Olocau-Gátova	6	Fair	No	No
CV-328	Diputación	Náquera-PortaCoeli	6	Fair	No	No
CV-331	Diputación	PortaCoeli–CV333	6	Fair	No	No
CV-315	Diputación	CV305–CV310	12	Good	Yes	No
CV-310	Diputación	Náquera-Bétera	12	Good	Yes	No
CV-310	Diputación	Serra-El Totxar	6	Good	No	No
CV-333	G.V.	Bétera-CV25	8–10	Good	Yes	Yes

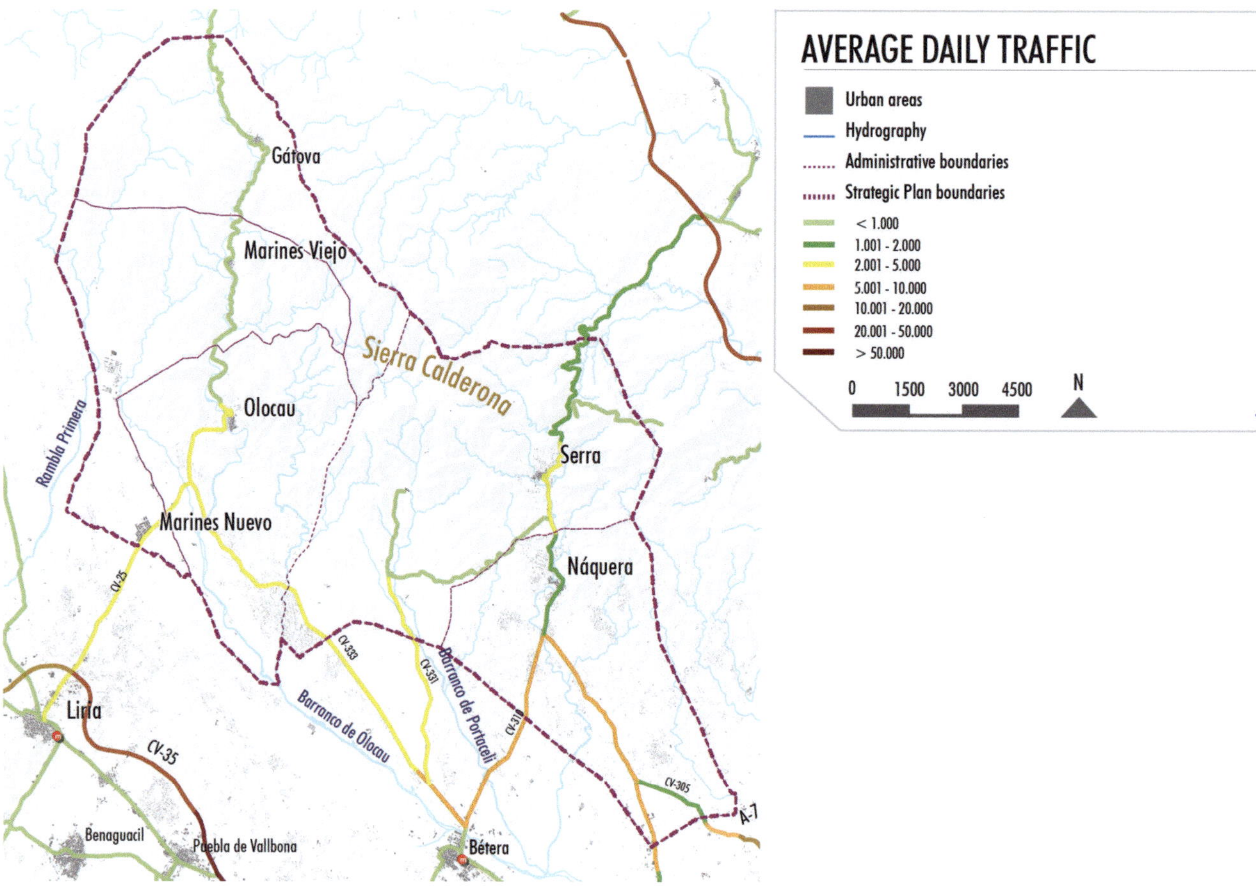

Fig. 3.44 Average daily traffic. *Source* CITMA and Diputación de Valencia

- **Isochrones maps**: The degree of connectivity of the Sierra Calderona towns has been represented with a set of isochrones maps (Fig. 3.46) displaying the accessible areas within the metropolitan area of Valencia in time intervals of 10 min. These diagrammatic maps highlight the good connectivity of the towns of Marines Nuevo, Olocau, Serra and Náquera, in contrast to Marines Viejo and especially Gátova.
- **Signposting from regional transport infrastructure**: Despite the high interest of the area and the intention to promote it, there is currently no signposting marking Sierra Calderona from the main national or regional traffic infrastructures (A23, CV35, A7 or V21). In order to address this deficiency, the spots where such signs should be located have been identified (Fig. 3.47).

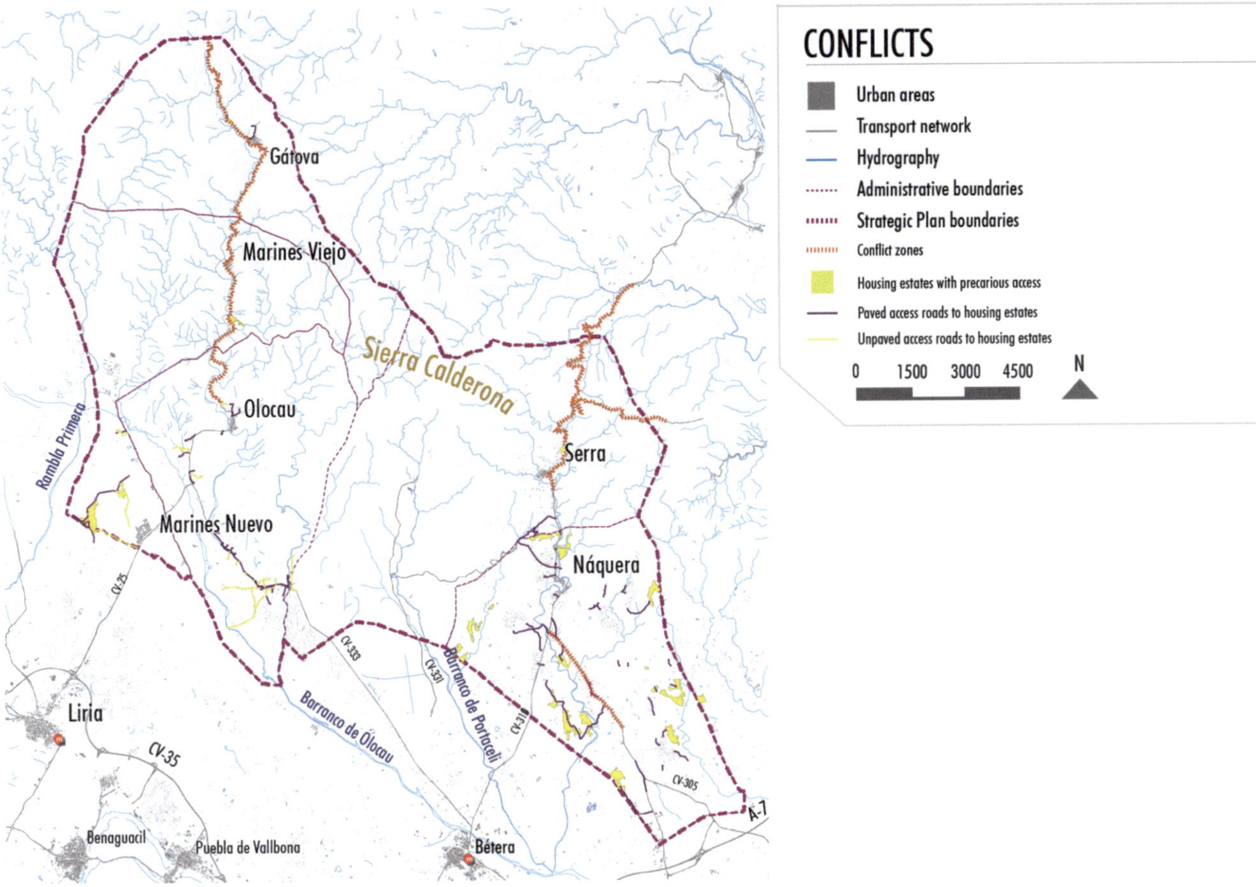

Fig. 3.45 Conflicts in transport infrastructure. *Source* Own elaboration

Table 3.27 Access to scattered settlements. *Source* Own elaboration

Settlement name	Settlement kind	Condition accesses
Urbanización Los Ríos	Housing estate attached to urban centre	Insufficient
Cañada de Les Cerveres	Scattered settlement	Insufficient
Barranc Les Voltes	Scattered settlement	Insufficient
Barranc del Sirer	Scattered settlement	Insufficient
Bassa Blanca	Scattered settlement	Insufficient
Cabeç Bord	Scattered settlement	Insufficient
Massapedro	Scattered settlement	Insufficient

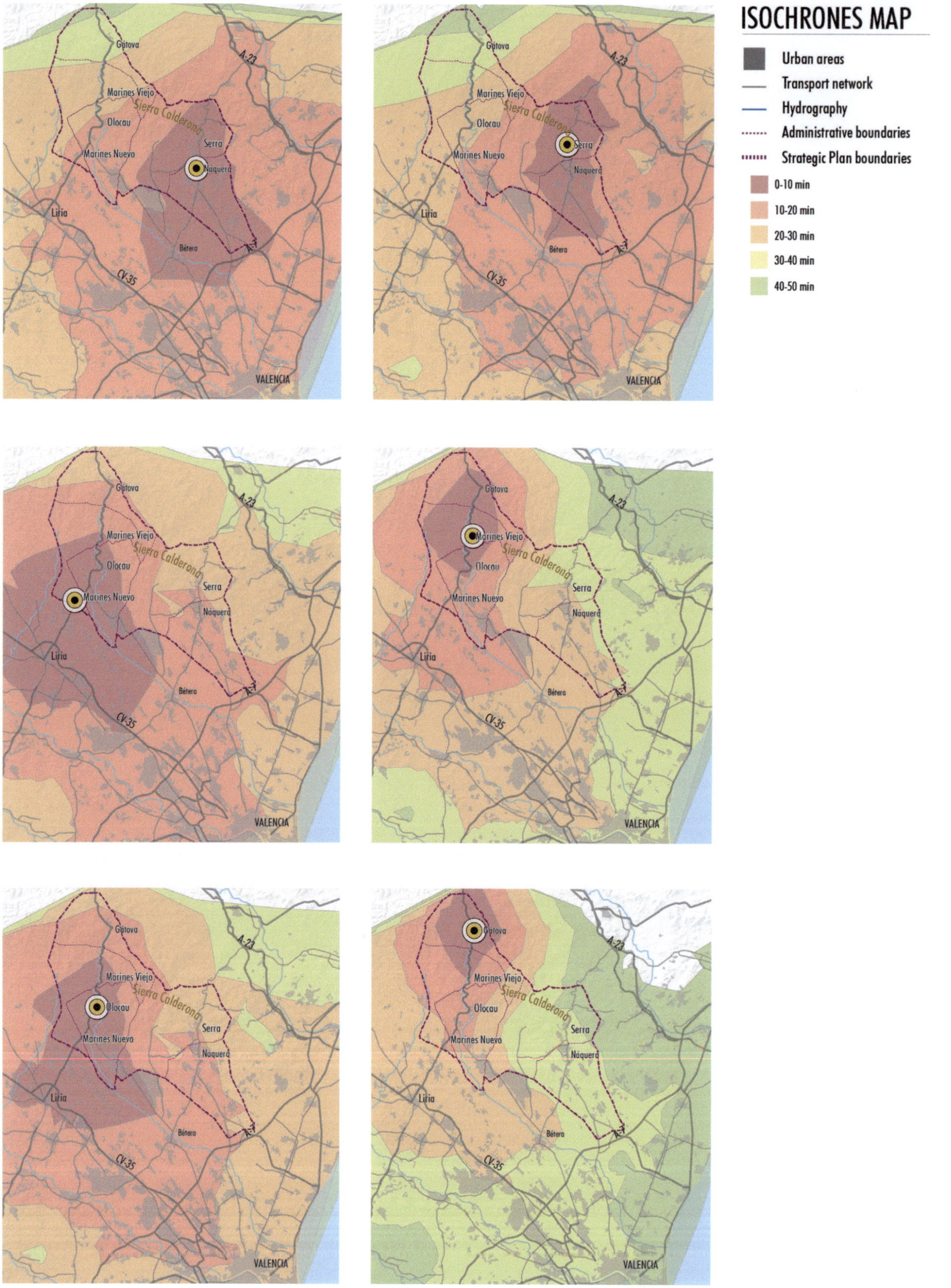

Fig. 3.46 Isochrones maps, travel time by car from the different urban centres. *Source* Own elaboration

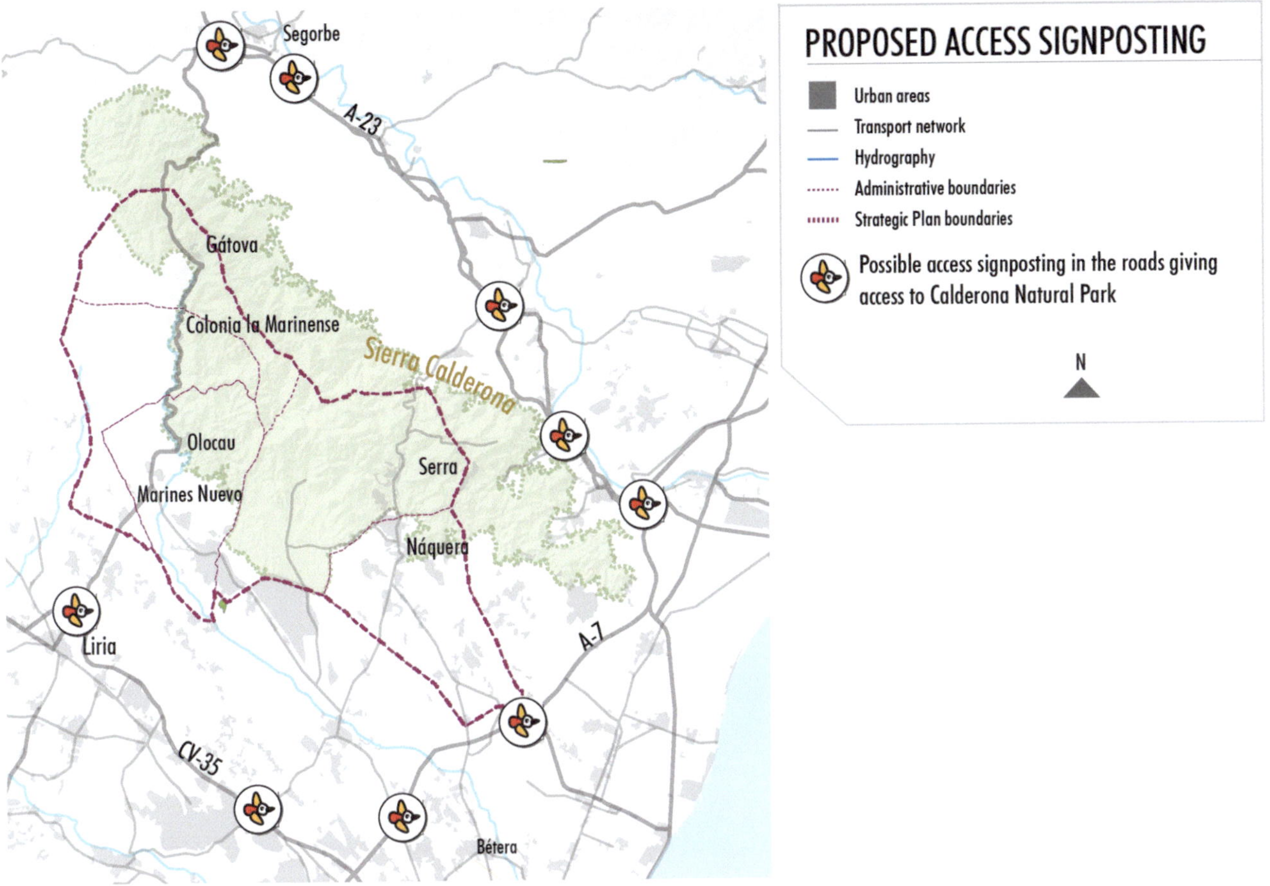

Fig. 3.47 Signposting from regional transport infrastructure. *Source* Own elaboration

3.5.3 Public Transport and Light Mobility

The following variables have been studied:

- **Public transport**: Two towns in the vicinity of the area covered by the Strategic Plan are the terminal stops of the railway line starting from Valencia (line 1—Metro Valencia). In Liria, the railway connects with the CV25 road and in Bétera with the CV310 road. Additionally, the railway connecting the airport (Manises) with Rafelbunyol via the city of Valencia (line 3—Metro Valencia) has a stop in Museros. Currently, there are three private bus services connecting these stations with the area covered by the Strategic Plan: the bus service 2 (Liria- Gátova), with 1–2 daily departures; the bus service 230 (Bétera-Serra), whose frequency varies between 2 and 13 daily departures; and the bus service 201 (Museros-Náquera). Due to the scattered presence of housing estates and their low density, it is difficult to connect them with public transport to the main regional administrative, commercial and economic hubs (Fig. 3.48).
- **Bicycle lanes**: The area is intensely used by both road and mountain cyclists. In the southern part, there is a bicycle lane network for road cyclists, while in the north, they have to share the road with motor vehicles. In the inner areas, the growing network is being progressively connected to roads and settlements. Unfortunately, the visual quality of their surroundings is quite poor (Fig. 3.49).
- **Pedestrian paths and rural roads**: In the past, the agricultural activity in the southern part of the region generated a dense network of rural roads. Nowadays, their use has increased, since they give access to housing estates and scattered settlements. Some sections of this network have a high potential for cyclist or pedestrian use, but they require improvement because of the precarious conditions of their surface and their discontinuity. In the inner mountainous areas, the network of roads is less dense but very important for fire extinguishing and for the establishment of pedestrian and bicycle paths in the natural environment. The roads belonging to this network could be classified as:
  - Paved roads
  - Rural paved roads
  - Rural unpaved roads accessible to all kinds of vehicles
  - Rural unpaved roads accessible only to 4WD
  - Pathways
- **Parking spaces in villages and towns**
  - In Gátova, there is an important lack of parking spaces and the existing ones are not properly signposted from the CV25 road. The project to channelise the ravine would create new parking options.

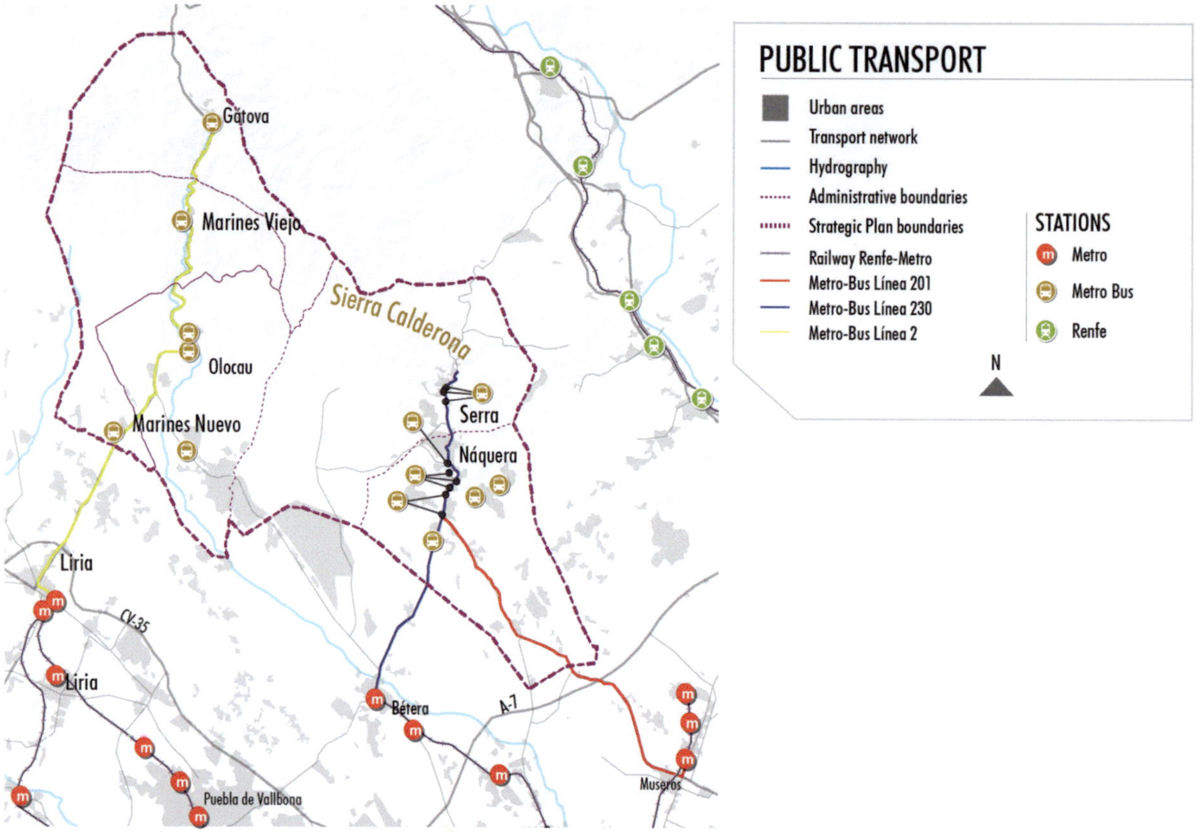

Fig. 3.48 Public transport. *Source* Own elaboration

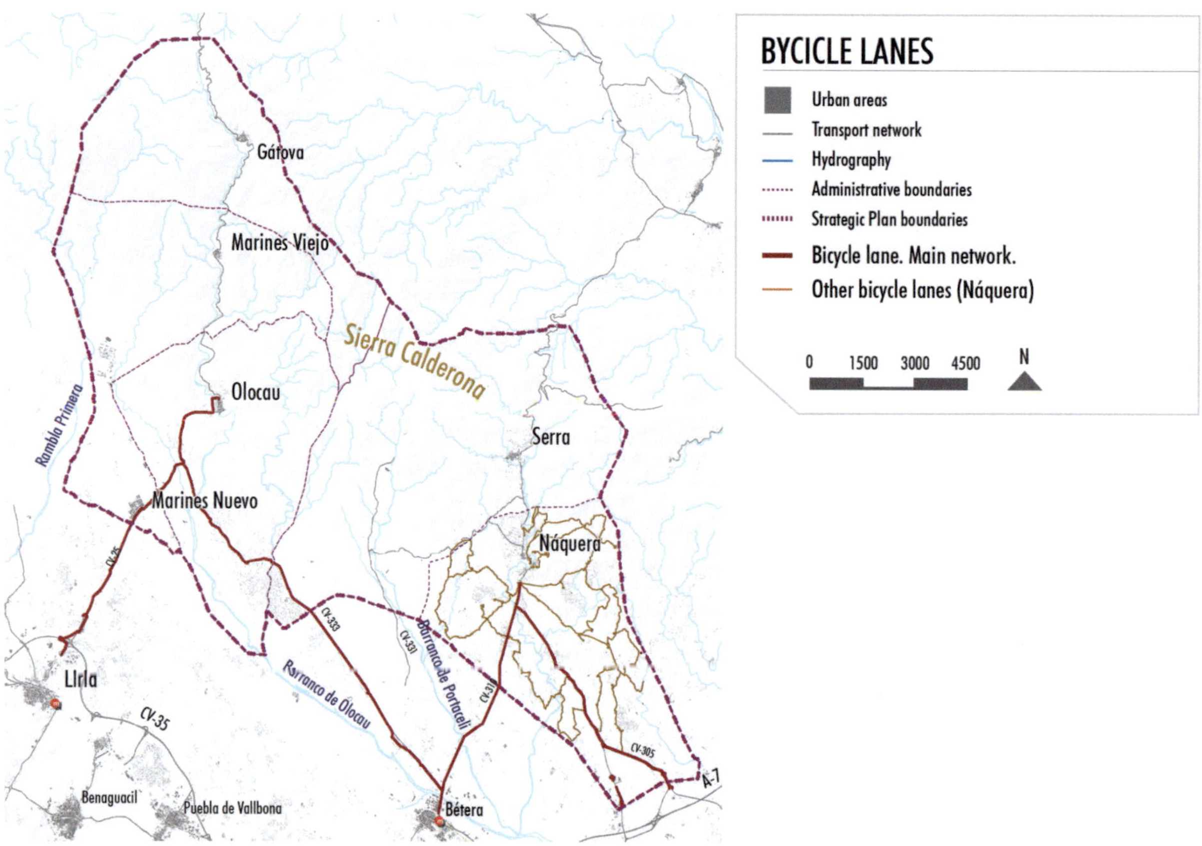

Fig. 3.49 Bicycle lanes. *Source* Own elaboration

- In Marines Viejo, they are difficult to find and are distributed in small parts of the village centre.
- In Marines Nuevo, they are properly integrated and distributed along all the streets.
- In Náquera, they are not present in the area of greatest interest, the historic centre, but concentrated on a massive scale in an area of high scenic interest at the edge of the ravine. They are not correctly signposted from the CV310 road.
- In Olocau, they are not properly signposted from the CV25 road. The main car park is located close to the sports complex, but it is functionally disconnected from the area of greatest interest in the historical centre.
- In Serra, they are not present in the area of greatest interest, the historic centre, and are not correctly signposted from the CV310 road.

* **Parking in housing estates**: There is often a lack of parking options around common services, unless they are close to an already existing restaurant. In some estates, the streets are very wide, even oversized for both circulation and parking.
* **Parking in rural areas**: Usually, there is sufficient space available, although areas like "Font de Marianet" in Serra have insufficient parking spaces for the current demand. In recreational areas, like "Pla del Lluc", there are enough spaces, but their visual impact is very strong (Fig. 3.50 and Table 3.28).

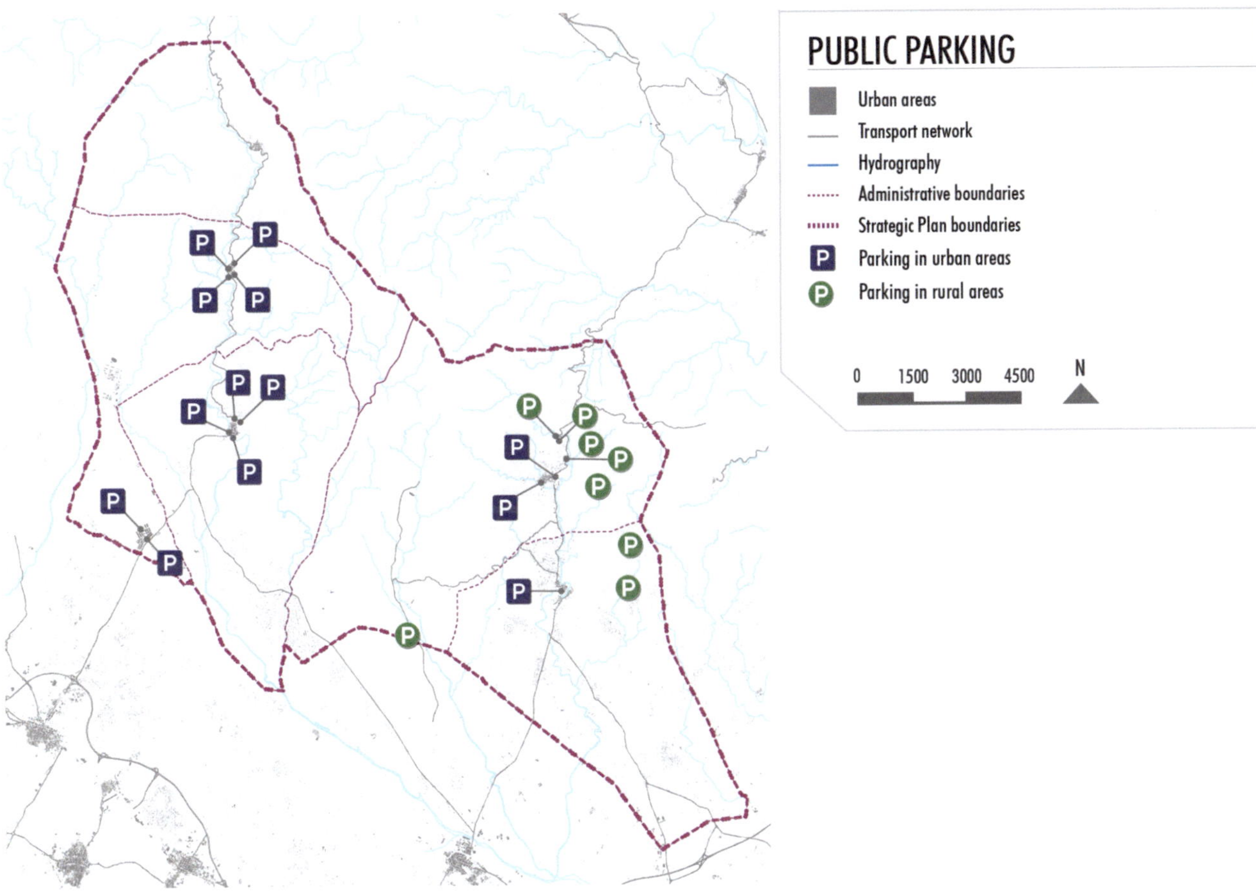

Fig. 3.50 Public parking. *Source* Own elaboration

Table 3.28 Parking spaces in urban areas (1 space/20 m^2). *Source* Own elaboration

Municipality	No. of public parking spaces
NÁQUERA	30
SERRA	20
OLOCAU	163
MARINES NUEVO	28
MARINES VIEJO	47
GÁTOVA	0

3.5.4 Intermodal Transport

In order to evaluate the potential of a more sustainable mobility model for the Sierra Calderona, a study of the estimated costs and the time consumed when travelling from the local towns and some housing estates to the Universidad Politécnica de Valencia, a health centre, the regional Hospital and the closest shopping mall have been made using three transport combinations:

(a) only private car
(b) public transport (bus and train)
(c) intermodal system (car/bicycle and bus/train)

The conclusions are:

- The use of a private vehicle is 30% more expensive than the use of car-train or bus-train, but it is faster, 100% and 300% faster, respectively.
- The buses do not give service to the housing estates, which account for a large number of dwellings and dwellers, especially during the summer.
- In order to enhance intermodal transport in the combination "car-train", it would be necessary to create parking areas in the railway stations of Liria and Bétera, as well as to reduce the prices of the train tickets.
- Intermodal transport based on the combined use of "bicycle-train" could be possible in the area of Náquera and Serra, as the average distance between the village centres or housing estates and the station of Bétera is 8–12 km. To reach Liria's railway station by bicycle only seems feasible from Marines Nuevo and Olocau, located between 7 and 12 km from the station. Large housing estates like Pedralbilla and Torre Portaceli currently have easier and faster access to Bétera than to Liria, although a new more direct bicycle lane to Liria could be arranged. This kind of intermodal transport could also be enhanced by allowing users to carry their bicycles onto trains or setting up bicycle parking spaces at the stations. The possibility to carry bicycles onto trains could, at the same time, promote an alternative way to visit Sierra Calderona. Currently, it is only possible to carry bicycles on trains during weekends and just at certain times during the weekdays (Table 3.29, 3.30, 3.31, 3.32 and 3.33).

Table 3.29 Public transport (frequency and duration of trip). Connection train Valencia-Liria (line 2—Metro Valencia) with bus Liria-Gátova. *Source* Own elaboration

One-way			Return	
Mean of transport	Stops	Working day	Stops	Working day
Bus	Gátova	7:10	Ángel Guimerá (Valencia)	17:13
	Marines	7:44	Liria	17:53
	Liria	7:58	Liria	18:30
Metro	Liria	8:08	Marines	19:00
	Ángel Guimerá (Valencia)	8:48	Gátova	19:10
		Working day	THURSDAY	OTHER
Bus	No. of services/day	2	2	1
		Working day	SATURDAY	SUNDAY
Metro	No. of services/day	17	17	6

Table 3.30 Public transport (frequency and duration of trip). Connection train Valencia-Bétera (line 1—Metro Valencia) with bus Bétera-Náquera-Serra (line 230). *Source* Own elaboration

One-way			Return	
Mean of transport	Stops	Mean of transport	Stops	Working day
BUS	Serra	6:55	8:35	11:05
	Náquera	7:02	8:43	11:13
	Bétera	7:20	8:53	11:22
METRO	Bétera	7:36	8:56	11:36
	Ángel Guimerá (Valencia)	8:03	9:23	12:03
BUS	No. of services/day	13	3	2
METRO	No. of services/day	17	17	6

Table 3.31 Private car (comparative costs and time consumed). *Source* Own elaboration

Private car		To														
		University			Healthcare centre primary school			Secondary school			Hospital			Shopping area		
		Time (min)	Distance (Km)	Cost (€)	Time (min)	Distance (Km)	Cost (€)	Time (min)	Distance (Km)	Cost (€)	Time (min)	Distance (Km)	Cost (€)	Time (min)	Distance (Km)	Cost (€)
From	Urbanización Monteamor (Náquera)	40	32	6.08	9	2.6	0.494	19	11	2.09	38	32	6.08	34	27.7	5.263
	Urb. Pedralbilla (Olocau)	38	32.4	6.16	11	7.3	1.39	13	10.4	1.98	36	31.3	5.95	23	21.5	4.09
	Marines viejo	49	46.8	8.89	11	6.5	1.24	26	17.1	3.25	47	45.7	8.68	29	26	4.94
	Gátova	55	51.4	9.77	2	0.5	0.10	32	21.6	4.10	53	50.3	9.56	35	30.5	5.80
	Marines	34	36	6.84	1	0.5	0.10	11	6.3	1.20	33	34.9	6.63	15	15.2	2.89
	Serra	40	32.9	6.25	5	0.5	0.10	17	11.8	2.24	36	32.8	6.23	32	28.2	5.36

Table 3.32 Public transport (comparative costs and time consumed). *Source* Own elaboration

Public transport		To														
		University			Healthcare centre primary school			Secondary school			Hospital			Shopping area		
		Time (min)	Distance (Km)	Cost (€)	Time (min)	Distance (Km)	Cost (€)	Time (min)	Distance (Km)	Cost (€)	Time (min)	Distance (Km)	Cost (€)	Time (min)	Distance (Km)	Cost (€)
FROM	Urbanización Monteamor (Náquera)	110	On foot + bus + metro	4.05	27	a pie	0.00	38	On foot + bus	1.25	130		4.05	77		4.05
	Urb Pedralbilla (Olocau)	*It does not have a stop (10 km to the next stop)*						*It does not have a stop (10 km to the next stop)*			*It does not have a stop (10 km to the next stop)*			*It does not have a stop (10 km to the next stop)*		
	Marines viejo	110	Bus + metro	4.9	10	Bus	1.50	30	Bus	2	73		4.8	46	Bus + metro	3.50
	Gátova	117	Bus + metro	5.1	4	On foot	0.00	48	Bus	2.3	86		5.1	64	Bus + metro	3.80
	Marines	96	Bus + metro	4.3	7	On foot	0.00	15	Bus	1.5	53		4.3	31	Bus + metro	3.00
	Serra	100	Bus + metro	4.05	5	On foot	0.00	19	Bus	1.25	110		4.05	62	Bus + metro	4.05

Table 3.33 Intermodal transport (comparative costs and time consumed). *Source* Own elaboration

Intermodality (car + metro)		To														
		University			Healthcare centre primary school			Secondary school			Hospital			Shopping area		
		Time (min)	Distance (Km)	Cost (€)	Time (min)	Distance (Km)	Cost (€)	Time (min)	Distance (Km)	Cost (€)	Time (min)	Distance (Km)	Cost (€)	Time (min)	Distance (Km)	Cost (€)
From	Urbanización Monteamor (Náquera)	85	Car + metro	4.70	–		–	21	car + bus	1.5	58	v + metro	4.70	44	Car + metro	4.7
	Urb. Pedralbilla (Olocau)	92	Car + metro	5.46	–		–	–		–	52	Car + metro	4.81	–		–
	Marines viejo	101	Car + metro	6.30	–		–	–		–	67	Car + metro	6.30	–		–
	Gátova	107	Car + metro	7.15	–		–	–		–	73	Car + metro	7.15	–		–
	Marines	87	Car + metro	4.13	–		–	–		–	53	Car + metro	4.13	–		–
	Serra	82	Car + metro	5.08	–		–	19	bus	1.25	55	Car + metro	5.08	41	Car + metro	5.08

3.5.5 Public Participation Plan

See Table 3.34.

Table 3.34 Survey results of Public Participation Plan_Phase 1: infrastructures and mobility. *Source* Own elaboration

Infrastructure and mobility	Total (%)	GÁTOVA (%)	OLOCAU (%)	MARINES (%)	NÁQUERA (%	SERRA %
Conflicts						
Poor quality public transport: irregular/low frequency	39.2	35.7	28.0	21.7	53.2	69.4
Poor roads	21.6	50.0	8.0	17.4	10.6	52.8
None	16.4	21.4	32.0	21.7	23.4	2.8
Bad coexistence pedestrian-bicycles-vehicles	15.2	0.0	12.0	4.3	21.3	33.3
Lack of signposting	13.5	21.4	24.0	17.4	6.4	19.4
Difficult accessibility to agricultural crops	12.9	35.7	12.0	4.3	8.5	25.0
Difficult mobility within housing estates	7.6	0.0	4.0	13.0	8.5	5.6
Road surface in housing estates in bad conditions	7.6	7.1	12.0	26.1	2.1	5.6
Insufficient bicycle lanes	7.6	0.0	12.0	0.0	10.6	13.9
Poor quality landscape from infrastructure	6.4	21.4	8.0	8.7	2.1	8.3
Do not know/no answer	15.8	0.0	8.0	4.3	2.1	0.0
Opportunities						
To improve public transport	20.5	28.6	16.0	4.3	31.9	27.8
To improve road accesses	20.5	64.3	12.0	30.4	8.5	33.3
To improve the bicycle lane network	11.7	0.0	8.0	0.0	17.0	27.8
To improve and extend the path network	11.1	14.3	20.0	26.1	4.3	11.1
To improve signposting in forest areas	11.1	28.6	12.0	21.7	4.3	11.1
To improve signposting in roads	9.9	28.6	20.0	13.0	0.0	13.9
Do not know/no answer	50.3	7.1	48.0	39.1	59.6	33.3

3.5.6 Synthesis and Diagnosis

Infrastructure:

The main roads CV35 and AP7 connect the surrounding regions with the area covered by the Sierra Calderona Strategic Plan; they give access to the roads forming the south–north axis across Sierra Calderona (CV25 and CV310) and to those running east-west (CV333 and CV305).

Due to the abrupt nature of the territory crossed by the roads CV25 and CV310, these roads are steep and winding, requiring slow driving. This fact is compensated by the high scenic value of the surrounding landscape. On the contrary, the roads CV333 and CV305 are located in plain areas with little scenic value due to the precarious treatment of their edges and of the housing estates and adjacent facilities along them.

In general, the Sierra Calderona Natural Park is insufficiently signposted along the motorways and main roads surrounding the region and giving access to it.

Altogether, it can be concluded that the area covered by the Strategic Plan is fairly well connected, as can be appreciated when looking at the average times to access the city of Valencia, Liria and Bétera. Nevertheless, those times increase considerably from the village centres of Marines Viejo and Gátova. However, the possible opening of a tunnel in the road CV50 under the Sierra Calderona would radically change the mobility pattern and the overall functioning of the area.

Light mobility: The network of rural roads, tracks and forest trails is extensive but discontinuous and not effectively connected with the urban settlements or roads.

Signposting is scarce and navigation difficult, hindering visitors' access to some places of high scenic and cultural value.

The bicycle lane network is extensive and continuous towards Valencia and Liria, although its visual environment is often quite poor. It lacks shading trees and seems to have been designed for external visitors rather than for locals moving between the housing estates and the local towns.

Public transport and intermodal transport: The area covered by the Strategic Plan does not have any railway station, and the closest ones are located at Liria, Bétera and Museros. Bus services connecting these stations with the studied area are irregular.

A first feasibility study of the reinforcement of public and intermodal transport systems at the Sierra Calderona shows the following:

- The existing network of low- or medium-density housing estates hinders the implementation of public transportation systems and encourages the use of private vehicles. This is also affected by the fact that housing estate dwellers do not use local schools and shopping facilities located in the nearest towns.
- The fact that the people work (industrial estates, shopping malls, etc.) or study in places outside the limits of the Plan's area, and that the connection with the public transport is not effective, promotes the use of private means of transport.
- To travel to Valencia by public transport or by commuting, saves money, around 33%, but requires a much larger investment of time, between two and three times, making these options unattractive. Under these circumstances, and in order to enhance the use of public transport or intermodal systems, the price of public transport should be decreased, together with the travel times to the railway stations.

3.6 Cultural Heritage

3.6.1 Generalities

The exceptional architectural, archaeological and ethnographic heritage of the south-central area of Sierra Calderona is specially concentrated in the three valleys crossing the area from north to south (ravines of Gátova-Olocau, Portaceli and Serra-Náquera). Because of the concentration of cultural assets and of its good accessibility (roads CV25, CV311 and CV310, respectively), these valleys have a great potential to become the cultural axes of the region.

In spite of its scattered distribution along those valleys, the existing cultural heritage is particularly concentrated around historical villages and towns or in vantage points with extensive views over the whole territory. Their state of preservation and accessibility are uneven. In some notorious cases, they have been well restored and adapted to new uses; whilst in some others, they are in a precarious state or remain virtually inaccessible.

In the next points of this subchapter, the existing heritage will be described and arranged according to their character and their state of preservation. That information will be used in the following stages of this Plan to generate an integrated network of natural and cultural sites for both locals and visitors.

3.6.2 Cultural Heritage

3.6.2.1 Types of Cultural Heritage

The existing cultural heritage has been classified according to the following groups: thirteen of the sites listed in this subchapter are *Bienes de Interés Cultural* or "BICs" (in English: Sites of Cultural Interest) and include seven towers, four castles, one archaeological site, one Carthusian monastery, and ten have been listed as *Bienes de Relevancia Local* or "BRLs" (In English: Sites of Local Relevance) and include five churches, two chapels, one aqueduct, one via dolorosa. The vast majority belong to the category of religious or defensive architectural heritage, with two exceptions: an archaeological site at the Iberian settlement of *Puntal dels Llops*, and a water infrastructure, the aqueduct of "L'Arquet".

The ethnographic heritage is often more difficult to identify as it is linked to traditional ways of living at the Sierra Calderona and to singular customs and techniques. In the area covered by the Strategic Plan, the most important festivities are *Fiesta de la Cereza* (cherry festival) in Serra and the Pig-slaughter Day in Serra and Gátova. There are also local craftwork techniques related to stonework and masonry using rodeno red stone, and also basket weaving, etc., though many of them are at risk of disappearing.

In relation to the gastronomic heritage, the wide variety of rice dishes, stews, sweets, and some processed products like cold meats and olive oils must be highlighted. The possibility of promoting local cuisine, as a complement to the cultural, recreational and environmental offer, could be one of the keys to generating a broader and more attractive tourist offer (Tables 3.35, 3.36 and 3.37).

Table 3.35 Architectural heritage. *Source* Own elaboration

Architectural heritage				
Religious	Defensive	Rural	Hydraulic	Civil engineering
Chapels = 3	Castles = 4	Tanks = 4	Aqueduct = 1	Bridges masonry, bricks = 16
Churches = 5	Watch towers = 7	Drinking trough = 3	Irrigation dams = 16	Bridges concrete = 2
Cemeteries = 4	Walled settlement = 1	Well = 2	Ponds = 30	
Carthusian monastery = 1		Pens = 1	Channel = 1	
		Shelter = 1	Deposit = 3	
		Country houses = 2	Fountains = 6	
		Farm = 1	Laundry sites = 5	
			Siphons = 7	
			Well = 0	
			Spring = 0	

Table 3.36 Gastronomic heritage. *Source* Own elaboration

Gastronomic heritage		
Typical dishes	Typical sweets	Processed products
Arròs de pollastre i carxofes	Rollets amb anís i mistela	Oil
Arròs caldós de marisc	Coques cristines	Ham
Arròs amb fessols i naps	Coca de llanda	
Arròs al forn	Orelletes	Candied fruit
Caragols	Congrets	Cold meat
Esgarrat		
Olla amb bledes		
Sopa coberta		
Rovellons		
Olla de col		
Ajolaire		
Tarta de pimiento y tomate		

Table 3.37 Archaeological and ethnographic heritage. *Source* Own elaboration

Archaeological heritage	Ethnographic heritage			
Archaeological	Livestock farming	Crafts	Industrial	Festivities
"Poblados" (settlements) = 7	"Vereda" (narrow path) = 4	Basket weaving = 1	Ice house = 1	Gastronomic = 3
"Villas" (settlements) = 7	"Colada" (mountain track) = 4	Stone cutting = 4	Lime kiln = 3	Cultural = 2
"Asentamientos" (settlements) = 7	"Cordel" (cattle trail) = 3	Masonry = 1	Windmill = 2	Religious = 27
Necropolis = 1	"Azagador" (path for sheeps) = 1		Watermill = 1	Bullfighting = 3
Ovens = 2				
Castles = 1				
Watch towers = 5				
Caves = 1				
Islamic "alquería" (farmhouse) = 1				

3.6.2.2 Inventory of Heritage Sites

The existing heritage sites are listed and displayed in Figs. 3.51 and 3.52 and as listed in Tables 3.38, 3.39, 3.40, 3.41, 3.42, 3.43, 3.44, 3.45, 3.46, 3.47.

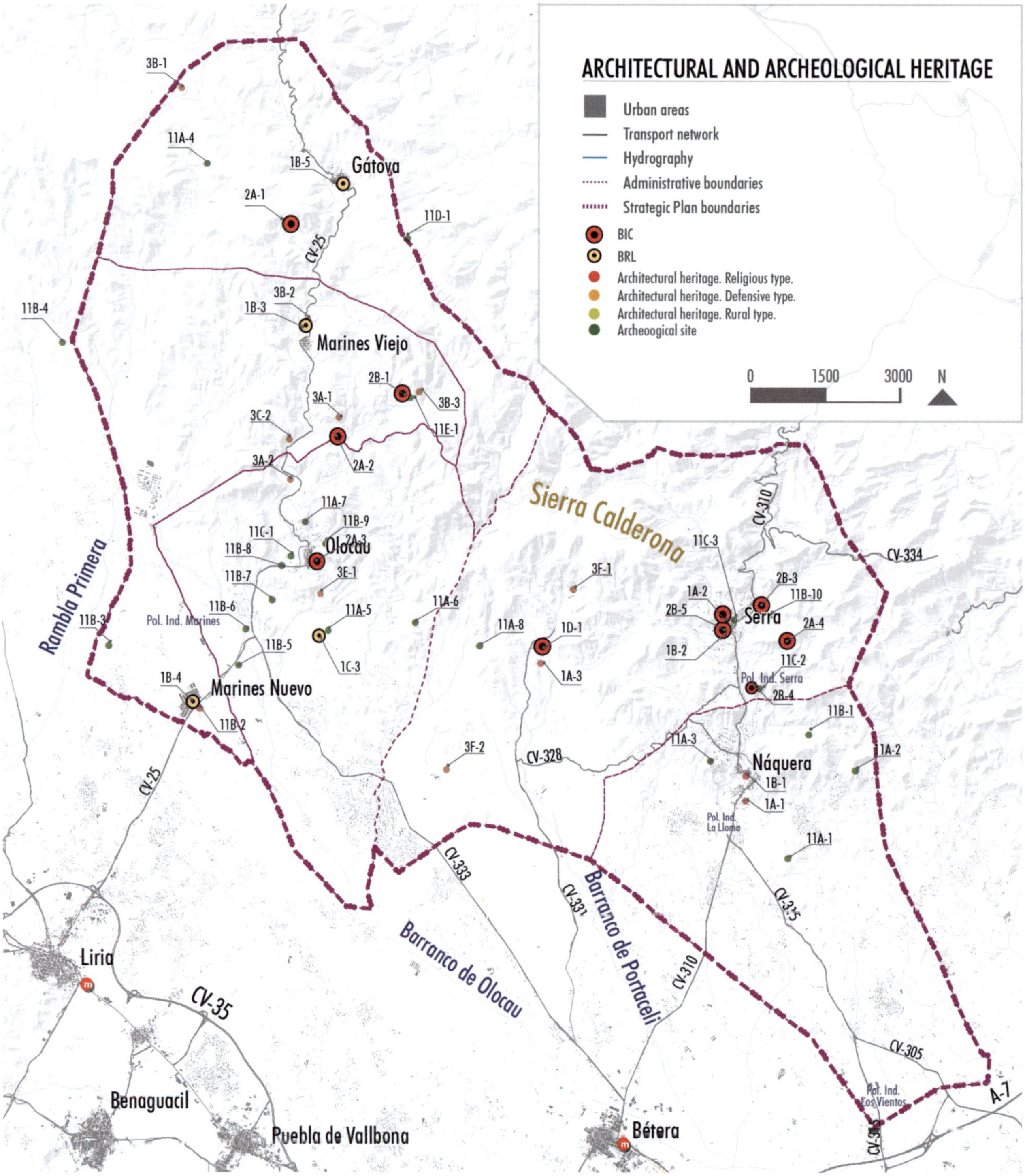

Fig. 3.51 Architectonic and archaeological heritage. Use codes from Tables 3.38, 3.39, 3.40 and 4.43 to identify sites. *Source* Own elaboration

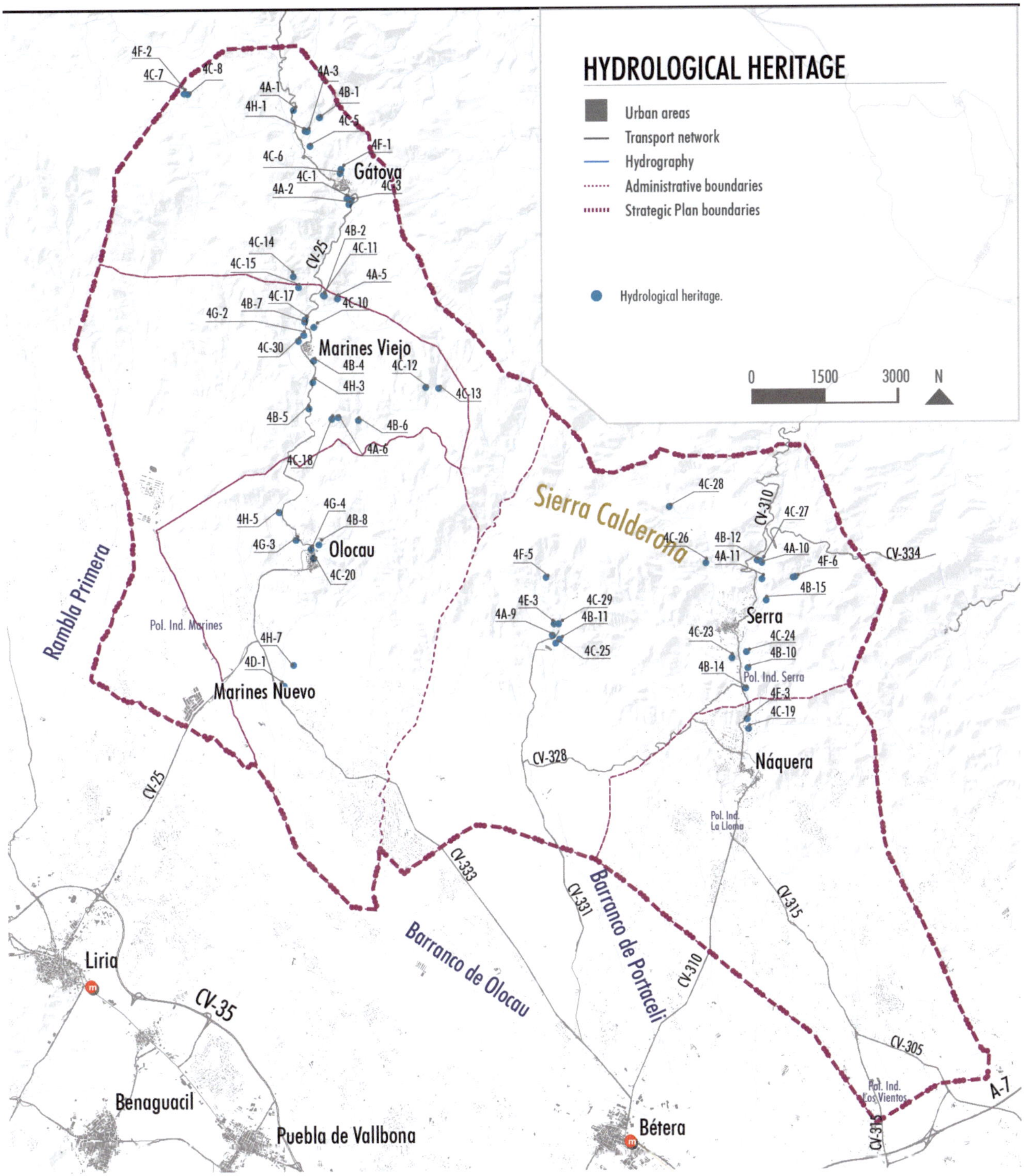

Fig. 3.52 Water (hydrological) heritage. Use codes from Table 3.41 to identify sites. *Source* Own elaboration

Table 3.38 Architectural heritage. Religious type. *Source* Own elaboration

Architectural heritage						
Name	Code in map	Type	Subtype	Municipality	Protection	Conservation status
Ermita de San Francisco de Asís y Calvario	1a-1	RELIGIOUS	CHAPEL	NÁQUERA	BRL	5
Ermita de San José y La Cruz y Calvario	1a-2		CHAPEL	SERRA	BRL	5
Ermita de Santa Bárbara	1a-3		CHAPEL	SERRA		Poor, boarded up
Iglesia Parroquial de Ntra. Señora de la Encarnación	1b-1		CHURCH	NÁQUERA	BRL	5
Iglesia Parroquial de Ntra. Señora de los Ángeles	1b-2		CHURCH	SERRA	BRL	
Iglesia del Santísimo Cristo del Perdón	1b-3		CHURCH	MARINES VIEJO	BRL	4
Iglesia Parroquial del Santísimo Cristo de las Mercedes	1b-4		CHURCH	MARINES NUEVO	BRL	5
Iglesia Parroquial Ntra Sra. De Los Angeles	1b-5		CHURCH	GÁTOVA	BRL	5
Cementerio Marines Viejo	1c-1		CEMETERY	MARINES VIEJO		
Cementerio Náquera	1c-2		CEMETERY	NÁQUERA		
Cementerio y Calvario Olocau	1c-3		CEMETERY	OLOCAU	BRL	5
Cementerio Serra	1c-4		CEMETERY	SERRA		
Convento Porta-Coeli	1d-1		CONVENT	SERRA	BIC	5

Table 3.39 Architectural heritage. Defensive type. *Source* Own elaboration

Architectural heritage						
Name	Code in map	Type	Subtype	Municipality	Protection	Conservation status
Castillo de Torrejón	2a-1	DEFENSIVE	CASTLE	GÁTOVA	BIC	1
Castillo del Real | Castillo Ali-Maimó	2a-2		CASTLE	OLOCAU	BIC	1
El Castell | Casa del Conde de Olocau	2a-3		CASTLE	SERRA	BIC	1–5
Castillo de Serra	2a-4		CASTLE	SERRA	BIC	1
Torre Olla	2b-1		WATCH TOWER	MARINES	BIC	2
Torre de la Ermita | Torre del Calvario	2b-2		WATCH TOWER	SERRA	BIC	2
Torre de Ría	2b-3		WATCH TOWER	SERRA	BIC	1
Torre de Satareña | La Torreta | Torre de Satarenya	2b-4		WATCH TOWER	SERRA	BIC	1
Torre del Señor de la Vila	2b-5		WATCH TOWER	SERRA	BIC	3
Torre de Torrejón	2b-6		WATCH TOWER	GATOVA	BIC	1
Poblado Amurallado Puntal dels Llops	2c-1		WALLED SETTLEMENT	OLOCAU	BIC	2

Table 3.40 Architectural heritage. Rural type. *Source* Own elaboration

Architectural heritage						
Name	Code in map	Type	Subtype	Municipality	Protection	Conservation status
Aljibe del Real	3a-1	RURAL	TANK	MARINES		
Aljub de la Sèquia d´Aigualeg	3a-2		TANK	OLOCAU		
Aljub Camí dels Frares	3a-3		TANK	OLOCAU		
Aljub de Sant Francesc	3a-4		TANK	NÁQUERA		
Abrevadero del Rebollo	3b-1		DRINKING TROUGH	GÁTOVA		
Abrevadero de la Huerta de Arriba	3b-2		DRINKING TROUGH	MARINES		
Abrevadero de Olla	3b-3		DRINKING TROUGH	MARINES		
Pozo de la Maimona	3c-1		WELL	MARINES		
Pozo Negro o Viejo de Aigualeg	3c-2		WELL	MARINES		
Corral de Moliner	3d-1		PEN	NÁQUERA		
Refugi-Caseta, Camí darrere la Sella	3e-1		SHELTER	OLOCAU		
Masía de la Pobleta	3f-1		COUNTRY HOUSE	SERRA		
Masía de la Torre	3f-2		COUNTRY HOUSE	SERRA		
Granja de la Majoralia	3g-1		FARM	SERRA		

Table 3.41 Architectural heritage. Water Heritage type. *Source* Own elaboration

Architectural heritage						
Name	Code in map	Type	Subtype	Municipality	Protection	Conservation status
Acueducto de la Carrasquilla	4a-1	WATER HERITAGE	CUEDUCTO	GÁTOVA		
Acueducto de la Huerta de Pulpé	4a-2		AQUEDUCT	GÁTOVA		
Acueducto del Piñel 01	4a-3		AQUEDUCT	GÁTOVA		
Acueducto del Piñel 02	4a-4		AQUEDUCT	GÁTOVA		
Acueducto de la Acequia de la Huerta de Arriba	4a-5		AQUEDUCT	MARINES		
Acueducto del Real	4a-6		AQUEDUCT	MARINES		
Acueducto de la Acequia del Cuarto Azud del Barranco del Carraixet	4a-7		AQUEDUCT	MARINES		
Acueducto de L'Arquet	4a-8		AQUEDUCT	OLOCAU		Good
Acueducto de la Cartoixa de Porta-Coeli	4a-9		AQUEDUCT	SERRA		Good
Acueducto de la Acequia de Toixima 01	4a-10		AQUEDUCT	SERRA		
Acueducto de la Acequia de Toixima 02	4a-11		AQUEDUCT	SERRA		
Azud del Piñel	4b-1		IRRIGATION DAM	GÁTOVA		
Azud de la Huertecica	4b-2		IRRIGATION DAM	MARINES		
Azud de Marines Viejo, Barranco de Gátova	4b-3		IRRIGATION DAM	MARINES		
Azud del Barranco de Carraixet 01	4b-4		IRRIGATION DAM	MARINES		
Azud del Barranco de Carraixet 04	4b-5		IRRIGATION DAM	MARINES		
Azud del Real	4b-6		IRRIGATION DAM	MARINES		
Azud del Pantano	4b-7		IRRIGATION DAM	MARINES		
Azud de L'Horta de baix	4b-8		IRRIGATION DAM	OLOCAU		
Azud de l'Horta de Baix	4b-9		IRRIGATION DAM	OLOCAU		
Azud de la Providencia	4b-10		IRRIGATION DAM	OLOCAU		
Azud de la Cartuja de Porta-Coeli	4b-11		IRRIGATION DAM	SERRA		
Azud de Marianet	4b-12		IRRIGATION DAM	SERRA		
Azud del Pantà	4b-13		IRRIGATION DAM	SERRA		
Azud o Balsò de Ria	4b-14		IRRIGATION DAM	SERRA		
Balsa de la Huerta de Pulpé 01	4c-1		POND	GÁTOVA		
Balsa de la Huerta de Pulpé 02	4c-2		POND	GÁTOVA		
Balsa de la Huerta de Pulpé 03	4c-3		POND	GÁTOVA		
Balsa de la Huerta del Caide	4c-4		POND	GÁTOVA		
Balsa de la Tejería	4c-5		POND	GÁTOVA		
Balsa del Matadero de la Huerta Común	4c-6		POND	GÁTOVA		
Balsa del Rebollo 01	4c-7		POND	GÁTOVA		

Table 3.41 (continued)

Architectural heritage						
Name	Code in map	Type	Subtype	Municipality	Protection	Conservation status
Balsa del Rebollo 02	4c-8	WATER HERITAGE	POND	GÁTOVA		
Balsa del Rebollo 03	4c-9		POND	GÁTOVA		
Balsa de la Huerta de Arriba	4c-10		POND	MARINES		
Balsa de la Huertecica	4c-11		POND	MARINES		
Balsa de Olla 01	4c-12		POND	MARINES		
Balsa de Olla 02	4c-13		POND	MARINES		
Balsa de Rama 01	4c-14		POND	MARINES		
Balsa de Rama 02	4c-15		POND	MARINES		
Balsa del Cuarto Azud del Barranco del Carraixet	4c-16		POND	MARINES		
Balsa del Molino	4c-17		POND	MARINES		
Balsa del Real	4c-18		POND	MARINES		
Bassa Gran	4c-19		POND	NÁQUERA		
Bassa de L'Horta de Baix	4c-20		POND	OLOCAU		
Basa de la Mara	4c-21		POND	OLOCAU		
Bassa de L'Horta del Molí	4c-22		POND	OLOCAU		
Bassa de L'Horta Nova	4c-23		POND	OLOCAU		
Bassa de l'Ombría	4c-24		POND	SERRA		
Bassa de la Cartoixa de Porta-Coeli	4c-25		POND	SERRA		
Bassa de la Miseria o de Santiago	4c-26		POND	SERRA		
Bassa del Molí	4c-27		POND	SERRA		
Bassa del Mugrón	4c-28		POND	SERRA		
Bassa Nova de la Cartoixa	4c-29		POND	SERRA		
Balsilla de la Huerta de Abajo	4c-30		POND	MARINES		
Canal principal de riego del embalse de Benagéber	4d-1		CHANNEL	OLOCAU		
Dipòsit municipal d´Aigua Nou	4e-1		DEPOSIT	OLOCAU		
Dipòsit municipal d´Aigua Vell	4e-2		DEPOSIT	OLOCAU		
Dipòsit municipal de PortaCoelli	4e-3		DEPOSIT	SERRA		
Fuente de San Isidro Labrador	4f-1		FOUNTAIN	GÁTOVA		5
Fuente del Rebollo	4f-2		FOUNTAIN	GÁTOVA		Good
Font del Sifò del Barranc de Nàquera	4f-3		FOUNTAIN	NÁQUERA		
Font de Sant Antoni	4f-4		FOUNTAIN	SERRA		
Font del Marge	4f-5		FOUNTAIN	SERRA		
Fon de l´Ombria	4f-6		FOUNTAIN	SERRA		4
Lavadero municipal de Gátova	4g-1		LAUNDRY SITE	GÁTOVA		5
Lavadero de Marines Viejo	4g-2		LAUNDRY SITE	MARINES		
Llavador de la Bassa de la Mara (cubierto por el agua)	4g-3		LAUNDRY SITE	OLOCAU		
Llavador municipal Olocau	4g-4		LAUNDRY SITE	OLOCAU		
Llavador municipal Serra	4g-5		LAUNDRY SITE	SERRA		

Table 3.41 (continued)

Architectural heritage						
Name	Code in map	Type	Subtype	Municipality	Protection	Conservation status
Sifón de la acequia de la mina (entrada y salida)	4h-1	WATER HERITAGE	SIPHONS	GÁTOVA		
Sifón de la Acequia de la Huerta de Arriba	4h-2		SIPHONS	MARINES		
Sifón de la Acequia del Cuarto Azud del Barranco del Carraixet	4h-3		SIPHONS	MARINES		
Sifón de la acequia d´aigualeg 01 (entrada y salida)	4h-4		SIPHONS	OLOCAU		
Sifón de la acequia d´aigualeg 02 (entrada y salida)	4h-5		SIPHONS	OLOCAU		
Sifón CV-333 (01)	4h-6		SIPHONS	OLOCAU		
Sifón CV-333 (02)	4h-7		SIPHONS	OLOCAU		

Table 3.42 Architectural heritage. Civil engineering type. *Source* Own elaboration

Architectural heritage						
Name	Code	Type	Subtype	Municipality	Protection	Conservation status
Puente de la CV-25 sobre el Barranco de Gátova	5a-1	CIVIL ENGINEERING	BRIDGE MASONRY OR BRICK	MARINES		
Puente de la CV-25 sobre el Barranco de la Hoya	5a-2		BRIDGE MASONRY OR BRICK	MARINES		
Puente de la CV-25 sobre el Barranco del Tello	5a-3		BRIDGE MASONRY OR BRICK	MARINES		
Pont de la Sabata	5a-4		BRIDGE MASONRY OR BRICK	NÁQUERA		
Pont del Peucal, Barranc de la Creueta	5a-5		BRIDGE MASONRY OR BRICK	NÁQUERA		
Pont sobre el Barranc de L´Or	5a-6		BRIDGE MASONRY OR BRICK	NÁQUERA		
Pont sobre el Barranc de Nàquera	5a-7		BRIDGE MASONRY OR BRICK	NÁQUERA		
Pont sobre el Barranc del Moliner	5a-8		BRIDGE MASONRY OR BRICK	NÁQUERA		
Pont sobre el Barranc d´Olocau, CV-25 km 10,2	5a-9		BRIDGE MASONRY OR BRICK	OLOCAU		
Pont sobre el Barranc d´Olocau, CV-333 km 13,1	5a-10		BRIDGE MASONRY OR BRICK	OLOCAU		
Pont sobre el Barranc dels Lladres	5a-11		BRIDGE MASONRY OR BRICK	OLOCAU		
Pont de l´Horta Nova	5a-12		BRIDGE MASONRY OR BRICK	SERRA		
Pont sobre el Barranc de L´Horteta, CV-310 km 24,2	5a-13		BRIDGE MASONRY OR BRICK	SERRA		
Pont sobre el Barranc de L´Horteta, CV-310 km 24,5	5a-14		BRIDGE MASONRY OR BRICK	SERRA		
Pont sobre el Barranc de Pedralvilla	5a-15		BRIDGE MASONRY OR BRICK	SERRA		
Pont sobre el Barranc dels Castellets	5a-16		BRIDGE MASONRY OR BRICK	SERRA		
Pont sobre el Barranc de Safra	5b-1		CONCRETE BRIDGE	OLOCAU		
Pont sobre el Barranc del Sirer	5b-2		CONCRETE BRIDGE	SERRA		

Table 3.43 Archaeological heritage. *Source* Own elaboration

Architectural heritage						
Name	Code in map	Type	Subtype	Municipality	Protection	Conservation status
Cerro de la Patada	11a-1	ARCHAEOLOGICAL	BRONZE	NÁQUERA		
Puntal dels Moros	11a-2		BRONZE	NÁQUERA		
Els Trencalls	11a-3		BRONZE	NÁQUERA		
Marmal, de Abajo	11a-4		BRONZE	GÁTOVA		
Les Solaniques	11a-5		BRONZE	OLOCAU		
Puntal Blanc	11a-6		BRONZE	OLOCAU		
Penya Roja	11a-7		BRONZE	OLOCAU		
La Cova de Soterrany	11a-8		BRONZE	SERRA		
Vinyes	11b-1		ROMAN	NÁQUERA		
Mas de Maimona	11b-2		ROMAN	MARINES		
Hospitalet	11b-3		ROMAN	MARINES		
La Carrasqueta	11b-4		ROMAN	MARINES		
El Pitxerí	11b-5		ROMAN	OLOCAU		
El Tollo	11b-6		ROMAN	OLOCAU		
La Cella	11b-7		ROMAN	OLOCAU		
La Cargadora o Pedra	11b-8		ROMAN	OLOCAU		
L'Arquet	11b-9		ROMAN	OLOCAU		
Eretes de Ría	11b-10		ROMAN	SERRA		
La Peixa	11c-1		ISLAMIC	SERRA		
Torre de Santarenya	11c-2		ISLAMIC	SERRA		
Torre de l'Ermita	11c-3		ISLAMIC	SERRA		
Torre de Ría	11c-4		ISLAMIC	SERRA		
Molino de Sacañet	11d-1		MEDIEVAL	GÁTOVA		
BIC Torre Olla	11e-1		ISLAMIC-MEDIEVAL	MARINES		
Castell de Serra	11e-2		ISLAMIC-MEDIEVAL	SERRA		
Torre del Senyor	11e-3		ISLAMIC-MEDIEVAL	SERRA		
Cartuja de Porta-Coeli	11e-4		MEDIEVAL-MODERN	SERRA		

Table 3.44 Ethnographic heritage. Traditional livestock farming techniques. *Source* Own elaboration

Ethnographic heritage						
Name	Code	Type	Subtype	Municipality	Protection	Conservation status
Vereda de Alto Romero	7a-1	TRADITIONAL LIVESTOCK FARMING TECHNIQUES	VEREDA (narrow path)	MARINES		
Vereda de Olocau	7a-2		VEREDA (narrow path)	OLOCAU		
Vereda de Pichon Serra	7a-3		VEREDA (narrow path)	SERRA		
Vereda de Olocau	7a-4		VEREDA (narrow path)	SERRA		
Colada de Masapedro	7b-1		COLADA (mountain track)	NÀQUERA		
Colada de Serra	7b-2		COLADA (mountain track)	NÀQUERA		
Colada de Sabato	7b-3		COLADA (mountain track)	SERRA		
Cordel de la Garrofera	7c-1		CORDEL (cattle trail)	MARINES		
Cordel de Aragón	7c-2		CORDEL (cattle trail)	NÁQUERA		
Cordel de la Garrofera	7c-3		CORDEL (cattle trail)	OLOCAU		
Azagador	7d-1		AZAGADOR (path for sheeps)	SERRA		

Table 3.45 Ethnographic heritage. Festivals and events. *Source* Own elaboration

Ethnographic heritage						
Name	Code	Type	Subtype	Municipality	Protection	Conservation status
Matanza del cerdo de Gátova		FESTIVALS AND EVENTS	GASTRONOMIC	GÁTOVA		
Matanza del cerdo de Serra			GASTRONOMIC	SERRA		
Fiesta de la Cereza			GASTRONOMIC	SERRA		
Ciclos de Concierto de Olocau			CULTURAL	OLOCAU		
Moros y Cristianos Gátova			CULTURAL	GÁTOVA		
Moros y Cristianos Serra			CULTURAL	SERRA		
Fototreking Objetivo Sierra Calderona			CULTURAL	SIERRA CALDERONA		
Cabalgata de Reyes			RELIGIOUS	TODOS		
San Antonio Abad			RELIGIOUS	TODOS		
Fallas, San José			RELIGIOUS	TODOS		
Semana Santa/ Pascua			RELIGIOUS	TODOS		
San Vicente Ferrer			RELIGIOUS	TODOS		
Comuniones			RELIGIOUS	TODOS		
San Isidro			RELIGIOUS	TODOS		
Corpus Christi			RELIGIOUS	TODOS		
San Antonio de Padua			RELIGIOUS	OLOCAU		
Festividad de la Purísima			RELIGIOUS	GÁTOVA		
Virgen de Agosto			RELIGIOUS	TODOS		
San Roque			RELIGIOUS	NÁQUERA		
San Roque			RELIGIOUS	OLOCAU		
Hijas de María			RELIGIOUS	OLOCAU		
San Rafael			RELIGIOUS	GÁTOVA		
Cristo de las Mercedes			RELIGIOUS	MARINES		
Inmaculada Concepcióm			RELIGIOUS	MARINES		
San Rafael			RELIGIOUS	MARINES		
Santa Cruzl			RELIGIOUS	MARINES		
San Roque			RELIGIOUS	MARINES		
San José			RELIGIOUS	SERRA		
Virgen de los Ángeles			RELIGIOUS	AGOSTO		
San Roque			RELIGIOUS	SERRA		
Ntra. Sra. Merced			RELIGIOUS	GÁTOVA		
San José			RELIGIOUS	GÁTOVA		
Cristo de la Misericordia			RELIGIOUS	GÁTOVA		
Virgen del Rosario			RELIGIOUS	OLOCAU		
Virgen del Rosario			RELIGIOUS	NÁQUERA		

(continued)

Table 3.45 (continued)

Ethnographic heritage						
Name	Code	Type	Subtype	Municipality	Protection	Conservation status
Niño Jesús			RELIGIOUS	NÁQUERA		
Encarnación			RELIGIOUS	NÁQUERA		
San Francisco de Asís			RELIGIOUS	NÁQUERA		
Santa Cecilia			RELIGIOUS	TODOS		
La Purísima			RELIGIOUS	TODOS		
Natividad del Señor			RELIGIOUS	TODOS		
Semana Taurina de Gátova			BULLFIGHTING	GÁTOVA		
Bous al Carrer de Náquera			BULLFIGHTING	NÁQUERA		
Bous al Carrer de Serra			BULLFIGHTING	SERRA		

Table 3.46 Ethnographic heritage. Craftwork techniques. *Source* Own elaboration

Ethnographic heritage						
Name	Code	Type	Subtype	Municipality	Protection	Conservation status
Basket weaving		CRAFTWORK TECHNIQUES	CRAFT	OLOCAU		
Stone cutting			CRAFT	SERRA		
Stone cutting			CRAFT	OLOCAU		
Stone cutting			CRAFT	NÁQUERA		
Stone cutting			CRAFT	GÁTOVA		
Masonry			CRAFT	OLOCAU		

Table 3.47 Ethnographic heritage. Traditional industrial techniques. *Source* Own elaboration

Ethnographic heritage						
Name	Code	Type	Subtype	Municipality	Protection	Conservation status
Ventisquer dels Rebalsadors		TRADITIONAL INDUSTRIAL TECHNIQUES	ICE HOUSE	SERRA		
Forn de Calç, Camí de la Font del Frare			LIME KILN	OLOCAU		Low
Forn de Calç, Camí a Porta-Coeli			LIME KILN	SERRA		
Forn de Calç, Cartoixa de Porta-Coeli			LIME KILN	SERRA		
Molino Nuevo o de Ceja			WINDMILL	GÁTOVA		5
Molino Viejo			WINDMILL	GÁTOVA		
Molino de Sánchez			WATERMILL	MARINES		

3.6.2.3 State of Preservation

In absence of a more detailed study, the state of preservation of the Sites of Cultural Interest has been analysed using existing information and different sources. The insufficient signposting to access some of the restored heritage sites must be highlighted, making their "discovery" and visiting difficult.

The studied area has a large number of heritage sites that are mainly concentrated next to the urban centres and the transport infrastructure. However, their access is often not very visible, and they are seldom associated with specific new functions.

3.6.3 Synthesis and Diagnosis

Cultural heritage: The study of the area covered by the Sierra Calderona Strategic Plan reveals the existence of a rich archaeological, architectural and ethnographic heritage, with relevant examples of religious and defensive architecture (towers and castles), as well as the Carthusian monastery complex of Portaceli. These sites are located mainly along the valleys of the ravines of Olocau and Náquera, giving the roads CV25 and CV310 a high potential as cultural axes.

State of preservation and use: The degree of preservation of the existing heritage is variable, as there are notable positive examples but also many elements in a severe state of abandonment. Signposting and access roads to some sites might be improved to attract more visitors.

3.7 Tourism and Public Use

3.7.1 Generalities

The purpose of this subchapter is to analyse the existing accommodation, dining and recreational services and facilities for both visitors and temporary residents in the area covered by the Sierra Calderona Strategic Plan. In addition, and in order to guide the future definition of a "Tourism Plan", the current subchapter analyses the profile of its visitors, the activities they participate in, and the economic impact of tourism on local economies.

The location of the Calderona mountain range on the edge of the metropolitan area of Valencia, with a population of 1.5 million inhabitants, together with its exceptional landscape, environmental and cultural values, bestows the area with a huge potential for recreational and tourism activities, though on the other hand, that can jeopardise some of the before-mentioned values.

It must be noted that sporadic visits, despite being important, currently have a moderate volume and a relatively low economic impact on the local communities. While the strong use of housing estates and, to a lesser extent, of compact towns, as secondary residences provide an important and constant volume of seasonal visitors, with a significant, but still undetermined effect on the local economy and with a strong impact on the territory.

All these economic aspects are studied in the subchapter "Local economy" and will guide the definition of objectives and strategies of a "Tourism Plan" that will raise the necessity to strengthen the social and economic relationships between casual and seasonal visitors with local communities, and also the importance of promoting, through an adequate supply of services and connections, the Sierra Calderona as a recreational, cultural and natural space, as well as a model for sustainable development.

3.7.2 Current Facilities

The following variables have been analysed:

(a) **Accommodation**: this consists mainly of apartments and hostels that are concentrated in one of the areas with less economic activity (Gátova), having a lower presence in Serra, Olocau and Marines Viejo. Existing hotels are quite basic, without clear provision for visitors with high purchasing power. The provision of campsites is also insufficient (Fig. 3.53 and Table 3.48).

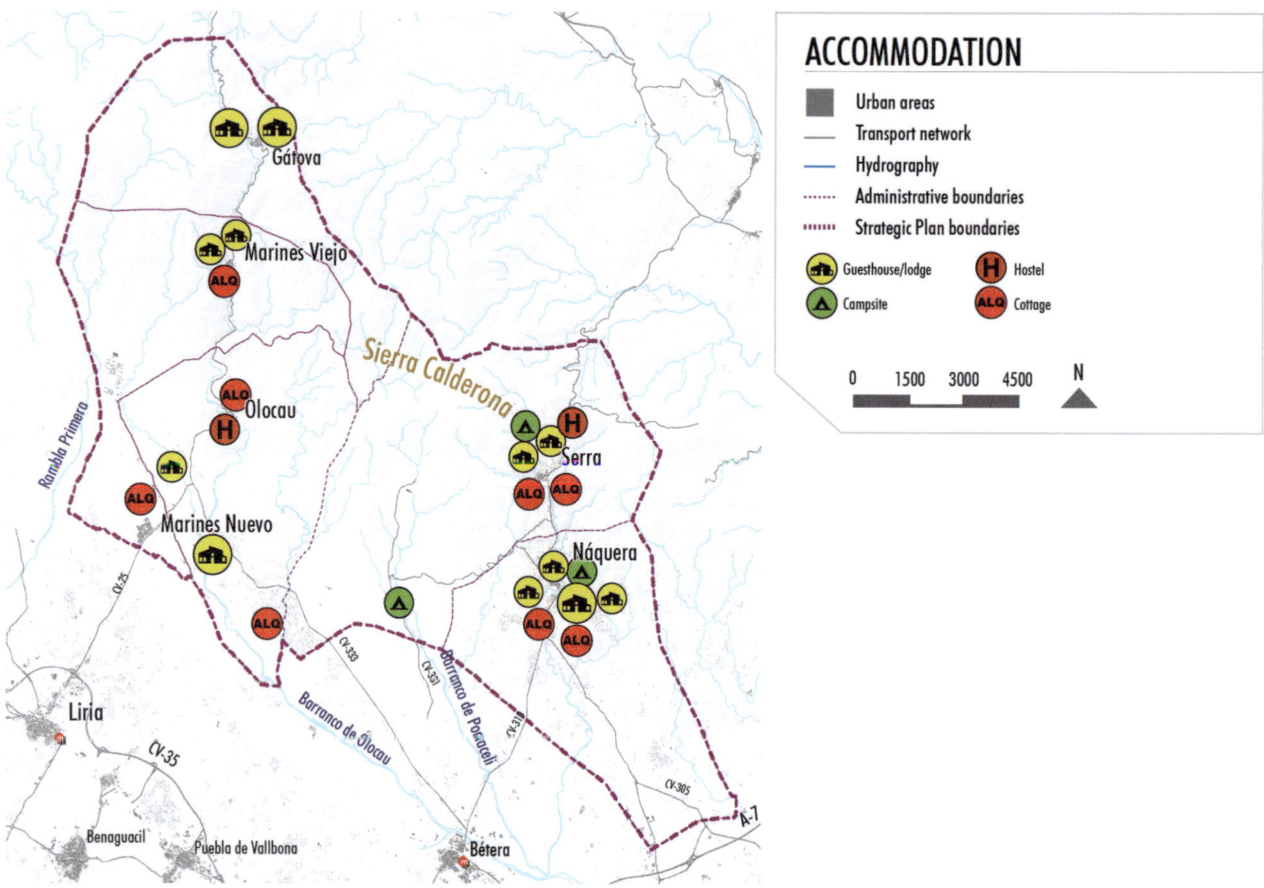

Fig. 3.53 Accommodation offer. *Source* Own elaboration

Table 3.48 Accommodation and dining offer. *Source* Agencia Valenciana de Turismo and own elaboration

	Accommodation								Dining			
	Cottages	Beds	Hostels	Beds	Guesthouses	Beds	Apartaments	Beds	To go	Bakeries	Restaurants	BAR/TAVERN/PUB
NÁQUERA	7	35	0	0	0	0	1	8	3	3	34	13
SERRA	2	15	1	30	0	0	1	21	1	3	8	8
MARINES	2	12	0	0	0	0	0	0	0	2	0	6
OLOCAU	0	0	1	10	2	280	0	0	0	0	3	4
GÁTOVA	12	46	0	0	0	0	0	0	0	1	2	4

(b) **Dining**: Restaurants are mostly concentrated in the housing estates located in the municipality of Náquera, where they have traditionally attracted numerous visitors and acquired a certain role as social centres. In the other municipalities, the dining possibilities are limited in both number and quality, despite the potential that historic centres might have to provide such services. The importance of bars must be noted, as they offer affordable prices for cyclists and hikers during the daytime (Fig. 3.54 and Table 3.48).

(c) **Other services**: There is a wide provision of equestrian centres that, curiously, are not located near to areas of special landscape interest, and there is an airfield, a strongly consolidated farm school, a youth education centre and a hospital for chronic treatments. This offer is complemented with beauty salons, hairdressers, garden centres, workshops, leisure centres, etc., mostly concentrated in Serra and Náquera, and aimed primarily at the permanent residents, despite the interest that they might also have for visitors. Sport facilities, included in the subchapter "Urban Environment and Wellbeing", as well as sport clubs and cultural or environmental associations also exist in the five municipalities (Table 3.49).

(d) Touristic activities: The following activities have been identified (Fig. 3.55):

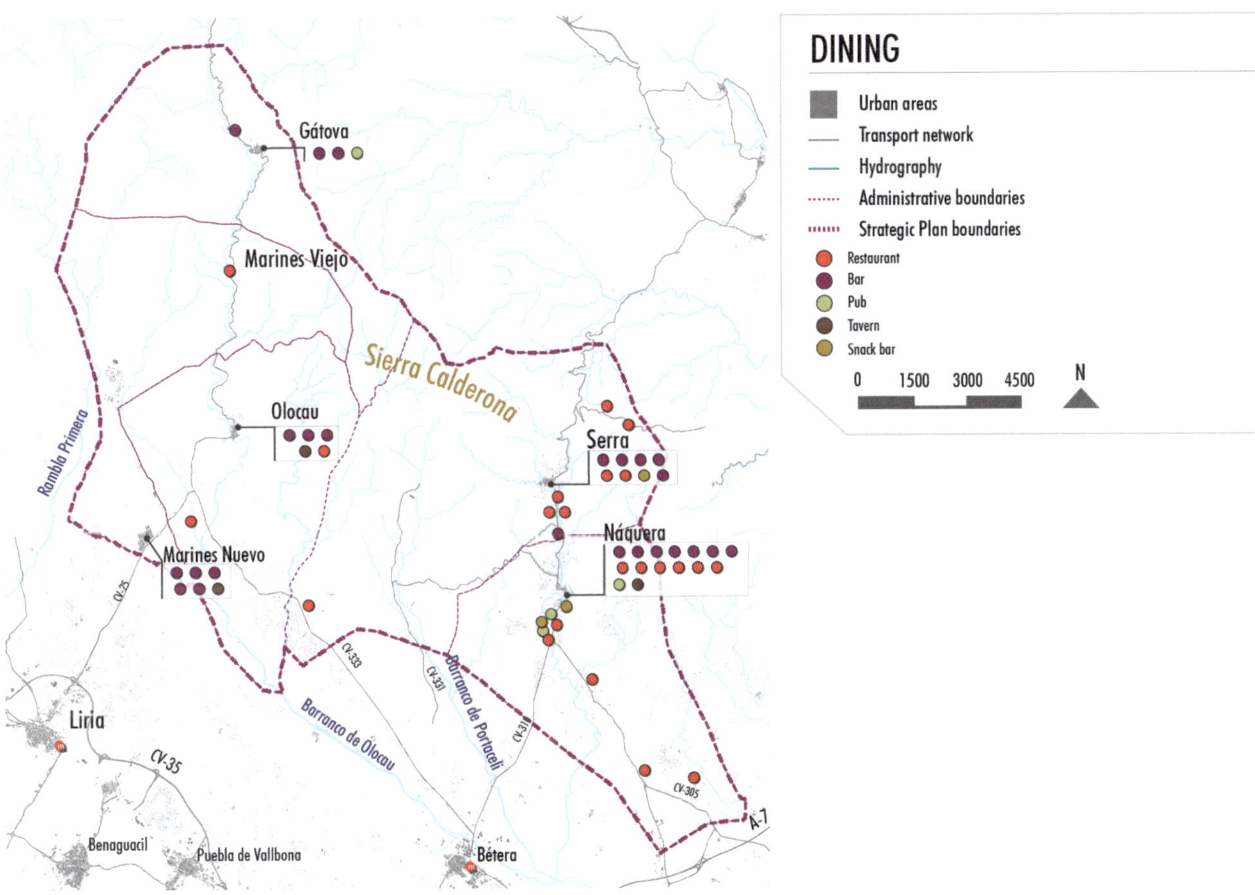

Fig. 3.54 Dining offer. *Source* Own elaboration

- **Itineraries in the natural environment**: The provision of itineraries in the woodlands (either for cars, cyclists or strictly for hikers) is located along dirt tracks. The existing network is extensive, but disconnected from towns that eventually could also function as service hubs, and it has discontinuities in some sections and it is not adequately signposted.
- **Recreational and picnic areas**: Recreational areas are concentrated in the area of Portaceli, with facilities and spaces suitable for large-scale use, in the fountains west of Gátova, where the recreational areas are small and often unknown to the external visitors, and north of Serra, in Aliguetes, which has a precarious access and is surprisingly disconnected from adjacent sites of great interest (Fig. 3.55). Picnic areas complement the recreational areas on a minor scale and are located mainly next to water springs or villages/towns. Within the central area of the Natural Park, there are neither recreational nor picnic areas.
- **Architectural, archaeological and ethnographic heritage**: The area has a rich cultural heritage, located both in villages or towns and rural areas (castles, Iberian settlements, convents, defensive buildings from the civil war, etc.). The reconstruction of the Iberian settlement in "Puntal dels Llops" and of "Casa de la Señoría", both in Olocau and of the tower situated in the historical centre of Serra, sets examples to be followed, though they may often go unnoticed to visitors because of the absence of signposts and the lack of services around them (Fig. 3.55).
- **Cycling routes**: Both road and mountain cycling are popular activities in the area. Road cycling makes use of roads, either sharing the carriageway with other vehicles or along adjacent specific cycle lanes. Mountain biking is developed along some specially marked routes in the forest paths and tracks network (Fig. 3.55).
- **Hunting**: See the subchapter about "Natural environment".
- **Events**: The municipalities included in the Strategic Plan accommodate some unique traditional festivities, i.e. "Fiesta de la Cereza (Cherry party)" in Serra

Table 3.49 Other services and activities. *Source* Own elaboration

Service or activity	Municipality	Amount
Academy	Náquera	1
	Serra	2
Coaches	Náquera	1
Library	Marines	1
	Náquera	1
	Olocau	1
	Serra	1
Beauty salon	Náquera	1
	Serra	1
Hardware	Náquera	1
Garden centre	Náquera	1
	Olocau	1
Nightlife	Náquera	1
Hairdresser	Náquera	2
	Serra	5
Taxi	Náquera	1
Theatre/cinema	Marines	1
	Serra	1
Workshop	Náquera	1
	Serra	1
Air field	Olocau	1
Equestrian centre	Náquera	3
	Marines	1
Botanic garden	Olocau	1
Bullfighting arena	Náquera	1
School farm	Olocau	1
Educational centre	Olocau	1

or the pig-slaughter in Serra and Gátova. Events of more recent origin, though still of high interest, are the musical evenings in "Casa de la Señoría" in Olocau and cultural activities in Marines, Olocau and Náquera. Any of these events, if properly promoted and collated with the rest of the touristic provision, could provide added value to the area.

- **Secondary residence**: In general, dwellings used as secondary residences are mainly located in housing estates, and the interaction between their residents, the local communities and the natural environment is very variable (Fig. 3.55).
- **Dining, accommodation and other services**: this has been partially described earlier in this subchapter (Fig. 3.55).

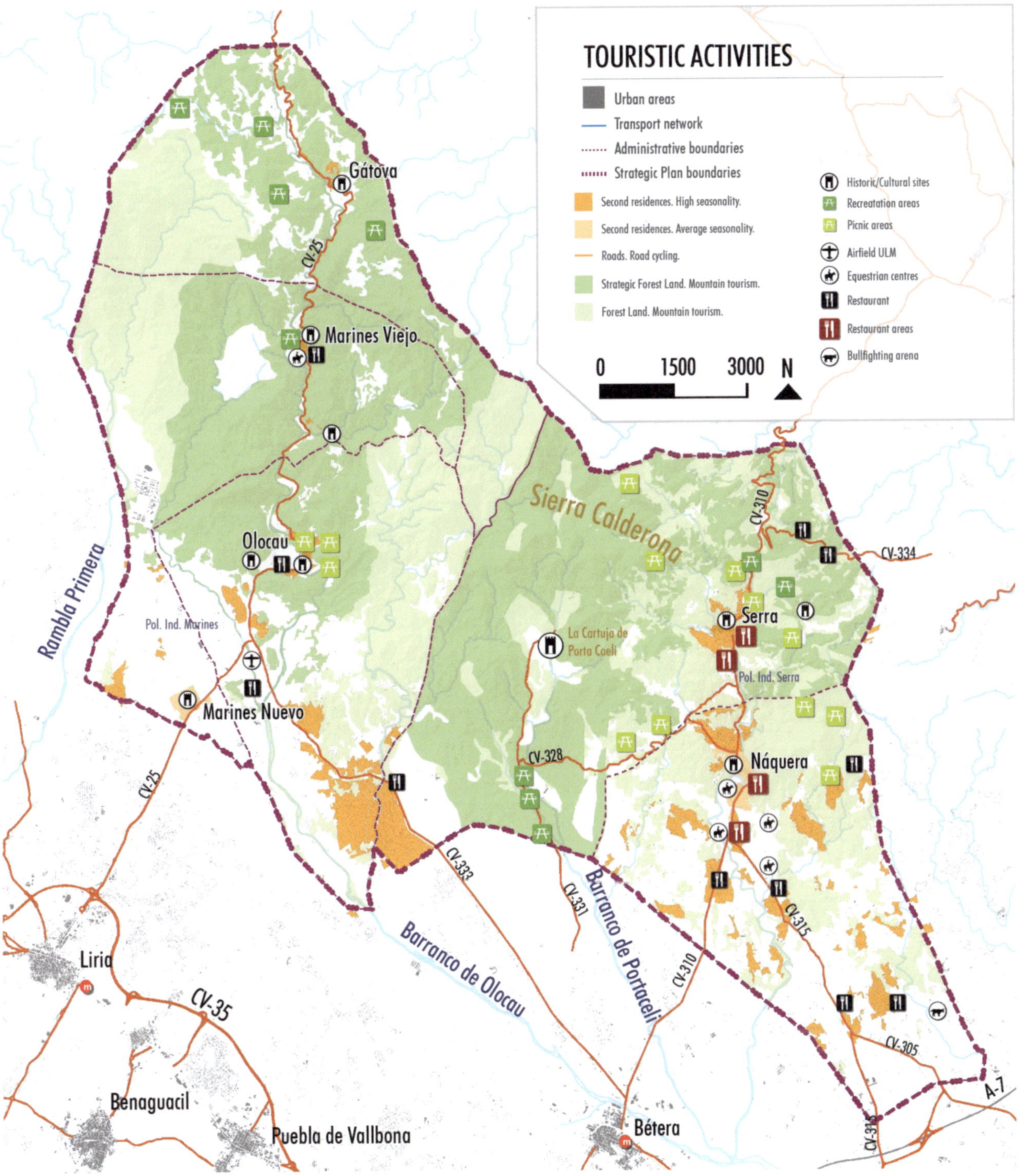

Fig. 3.55 Touristic activities. *Source* Own elaboration

3.7.3 Visitors

Number of visitors: Starting with the number of visitors estimated in the annual report 2009 of the Sierra Calderona Natural Park, an adjustment has been made in order to estimate the proportion that visits the area of the Natural Park covered by this Strategic Plan and to include those visitors who come to visit areas outside the National Park but which are included in the scope of the Strategic Plan. This indicative and approximate figure refers only to casual visitors. At the same time, the number of seasonal visitors (amount of people having a secondary residence in the area) has been estimated from indirect indicators, such as the increase in energy or water consumption, as well as of waste production. It has also been assumed that these seasonal visitors remain in the area an average of 25 days a year.

Profiles, typical visiting patterns and activities developed by casual visitors: The percentage of occasional visitors ascribable to each type of profile (i.e. hikers, users of recreational areas, cyclists) has been estimated, providing an indication of the volume of people performing different activities (Tables 3.50, 3.51).

Table 3.50 Number of visitor. *Source* Own elaboration from Annual Report 2009 of the Sierra Calderona natural park

Visits Natural Park SIERRA CALDERONA (2009)	JAN.	FEB.	MAR.	APR.	MAY	JUN.	JUL.	AUG.	SEPT.	OCT.	NOV.	DEC.	TOT.	Estimate whole area Strategic Plan (65% total area Natural Park)	Increment 12% counting total area Strategic Plan (areas outside Natural Park)
Notified visits		192	37	170	775	4	50				52	50	1,330	865	1,037
Drafting PUP	57	13	60	416		4							550	358	429
Hiking and sport competitions		45		500	600			70	750	3,100	1,160	2,060	8,285	5,385	6,462
Heritage interpretation	20	35	41	65	17	25							203	132	158
Voluntary work management			60										60	39	47
Estimate surveillance	3,430	1,180		3,400	8,400					100			16,510	10,732	12,878
Estimate establishments	4,725	4,275	4,275	4,275	4,275	4,275	4,275	4,275	4,275	4,275	4,275	4,275	51,750	33,638	40,365
Environmental education	173	350	485	40	133								1,181	768	921
Communication and promotion					3,000								3,000	1,950	2,340
Attended in office	15	17	12	16	13	6	13			150			242	157	189
TOTAL	8,420	6,107	4,970	8,882	17,213	4,314	4,338	4,345	5,025	7,625	5,487	6,385	83,111	54,022	64,827

Table 3.51 Number of visits by visitor profile. *Source* Own elaboration

Visitor profile	Main activity	%	No. of visits/year	Duration visit (days)
Occasional visitors (not owning/renting a house/flat in the area)	Group visiting recreational	27	17,503	1
	Hikers	30	19,448	1
	Group of sportsmen/women (mountain/road cyclist)	28	18,151	1
	Group visiting rural + urban areas	14	9,076	1
	Group staying overnight visiting the rural environment (*)	0.8	519	3
	Group staying overnight visiting the rural environment and urban centres (*)	0.2	130	3
	Total	100	64,827	
Visitor profile	Main activity	%	No. of persons	No. of days in a year/persons
Seasonal visitors (owning/renting a house/flat in the area)	In housing estates and urban centres	100	13,774	25

*overnight stays = 5% of the total in the area of socio-economic influence of the Natural Park
*accommodation vacancies = 5% of the total in the area of socio-economic influence of the Natural Park

3.7.4 Economic Impact of Tourism

The estimation of the daily spending of each occasional or seasonal visitor profile, together with the preliminary estimation of the number of visitors in each category, allows us to calculate the overall economic impact on the local economy of both types of tourists.

These figures, developed further in the subchapter "Local economy", are purely indicative. A more precise calculation would have required resources and time frames that exceeded those available for the drafting of the Sierra Calderona Strategic Plan. Nevertheless, they can provide valuable information to find out which aspects should be reinforced in the touristic provision and in further studies (Table 3.52).

Table 3.52 Average and total spending by type of visitor. *Source* Own elaboration

Visitor profile	Main activity	No. of visits/year	Duration visit (days)	Spending (€/Day)	Total spending (€)
Occasional visitor (not owning/renting a house/flat in the area)	Group visiting recreational areas	17,503	1	1.5	26,254.76 €
	Hikers	19,448	1	3	58,343.92 €
	Group of sportsmen/women (mountain/road cyclist)	18,151	1	6	108,908.65 €
	Group visiting rural + urban areas	9,076	1	12	108,908.65 €
	Group staying overnight visiting the rural environment (*)	519	2.53	50	65,604.50 €
	Group staying overnight visiting the rural environment and urban centres (*)	130	2.53	80	26,241.80 €
	Total	64,827			394,262.29 €
Visitor profile	Main activity	No. of persons	No. of days in a year/person	Spending (€/Day)	Total spending (€)
Seasonal visitor (owning/renting a house/flat in the area)	In housing estates and urban centres	25,000	20	15	7,500,000 €
	Total	25,000			7,500,000€

*overnight stays = 5% of the total in the area of socio-economic influence of the Natural Park
*accomodation vacancies = 5% of the total in the area of socio-economic influence of the Natural Park

3.7.5 Public Participation Plan

See Table 3.53.

Table 3.53 Results of Public Participation Plan. Phase 1. *Source* Own elaboration

Tourism	Total (%)	GÁTOVA (%)	OLOCAU (%)	MARINES (%)	NÁQUERA (%)	SERRA (%)
Conflicts						
Itineraries: poor signposting, difficult access, flack of maps and guides, lack of general information	38.6	21.4	52.0	34.8	46.8	52.8
Lack of cooperation in the promotion of tourism in the area	25.1	14.3	44.0	13.0	29.8	33.3
Insufficient conservation and promotion of the landscape resources (i.e. archaeological, architectonic, gastronomic heritage)	23.4	7.1	56.0	34.8	10.6	30.6
Insufficient accommodation/dining offer of high quality	19.3	14.3	28.0	13.0	25.5	22.2
Limited impact of tourism in the local economy	15.8	21.4	16.0	17.4	12.8	25.0
None	9.9	28.6	12.0	26.1	8.5	0.0
Lack of professionals and formation options	9.4	14.3	16.0	0.0	4.3	19.4
Lack of high-quality local products	7.6	14.3	16.0	0.0	2.1	13.9
Do not know/no answer	22.2	7.1	0.0	17.4	17.0	0.0
Opportunities						
To improve tourist information (information points maps, guides, signposting, …)	32.7	21.4	40.0	34.8	31.9	52.8
To promote tourism in order to support the local economy	30.4	64.3	40.0	52.2	19.1	33.3
To extend accommodation/dining offer of high quality	20.5	21.4	32.0	17.4	21.3	27.8
To properly plan tourism: avoiding crowds and urban sprawl, diversifying and improving services	16.4	0.0	60.0	17.4	8.5	13.9
To extend and diversify the network of itineraries for cyclists, hikers and horse riders	13.5	28.6	24.0	13.0	6.4	19.4
Do not know/no answer	36.3	21.4	12.0	8.7	42.6	25.0

3.7.6 Synthesis and Diagnosis

Volume of visitors and economic impact: Using data from the annual reports of the Sierra Calderona Natural Park and the high season census, it is estimated that the number of occasional visitors to the area is 64,000 people, and of seasonal visitors is 13,000 people. These visitors represent an income to the local economy estimated at €400,000 and €7,500,000, respectively, corroborating that sporadic or casual visitors make a small contribution to local economies when compared with the contribution of seasonal residents owning a secondary residence in the area.

Current facilities: There is a limited provision of regulated accommodation (hotels, hostels and apartments), as well as campsites. The dining provision is abundant in the area of Náquera, with consolidated establishments, but very limited in other municipalities. Fine dining restaurants are not present in the area. There is also a very limited provision of services and facilities for visitors, such as guided tours to heritage or natural sites, health and rehabilitation centres, guided outdoor activities. The existing services focus on the permanent population, despite the strong flow of seasonal and occasional visitors.

Potential services: The presence of exceptional environmental and landscape values, together with a network of attractive villages and heritage sites of high interest in the proximity of the metropolitan area of Valencia, gives the Calderona Mountain Range (Sierra Calderona) a high potential to enhance its touristic use. It would be possible to expand the provision of services for tourists in many sectors and to jointly promote such services in collaboration with public and private actors. According to the presented analysis, the main sectors where tourism could be developed are:

- Itineraries and activities in the natural environment.
- Recreational and picnic areas.
- Guided tours on archaeological, architectural and ethnographic heritage sites.
- Cycling itineraries.
- Specialised services and specialised events (elderly, health and beauty, high performance sports centres, cultural events associated with art, music, etc., equestrian activities).
- Dining and accommodation.

3.8 Landscape and Territorial Models

3.8.1 Generalities

Following the approach defined by the European Landscape Convention and the Valencian landscape legislation, this subchapter presents an analysis of the landscape, understood as the perceived dimension of any territory (urban, natural, agricultural, infrastructural, etc.), as well as in their areas of contact. The study assumes that high-quality and sustainable landscapes are the result of a positive interaction between humans and the environment, as well as that they promote prosperity and development in the short-, medium- and long-term.

In order to give a proactive role to the landscape in the management and development of the Sierra Calderona, this subchapter includes a preliminary landscape characterisation and assessment as well as the identification of conflicts and potentials concerning the landscape.

In addition, and in order to connect cross-cutting factors affecting several of the topics included in the overall analysis included in this subchapter, different spatial models that explain the current situation and its possible evolution are discussed.

3.8.2 Natural Landscapes

These are mainly located at forested areas. The spaces or elements of higher natural value are located in areas with dense and mature forests of cork oak and pine, as well as in the steepest topographies. Many of the valleys are traversed by the transport infrastructure corridors and present a high landscape quality, which gives those roads an added value as scenic routes. Among the elements with a special value are viewpoints, itineraries and individual rural buildings.

The main problems are situated in poorly integrated tracks or routes, some recreational areas and some isolated and highly visible low-quality buildings.

In general, the natural landscapes in the south-central area of the Sierra Calderona present a high or very high landscape quality that usually diminishes in those areas where the vegetation cover gets less dense or mature (Figs. 3.56 and 3.57).

Fig. 3.56 Natural landscape at Sierra Calderona. *Source* Bingmaps

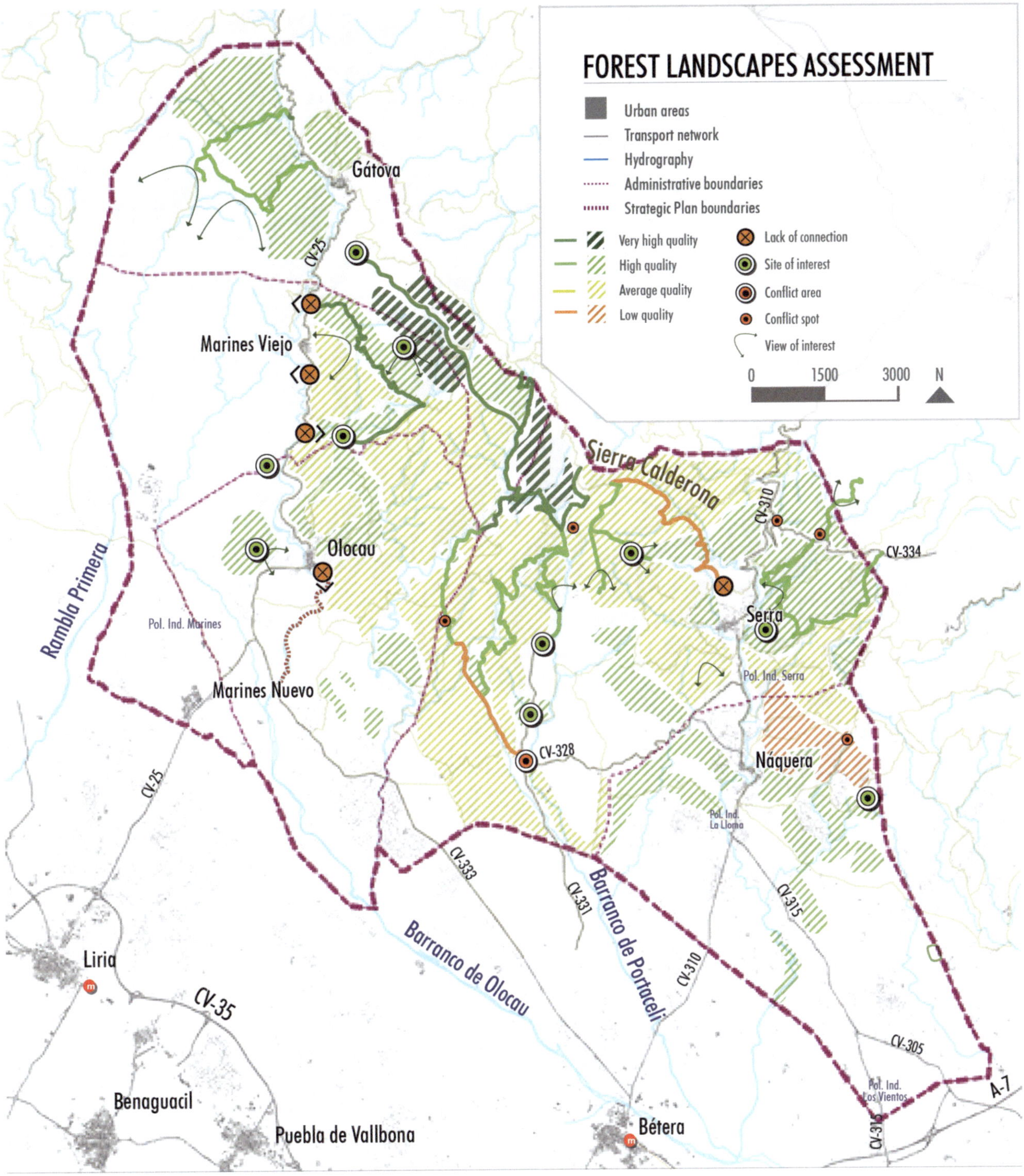

Fig. 3.57 Natural landscapes assessment. *Source* Own elaboration

3.8.3 Agricultural Landscapes

The agricultural areas situated in the mountains are scarce and mostly defined by scattered, terraced, rainfed plots which add value to the rural landscape mosaic, and which, if abandoned, are spontaneously overgrown by forest vegetation that maintains an adequate landscape quality.

The orchards and small vegetable gardens located next to the historical settlements and villages usually have a very high landscape value, although the presence of intrusive elements is frequent.

The agricultural areas located in the foothills and plains constitute a territorial matrix interrupted by a dense network of housing estates, roads and other infrastructure. The high level of abandonment of some plots, especially those dedicated to rainfed agriculture, together with the strong presence of intrusive constructions of low quality (i.e. poorly integrated buildings, fences or walls), generates a landscape of medium or low quality which slightly improves around irrigated plots of citrus trees (Figs. 3.58, 3.59, 3.60 and 3.61).

Fig. 3.58 Agricultural plots in plains. *Source* Bingmaps

Fig. 3.59 Orchards and vegetable gardens in mountain villages. *Source* Bingmaps

Fig. 3.60 Rainfed and terraced agriculture in mountainous areas. *Source* Bingmaps

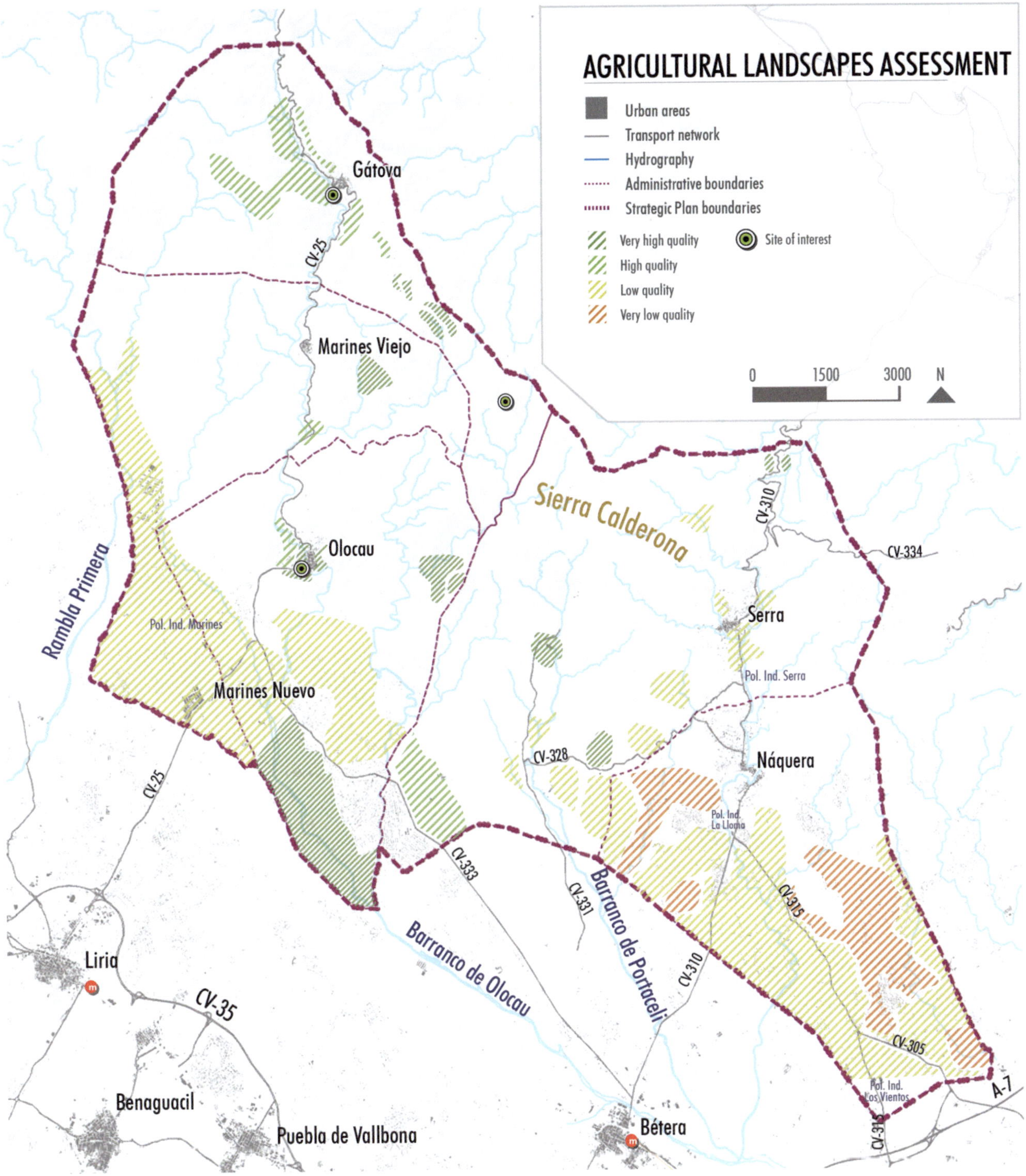

Fig. 3.61 Agricultural landscapes assessment. *Source* Own elaboration

3.8.4 Urban Landscapes

This section identifies and assesses the landscape quality of urban areas and urban edges. The evaluation of urban edges has considered the character and treatment of the border elements (i.e. façades, closing or retaining walls, fences, hedgerows). Inner urban landscapes have been assessed mainly by analysing the quality of public spaces (streets, green areas, squares, etc), the consistent use of building types and volumes, the treatment of façades or fences and the transition between adjacent urban fabrics.

The study was developed in villages and towns, housing estates, scattered residential settlements and industrial areas (Figs. 3.62, 3.63, 3.64 and 3.65).

Fig. 3.62 Industrial areas. *Source* Bingmaps

Fig. 3.63 Housing estates. *Source* Bingmaps

Fig. 3.64 Scattered settlements. *Source* Bingmaps

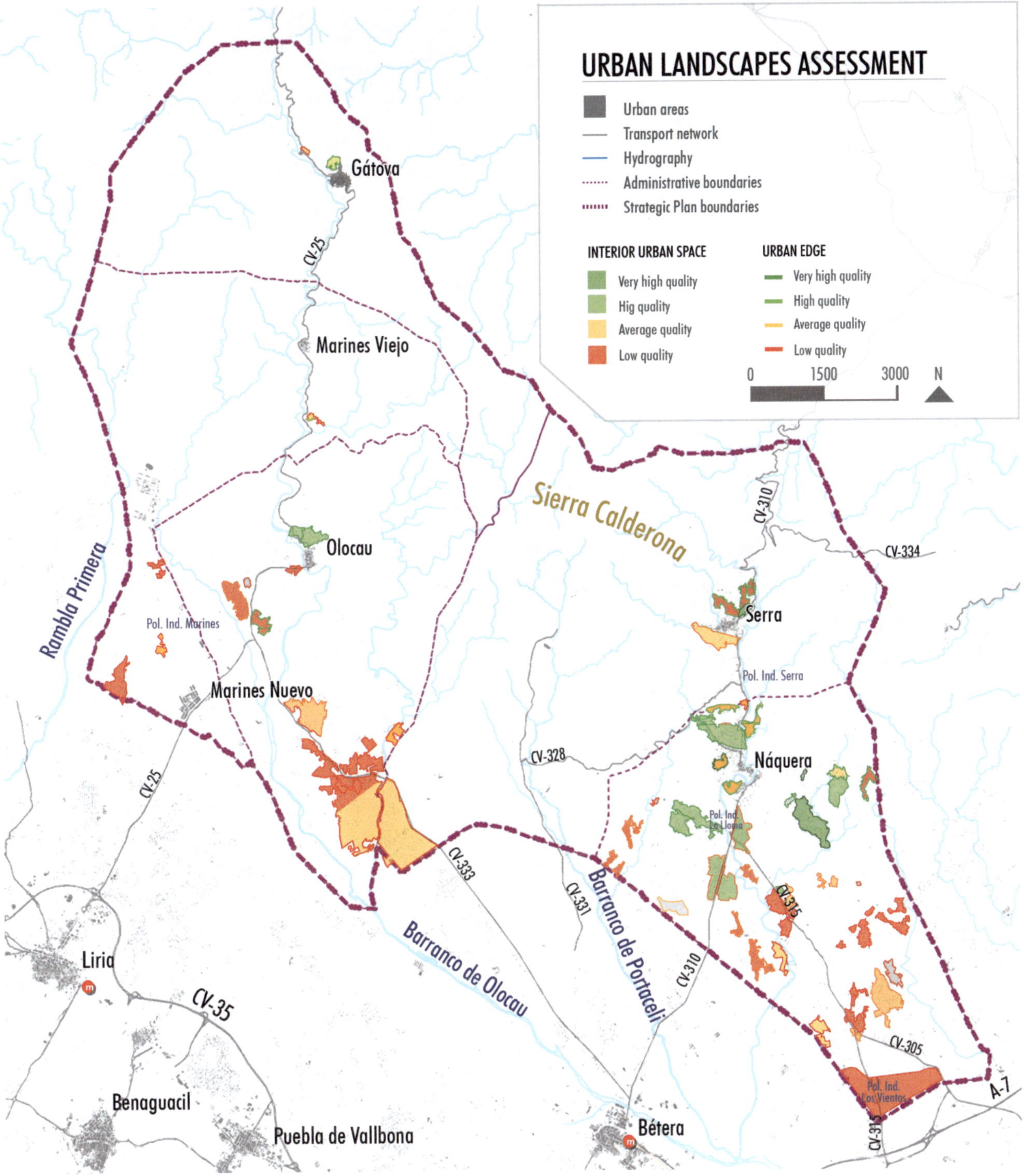

Fig. 3.65 Urban landscapes assessment. *Source* Own elaboration

3.8.4.1 Villages and Towns

The historical tissues of Sierra Calderona's villages and towns have a high or very high landscape value despite the occasional presence, in some of them, of intrusive and poorly integrated buildings. Marines Nuevo stands out for the quality and consistency of both its interior landscape and its edges, and it constitutes a remarkable example of a twentieth century colonisation settlement. In Serra and Náquera, the addition at their margins of new urban fabrics of either low quality or poorly articulated with the historical centres decreases their overall quality significantly. In general, the relationship between towns and their peripheral or inner ravines is quite weak, in spite of their potential to articulate the towns and their green areas, functioning as linear parks and regional connectors (Figs. 3.66, 3.67 and 3.68).

Fig. 3.66 Urban historic centre. *Source* Bingmaps

Fig. 3.67 Compact urban centre. *Source* Bingmaps

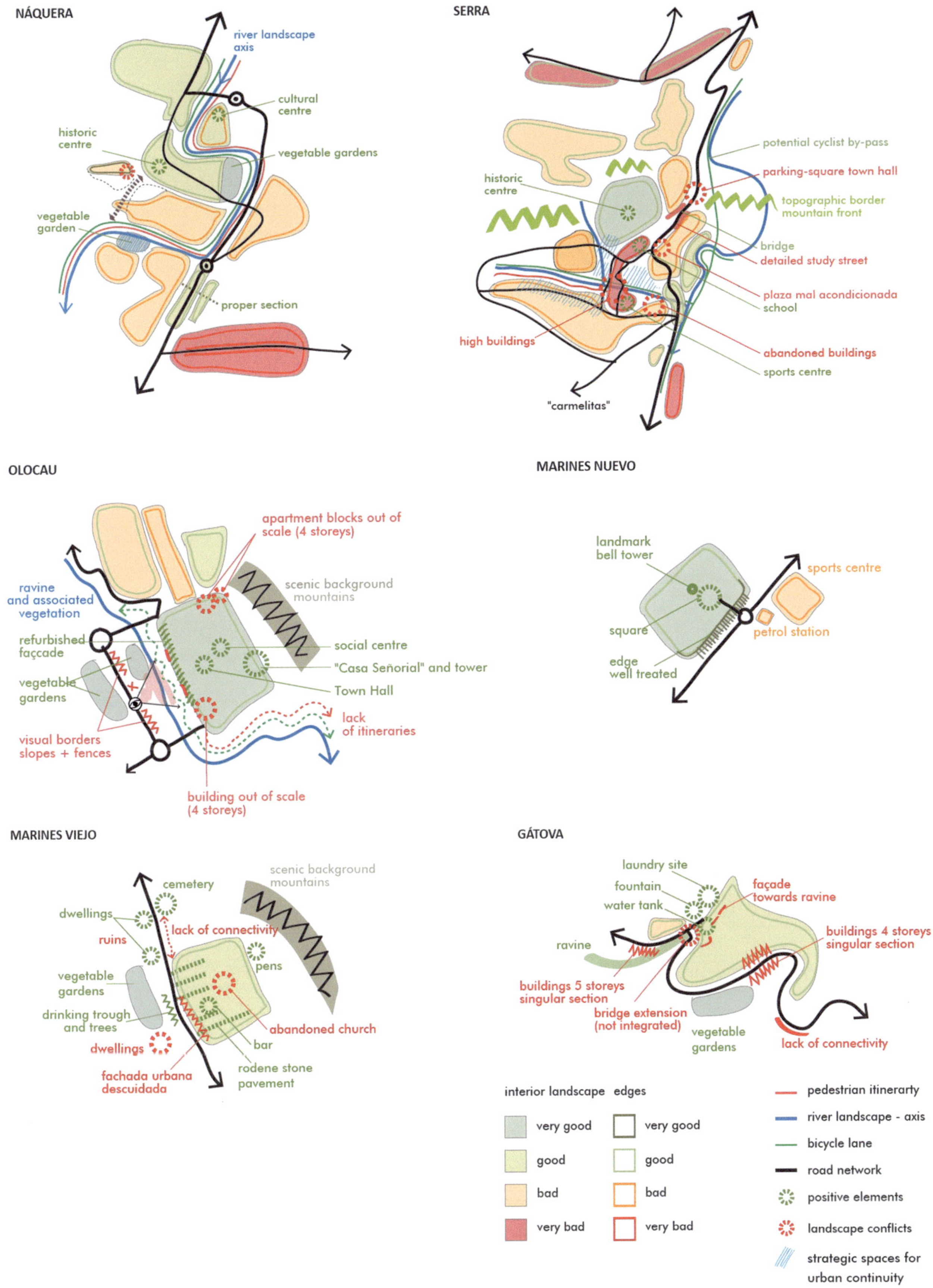

Fig. 3.68 Urban landscape. Urban centres of Náquera, Serra, Olocau, Marines Nuevo, Marines Viejo and Gátova. *Source* Own elaboration

3.8.4.2 Housing Estates and Scattered Settlements

The quality of both the edges and the inner landscape of these settlements varies considerably within the area of the Strategic Plan. In general, the settlements located in the mountainous areas have a higher quality, as they are often more consolidated, surrounded by forests and with a dense internal tree cover that facilitates their integration into the landscape. The settlements located in the plains often have a poorer treatment of their edges, as a result of both the abandonment of the surrounding agriculture fields and the frequent presence of poorly finished retaining walls or fences. Although the quality of their inner urban landscape varies, in general, the public space is poorly treated, i.e. very narrow sidewalks, lack of street trees, uneven treatment of fences and walls, non-existing or poor maintenance of green areas. This becomes more critical in almost all scattered settlements, regularised or not. The intersections between housing estates and transport infrastructures illustrate many of the problems already discussed, impairing the perception that drivers and cyclists get of the whole territory (Figs. 3.69, 3.70 and 3.71).

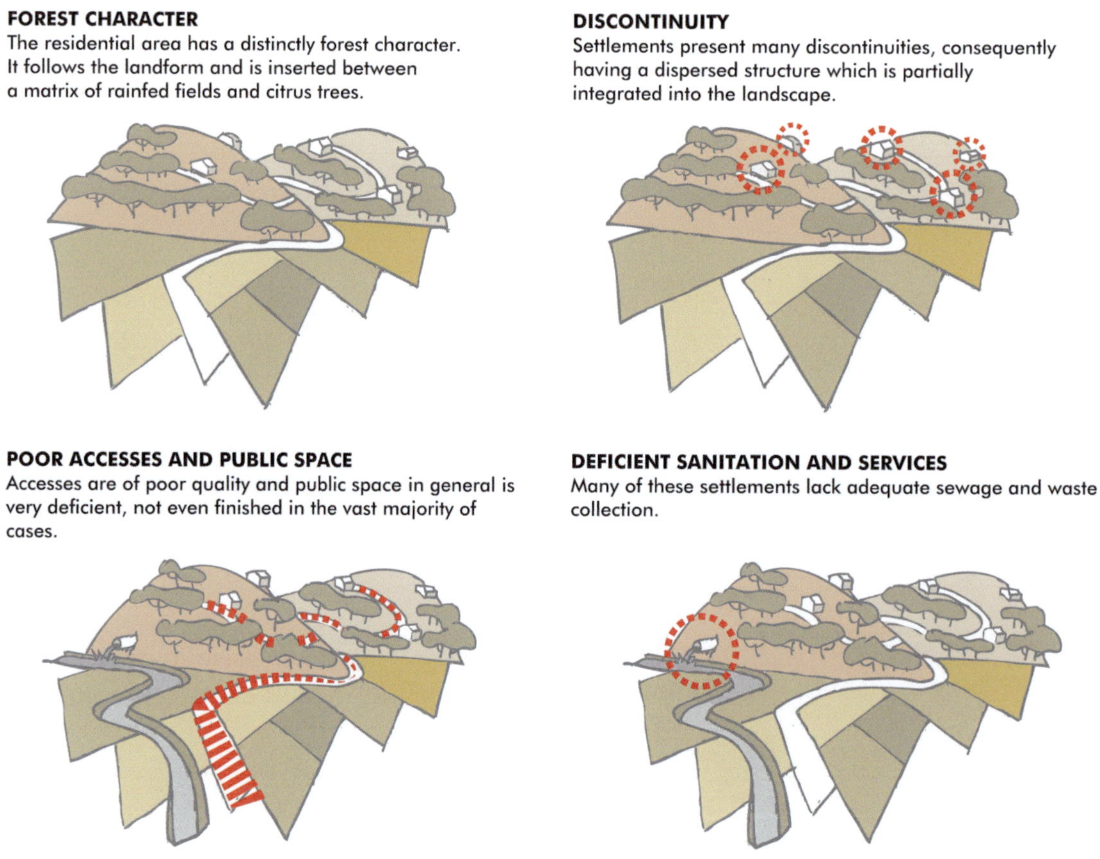

Fig. 3.69 Scattered settlements in mountainous areas: characteristics and problems. *Source* Own elaboration

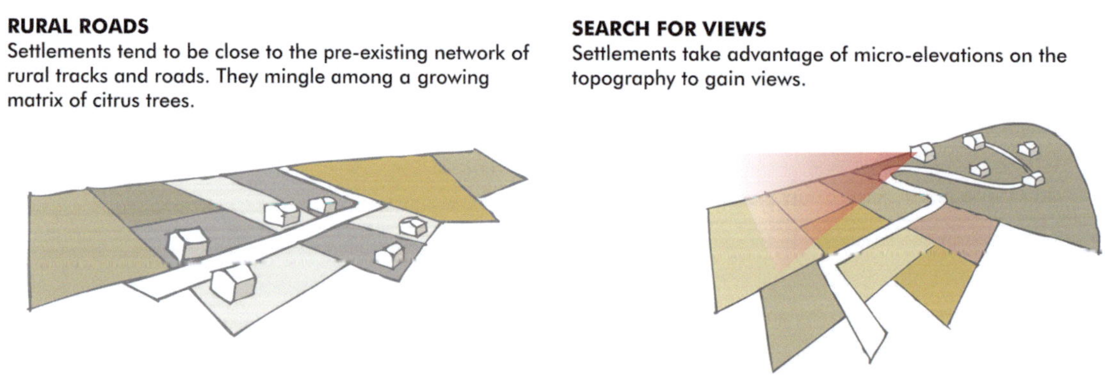

Fig. 3.70 Scattered settlements in low plains: location criteria. *Source* Own elaboration

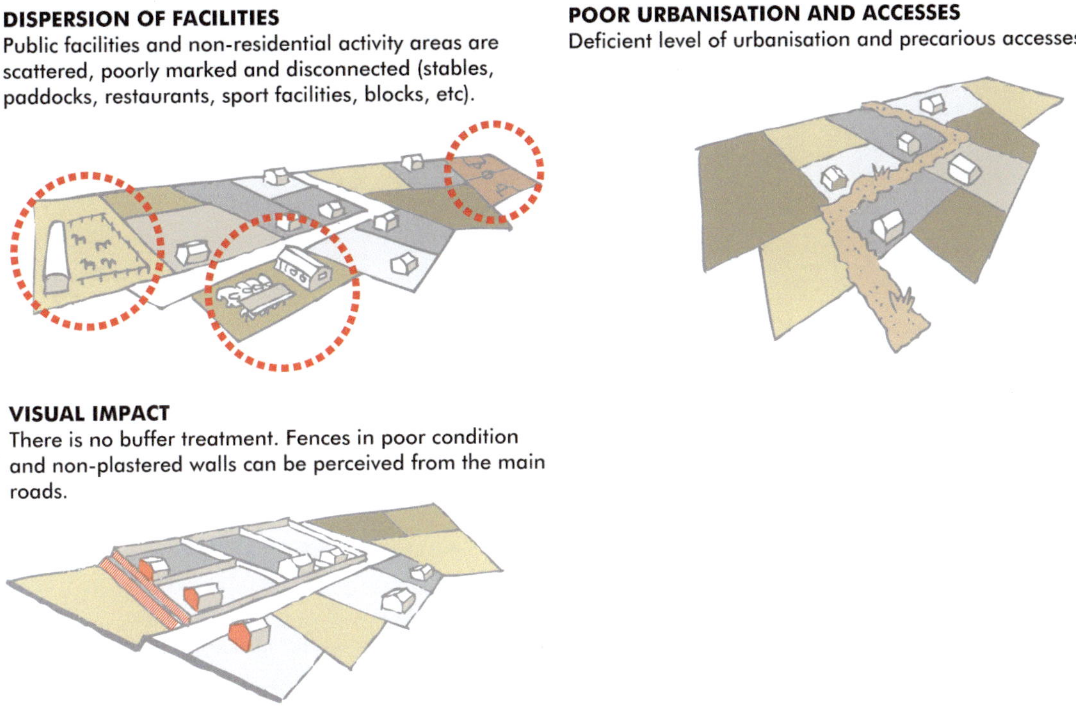

Fig. 3.71 Main problems in scattered settlements situated in low plains. *Source* Own elaboration

3.8.4.3 Industrial Areas

There are few and small scattered industrial areas, so, despite being close to roads, they do not cause an excessively negative effect. Industrial estates attached to urban centres are often embedded within them, though their incidence is not particularly serious.

The industrial estate *Los Vientos* is the only exception. Due to its large size and volume, it exerts a strong impact on the surrounding landscape. This negative impact is accompanied by a particularly deficient inner public space.

Throughout the area, there are no buffering green spaces between the industrial areas and the traffic infrastructure.

3.8.5 Landscape from Transport Infrastructures

The character and quality of the landscape as perceived from traffic infrastructures are highly defined by the landscape of the natural, agricultural or urban spaces adjacent to them. However, the infrastructures themselves often generate dynamics within the tissues they cross that modify that perception. This consideration, together with the fact that the roads constitute the main and most frequent viewpoint towards the landscape, gave rise to the necessity of specifically analysing the quality of the landscape along the edges of the main roads. The conclusion of that analysis has been that the improvement of the visual conditions along those edges would have a considerable and positive impact on the perceptions that locals and visitors can have of the surrounding territory. It would also be important that this positive experience started already beyond the limits of the area covered by the present Strategic Plan (Figs. 3.72 and 3.73).

On the basis of this analysis and in keeping with the results obtained in the Visual Plan of Comunidad Valenciana ("Plan Visual de la Comunidad Valenciana"), it is concluded that the perception of the quality of the landscape from the roads is considerably worse than the quality of the spaces around them. This is largely due to the poor treatment of the edges (hard shoulders and margins) and the low quality of the adjacent elements (walls, fences, buildings, etc.). However, there are already many road sections which do have a high scenic value, mostly located in mountainous areas, but also some that traverse agricultural plains or next to urban areas (Figs. 3.74 and 3.75).

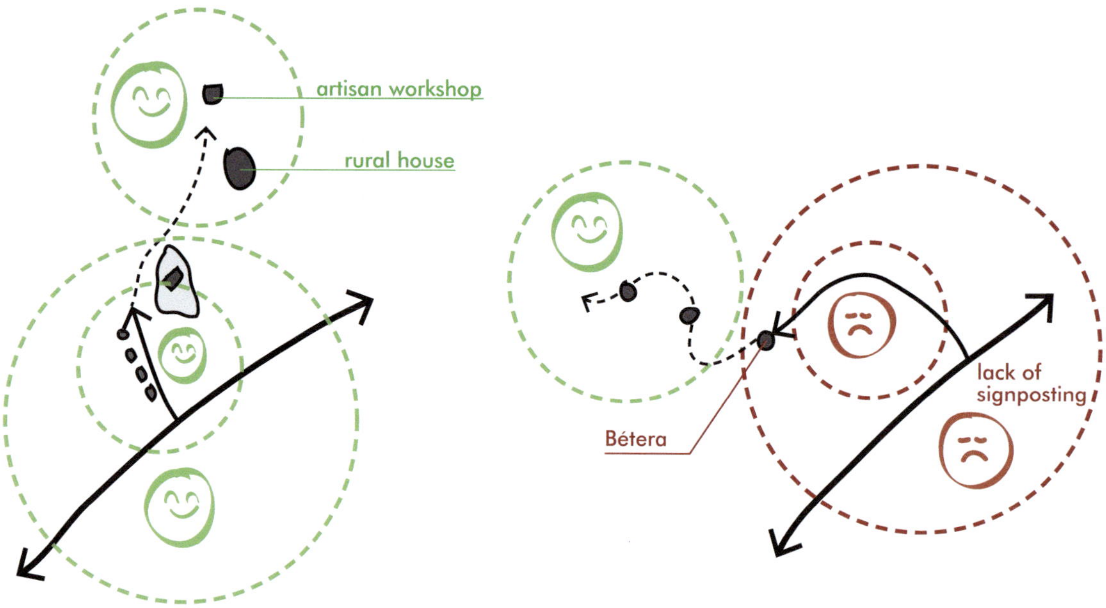

Fig. 3.72 A positive (left) and negative (right) experience of the landscape from the city towards the rural areas. *Source* Own elaboration

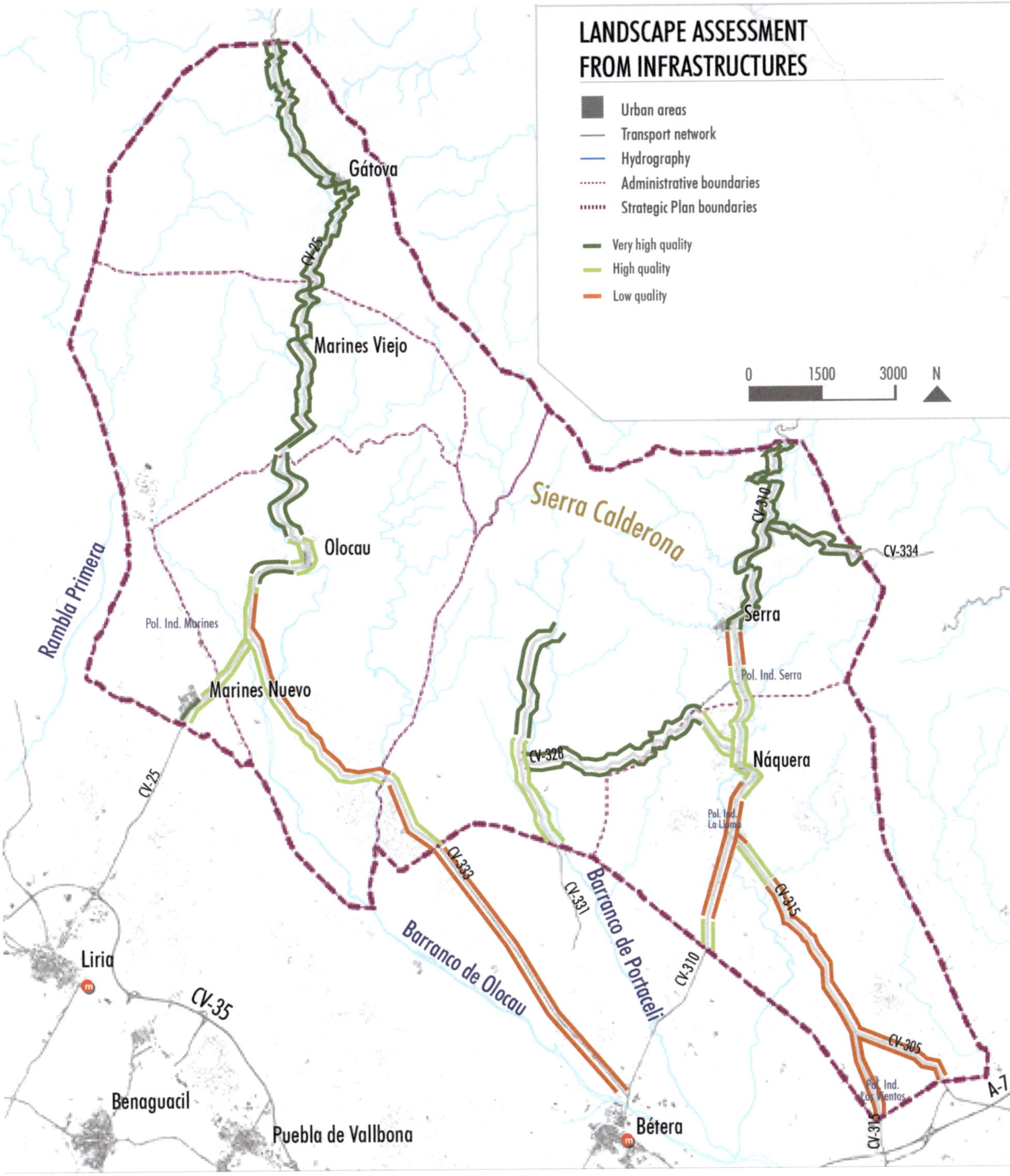

Fig. 3.73 Landscape quality assessment from the main transport infrastructures. *Source* Own elaboration

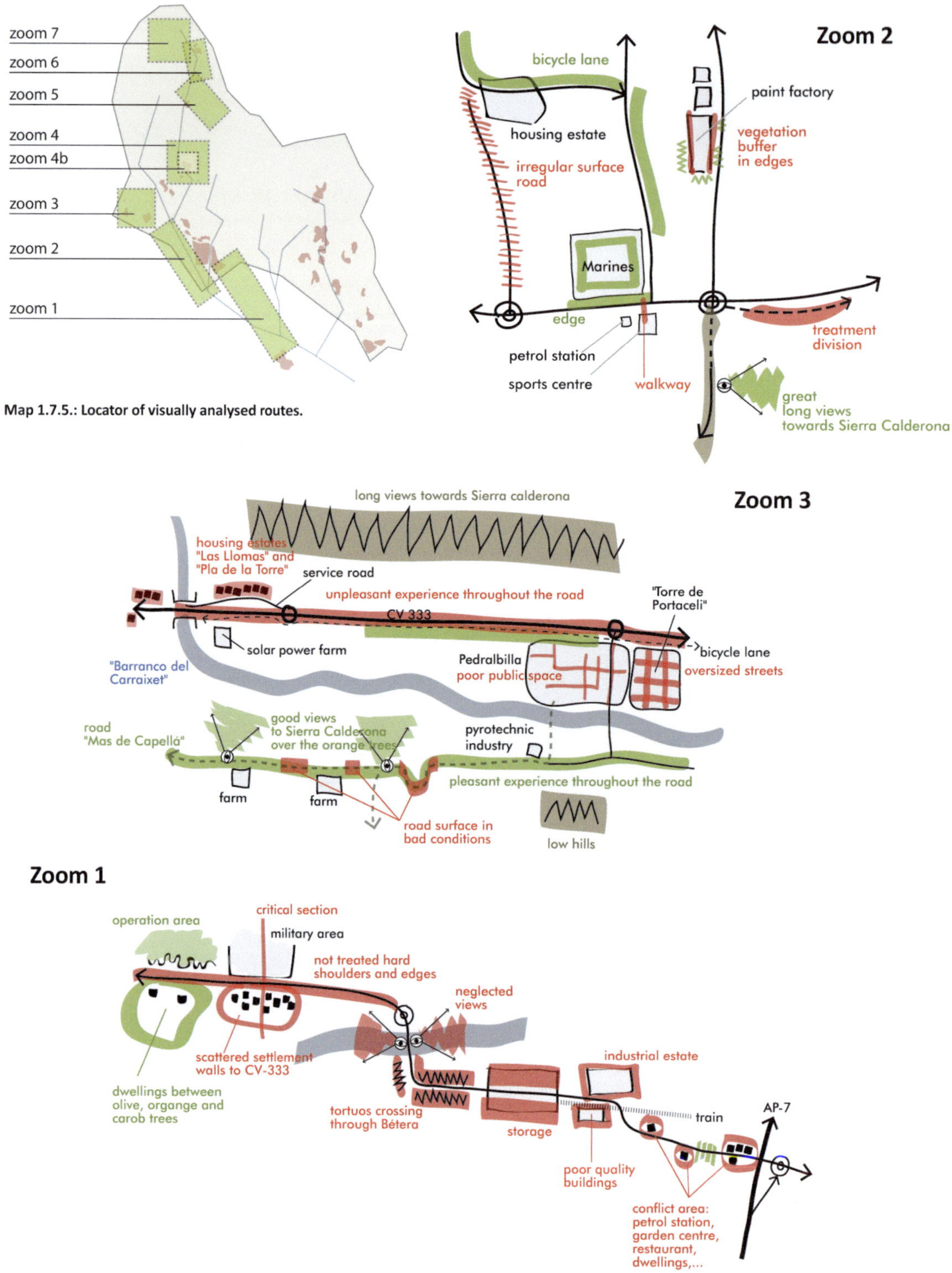

Fig. 3.74 Locator map of visually analysed routes and Zooms 1, 2, 3. *Source* Own elaboration

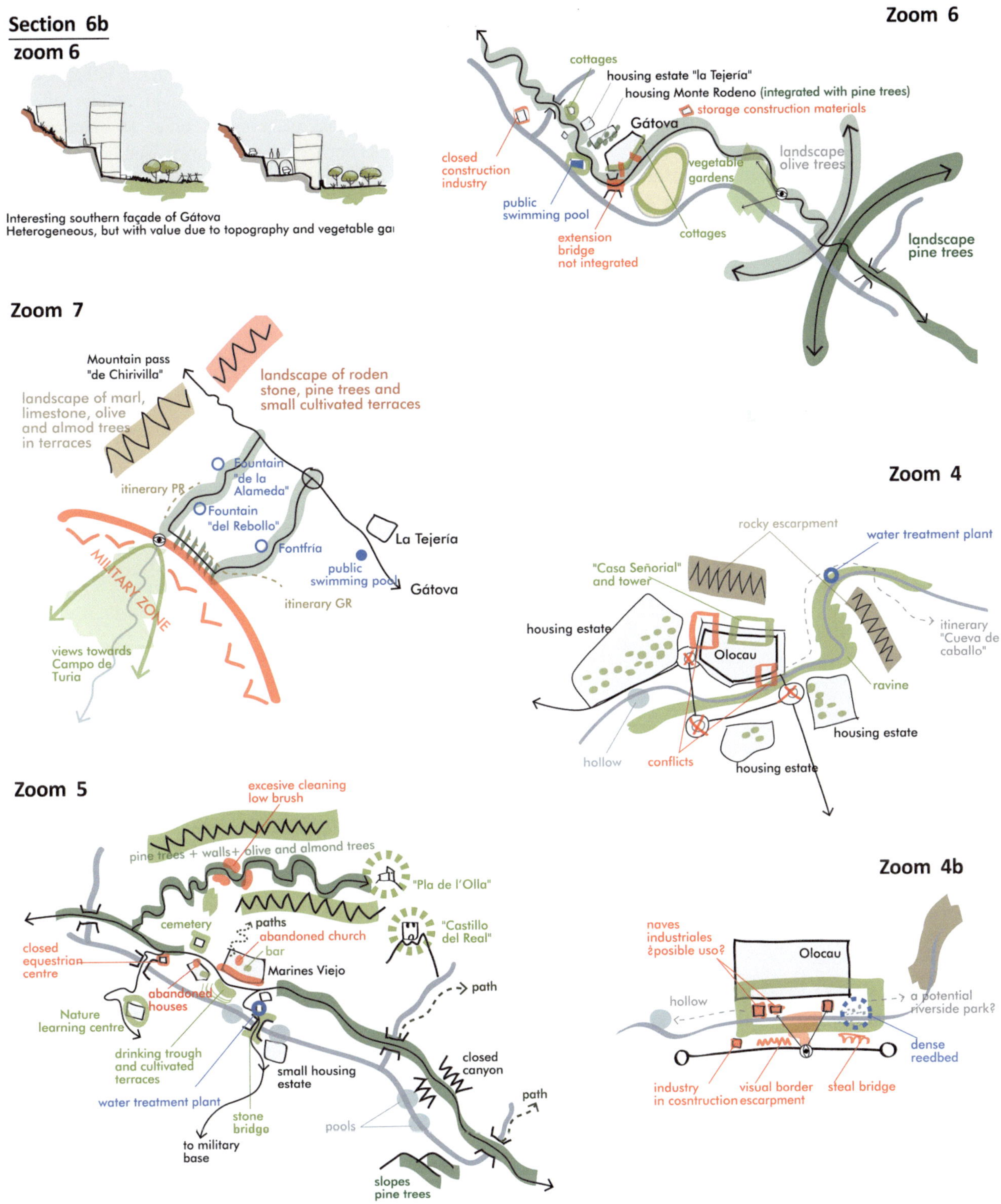

Fig. 3.75 Visually analysed routes and Zooms 4, 4b, 5, 6, 6b and 7 (see locator map in Fig. 3.74). *Source* Own elaboration

3.8.6 Territorial Models for Urbanisation

This section includes a conceptual reflection about the logics and possible changes in the structural configuration and spatial distribution of some land uses.

3.8.6.1 A Regional Territorial Model for the Evolution of Towns and Housing Estates

This examines the relationship between regional infrastructure corridors, urban land uses and the agricultural-forest matrix.

The unplanned evolution of the territory would lead to a scenario 0, where low–medium-density built areas would consolidate in the foothills, while the interior villages would be neglected. Other alternatives would require effective strategic and spatial planning in order to generate new functional links between compact towns and housing estates and avoid increasing and delocalised urban sprawl (Fig. 3.76).

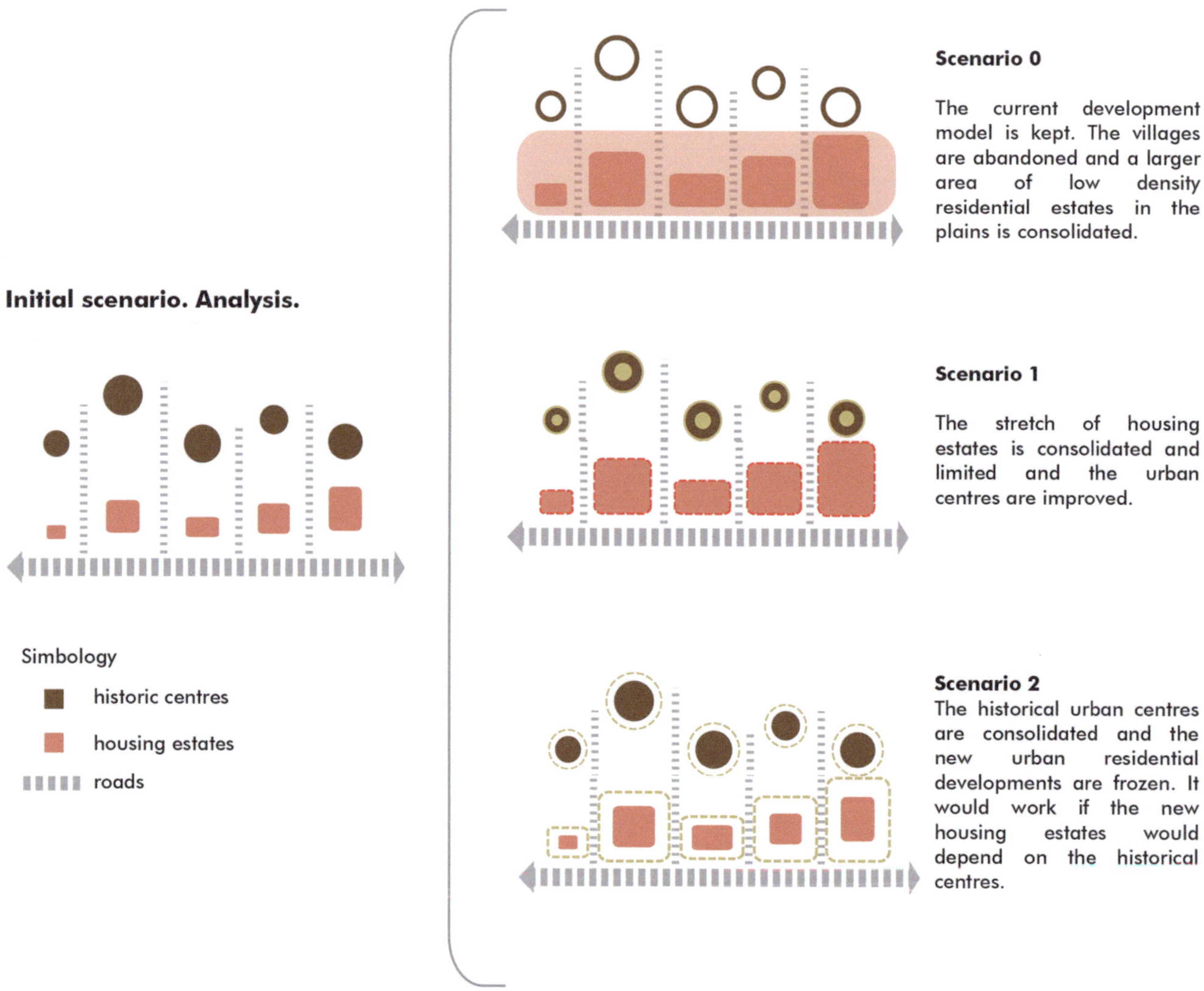

Fig. 3.76 Alternative models for the regional evolution of towns and housing estates. *Source* Own elaboration

3.8.6.2 Regularisation of Scattered Settlements

Regularisation of existing scattered settlements has often been based on the aggregation of new buildings that, allegedly, should have improved the conditions of the already existing housing stock. However, the application of this strategy has seldom been successful and has increased the number and size of housing estates on the agricultural plains and in natural areas (Figs. 3.77 and 3.78).

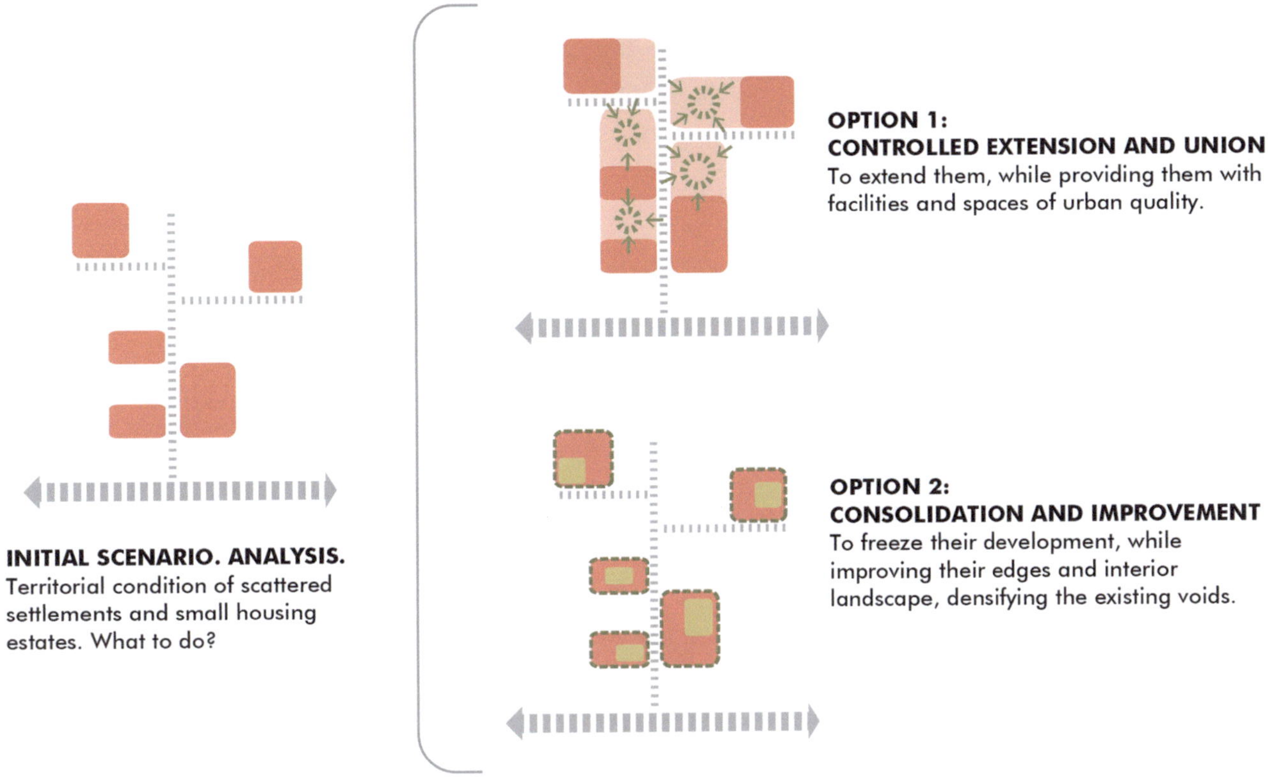

Fig. 3.77 Patterns and alternatives for the regularisation/integration of scattered settlements. *Source* Own elaboration

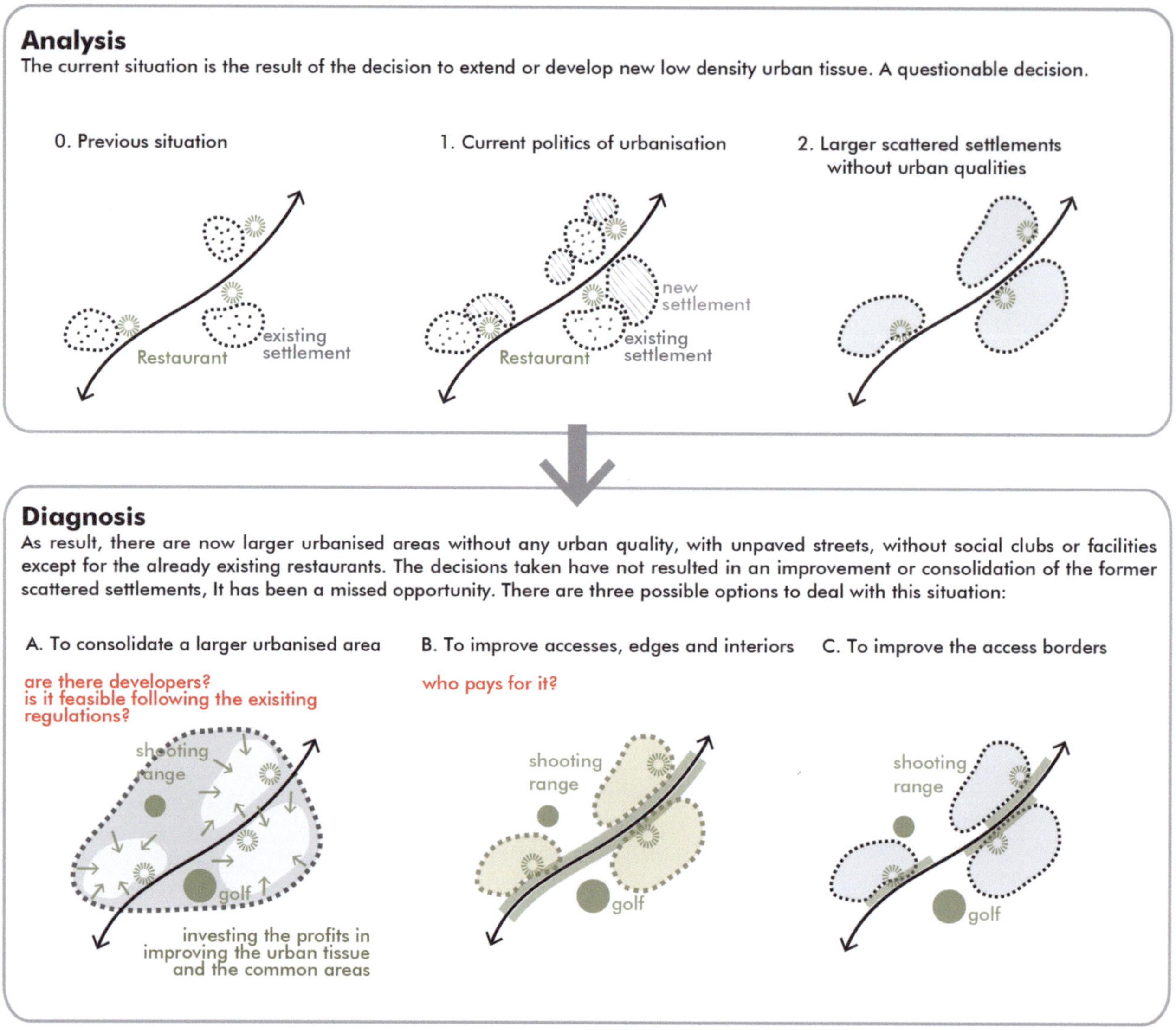

Fig. 3.78 Evolution of scattered settlements in plains. Current situation and alternative strategies for their improvement. *Source* Own elaboration

3.8.6.3 Integrated Network of Itineraries

Frequently, there is a disconnection between itineraries both on natural and agricultural areas, since they are usually defined as isolated routes without a common inter-municipal plan. Alternatively, itineraries should be unified in both areas, take advantage of existing continuous public domains, such as riverbeds and their easement areas and contribute to the general objectives defined for the Green Infrastructure of the Valencian Region (Fig. 3.79).

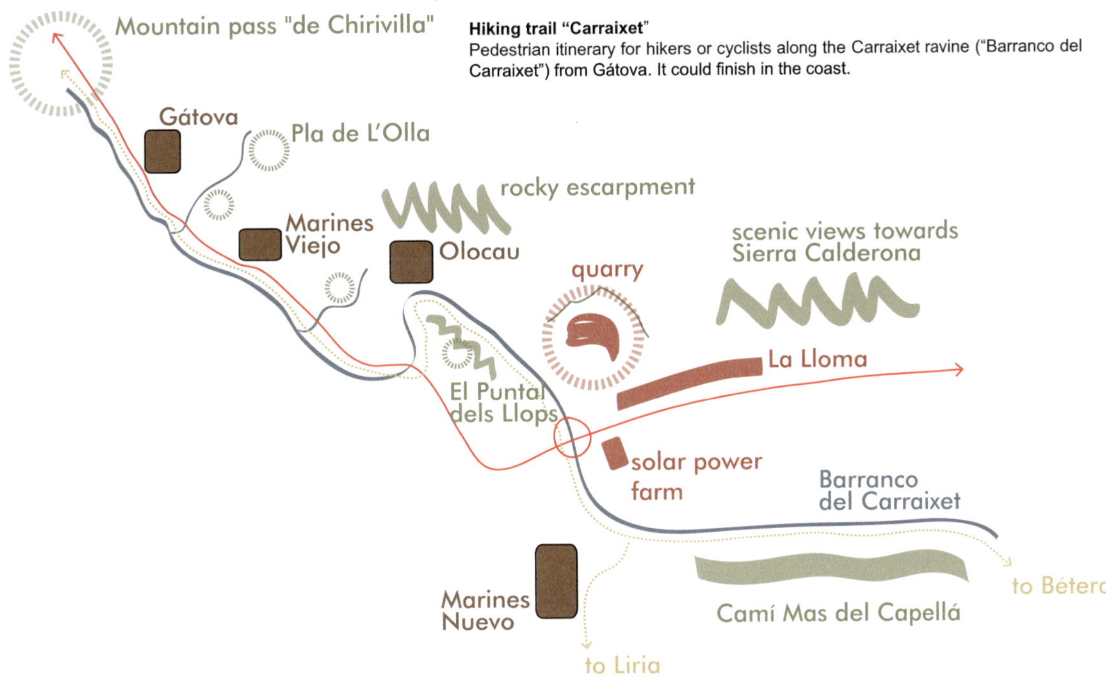

Fig. 3.79 Strategies to increase continuity and interest of itineraries in the rural environment. *Source* Own elaboration

3.8.7 Public Participation Plan

See Table 3.54.

Table 3.54 Results of the Public Participation Plan. Phase 1_landscape quality. *Source* Own elaboration

Landscape quality	Total (%)	GÁTOVA (%)	OLOCAU (%)	MARINES (%)	NÁQUERA (%)	SERRA (%)
Conflicts						
Paths, avenues and streets in bad conditions	15.8	7.1	28.0	0.0	19.1	27.8
Abandonment mountains	15.8	14.3	12.0	8.7	19.1	30.6
Poor landscape quality of accesses to urban centres and housing estates	15.2	7.1	20.0	8.7	12.8	33.3
Abandoned dwellings, often not finished	14.6	7.1	20.0	13.0	14.9	25.0
Identity loss of urban centres as "mountain villages"	14.0	7.1	8.0	0.0	23.4	27.8
Deficient landscape along roads	14.0	35.7	0.0	8.7	27.7	11.1

(continued)

Table 3.54 (continued)

Landscape quality	Total (%)	GÁTOVA (%)	OLOCAU (%)	MARINES (%)	NÁQUERA (%)	SERRA (%)
Ravines not maintained. Poor image.	12.3	0.0	52.0	8.7	4.3	11.1
Deficient landscape in urban centres	12.3	7.1	24.0	8.7	14.9	13.9
None	27.5	57.1	20.0	52.2	23.4	27.8
Do not know/no answer	17.0	0.0	0.0	0.0	4.3	5.6
Opportunities						
To improve the visual quality of the accesses to urban centres, housing estates, ravines, view points, …	16.4	14.3	48.0	13.0	4.3	25.0
To recuperate and use the existing heritage	14.6	7.1	4.0	13.0	8.5	44.4
To promote the identity of the "mountain villages"	12.9	21.4	12.0	0.0	14.9	25.0
To define regulations that will allow the creation of villages and housing estates with an identity	9.9	14.3	4.0	4.3	17.0	13.9
To improve the visual environment of the infrastructure	8.2	42.9	8.0	4.3	6.4	5.6
To improve the image of the urban centres	8.2	21.4	20.0	8.7	2.1	8.3
To improve and unify the image of the urban space	7.6	7.1	20.0	8.7	2.1	11.1
Do not know/no answer	54.4	21.4	40.0	69.6	53.2	36.1

3.8.8 Synthesis and Diagnosis

Landscape: In general, the forest and natural landscapes of the south-central area of Sierra Calderona have a high or very high landscape value that diminishes in areas with undeveloped vegetation or around forest tracks or large recreational areas. Traditional mountainous agricultural areas, orchards and small vegetable gardens around old settlements also have high to very high value, although rainfed and terraced fields are in severe regression, with the consequent diminishment on the diversity of the agroforestry mosaic that characterises many mountainous areas in the Mediterranean region.

Agricultural areas located in the foothills and plains define a territorial matrix, now interrupted by a dense network of settlements, roads and other infrastructure. In those plains, irrigated agricultural fields, especially of citrus fruit plantations, are gaining importance, while the high level of abandonment of non-irrigated crops, coupled with the strong presence of intrusive elements or of low-quality constructions, generates an overall medium- or low-quality landscape.

In relation to urban landscapes, historical villages and towns, and the town of Marines Nuevo in particular, have a high value, despite the occasional inclusion of buildings with intrusive volume or architecture. The situation gets significantly worse in the housing estates where, with the exception of some settlements located in the mountainous areas and surrounded by forest vegetation, both their inner landscape and fringes have been poorly treated.

Although industrial land is scarce, its location near to the accesses to towns worsens the overall quality of the landscape. The quality of the inner public spaces of the industrial estate "Los Vientos" is very poor. Throughout the area, there are no green buffer spaces between the industrial areas and traffic infrastructures.

Consequently, the quality of the landscape perceived from infrastructural corridors is largely defined by the landscape of the natural, agricultural or urban spaces surrounding them. However, it should be noted that the presence of inadequate elements in the immediate margins of some roads worsens the perception of the surrounding landscape, even though this, as a whole, might have a high quality.

Territorial models for urbanisation: At this point, it can be predicted that the spontaneous evolution of the territory might lead to a scenario in which low–medium-density built areas would be consolidated in the plains and foothills, while the inner villages would progressively lose functionality and relevance. The generation of alternatives to this phenomenon would require effective strategic and spatial planning in order to redefine the links between compact villages/towns and housing estates and avoid the progressive coalescence of the latter. The analysis of current transformations suggests three main future possible scenarios:

(1) Creation of a built continuum of low-density settlements on the plains and the foothills of Sierra Calderona. This alternative would rely on quick access to these areas from transport infrastructures and would result in loss of importance of the historical towns and villages.
(2) Organised consolidation of the low-density settlements located on the plains and foothills of Sierra Calderona and revitalisation of villages.
(3) Freezing the sprawl of low-density settlements while increasing their density and facilities, as well as strengthening/revitalising the villages by enhancing the provision of strategic services and renewing or upgrading the housing stock (in buildings requiring renovation, vacant lots or restrained urban growths, etc.).

Although the existing scattered settlements have often been regulated by the addition of new constructions that, allegedly, should improve the conditions of the already existing housing stock, the results of this strategy have rarely been satisfactory, having generally resulted in an increase in the amount and size of the housing estates in the agricultural and forest areas. Due to their high incidence, the following alternatives might be considered:

(1) To integrate scattered settlements into a more extensive urban fabric, in order to improve their current precarious conditions. This option would have a heavy territorial impact and would not be feasible in the current economic situation.
(2) To improve the interior landscape and edges of scattered settlements. This option would require their inhabitants to bear the brunt of the costs associated with these improvements.
(3) To improve just the edges and areas of greatest visual impact of the scattered developments.

3.9 Regional Sustainability

3.9.1 Generalities

The integration of socio-economic development with the preservation and enhancement of environmental and cultural values is one of the basic pillars of sustainable development. This concept has been used throughout this document to analyse and diagnose variables such as mobility, urban patterns and land-use patterns. However, this subchapter intends to objectively measure a number of variables related to the use and proper management of resources, energy and waste, as a way to identify problems and propose alternative solutions.

Three factors would be studied: water resources, energy consumption and production/management/utilisation of waste. Concurrently, and given the strategic importance that housing estates have on the sustainability level of the analysed territory, a detailed study of their services (i.e. sewage, garbage collection, water supply) has been conducted.

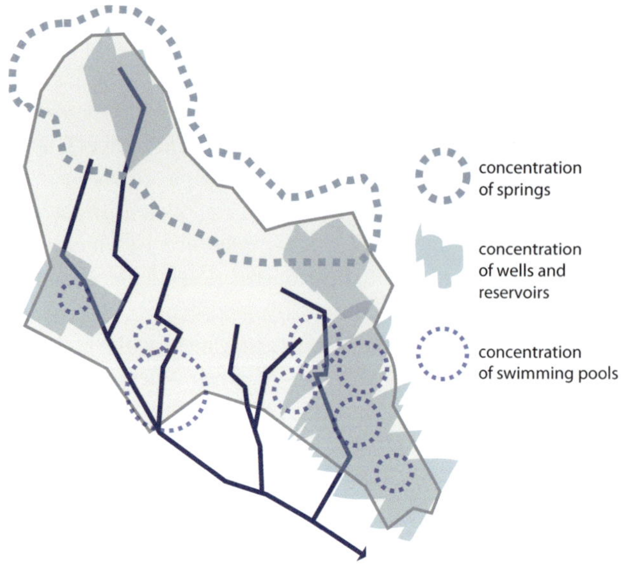

Fig. 3.81 Springs, wells, reservoirs and swimming pools. *Source* Own elaboration

3.9.2 Water

The following aspects have been analysed (Figs. 3.80, 3.81 and 3.82):

Natural Water network: This mainly consists of ravines and brooks that are often in poor condition along their urban or peri-urban stretches due to uncontrolled dumping of waste, presence of invasive vegetation, hard channelisation, etc. However, this network has an enormous potential to articulate the main network of itineraries in the natural environment, as well as the system of green spaces in urban areas, especially in Serra, Náquera and Olocau. The area included in the Strategic Plan also has numerous springs of excellent water that could be commercialised in Gátova and Serra, as long as this would not have an ecological impact.

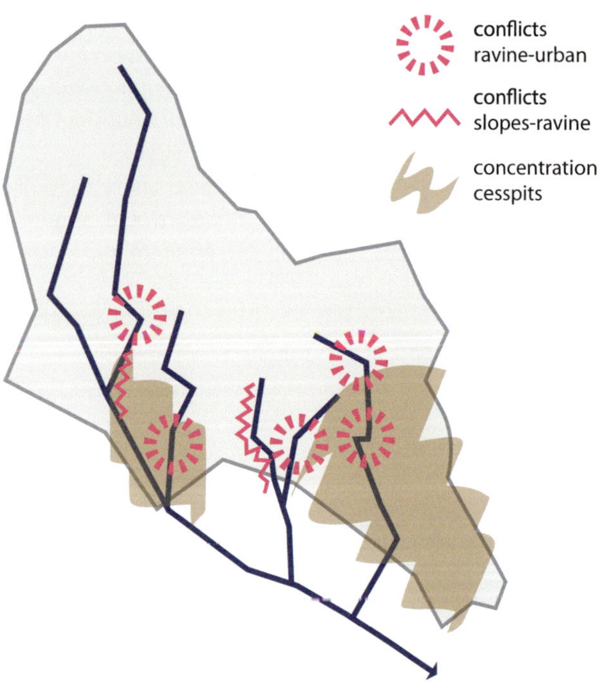

Fig. 3.80 Main ravines and main conflicts. *Source* Own elaboration

Artificial Water network: The area has many wells, both public and privately managed. In Náquera and Marines, due to the importance of irrigated agriculture in the plains of these municipalities, there are numerous irrigation ponds. In Marines, the channel of Campo del Turia irrigates the southern part of the municipality with water coming from the Benageber dam. Finally, because of its uniqueness and cultural value, the ancient system of ponds and channels irrigating the orchards located next to historic settlements, villages and towns should also be mentioned.

Drinking water supply: Water supply is mostly provided by water coming from privately managed wells. The network is generally old and is expected to have many leaks (the lack of metres in almost all the public wells prevents the possibility of obtaining a reliable record of the water lost between the wells and the points of consumption).

Waste water: There is a strong deficit of sewage infrastructure in most of the largest housing estates of Náquera and Olocau, which currently discharge their wastewater through cesspits (in Náquera in almost all housing estates, in Olocau in around half of them). As the vulnerability of the aquifers in the area is high, it is important to find an urgent solution to this infrastructural problem. In the villages and housing estates with sewage, the

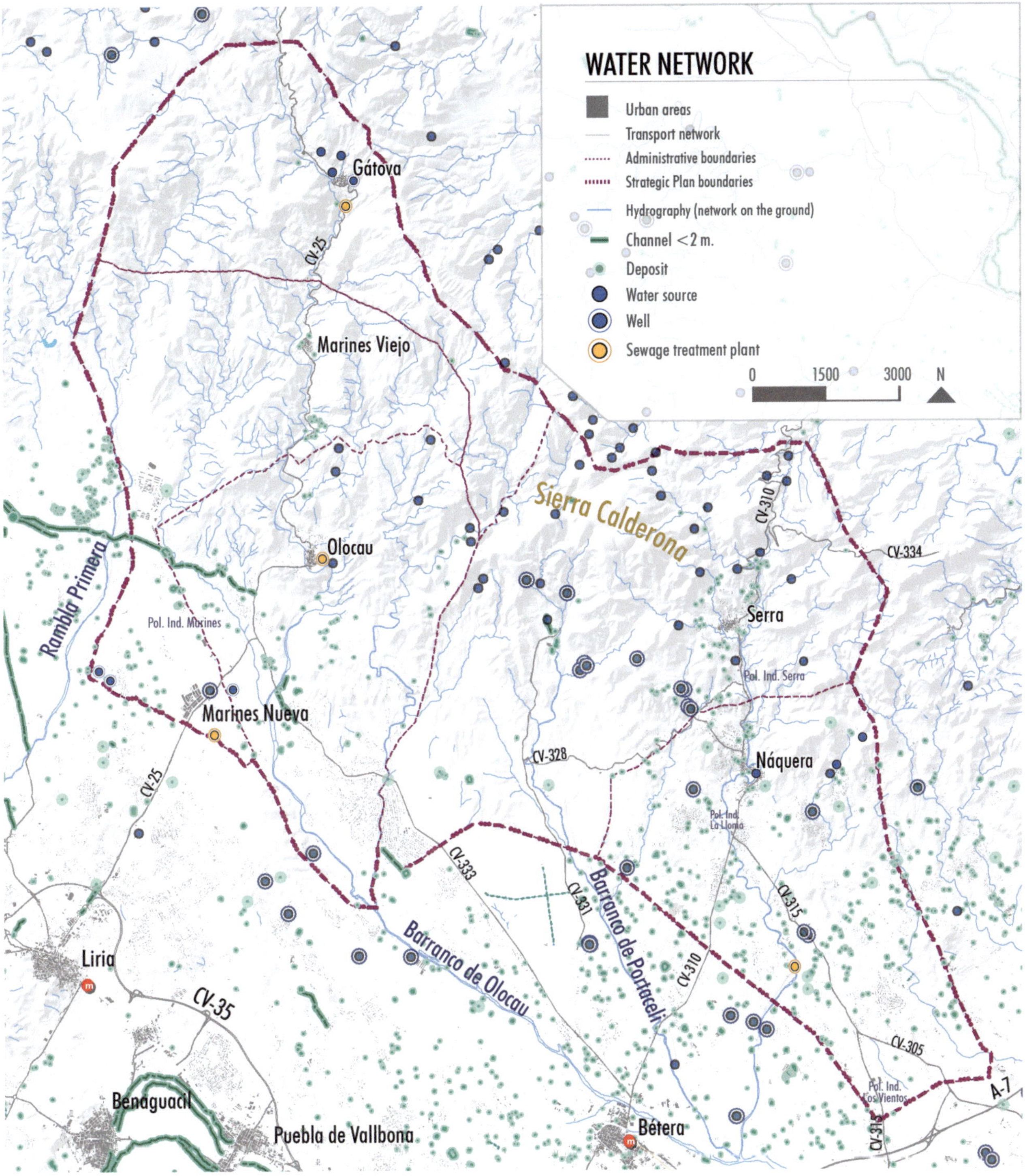

Fig. 3.82 Water network. *Source* Own elaboration

wastewater is treated in the existing sewage treatment plants, except in Marines Viejo, where despite having a sewage treatment plant, it remains inactive, with the subsequent discharge of wastewater into the Olocau ravine (Fig. 3.83).

Water consumption: This has been estimated for each municipality and for each activity sector (agriculture, industry and domestic use) for the whole area included in the Sierra Calderona Strategic Plan. This estimation has been made by using generic ratios that have been adjusted depending on the weight that each activity sector has in each municipality. The average consumption per capita for domestic use slightly exceeds the Spanish average. The

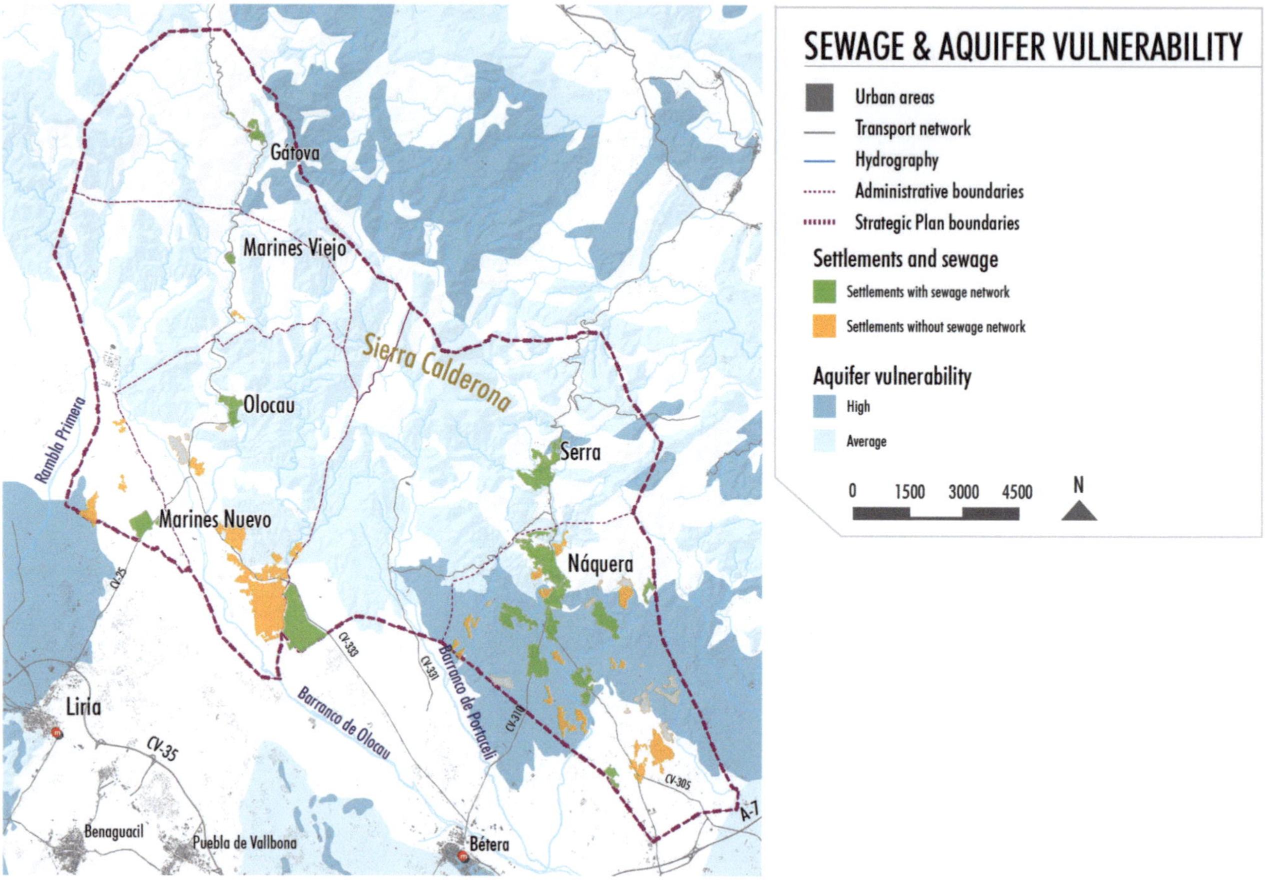

Fig. 3.83 Sewage and aquifers vulnerability. *Source* Own elaboration

water consumption differs between the residents living in the compact villages and towns and those living in housing estates, as the maintenance of gardens and private pools require more water and the extensive water networks associated with low-density settlements increase the probability of leakage. Finally, the seasonal fluctuation of water demand must be noted, which can be multiplied by as much as eightfold in some residential areas during the high season. This seasonality affects not only water consumption, but also energy consumption and waste production (Figs. 3.84, 3.85 and Tables 3.55 and 3.56).

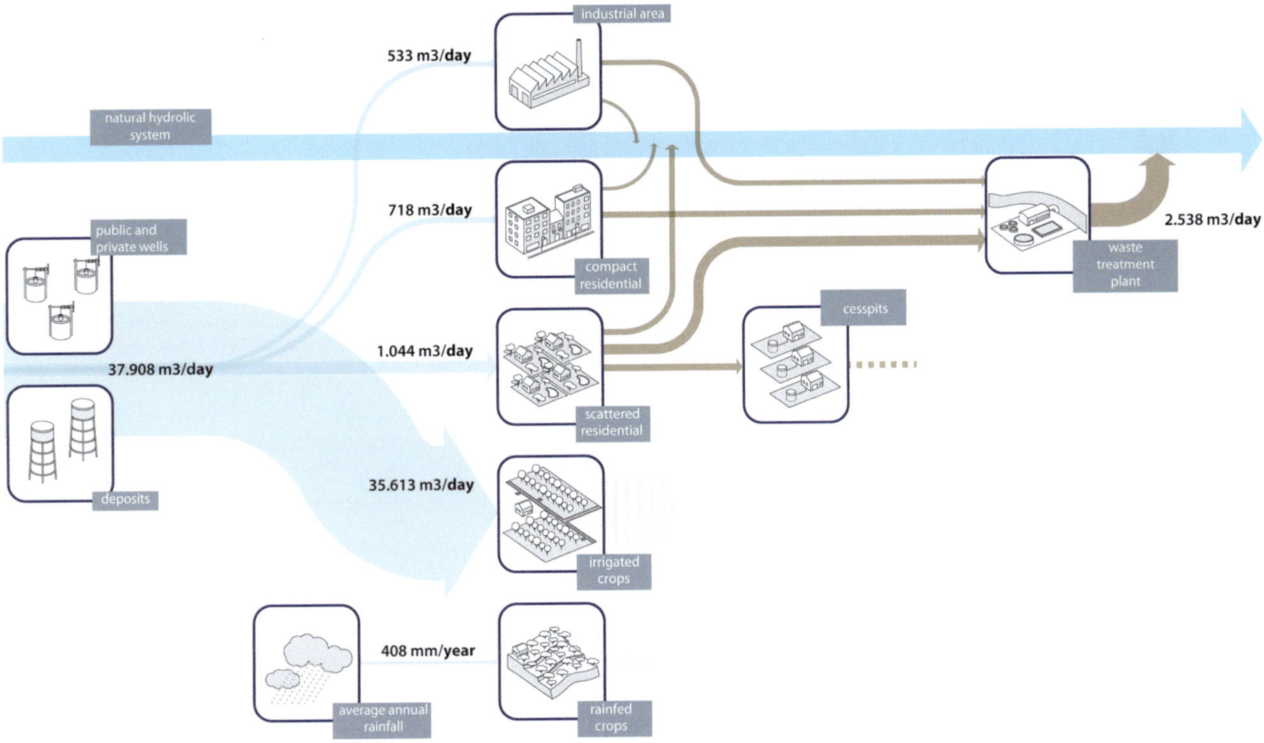

Fig. 3.84 Water. Estimated annual flows. *Source* Own elaboration

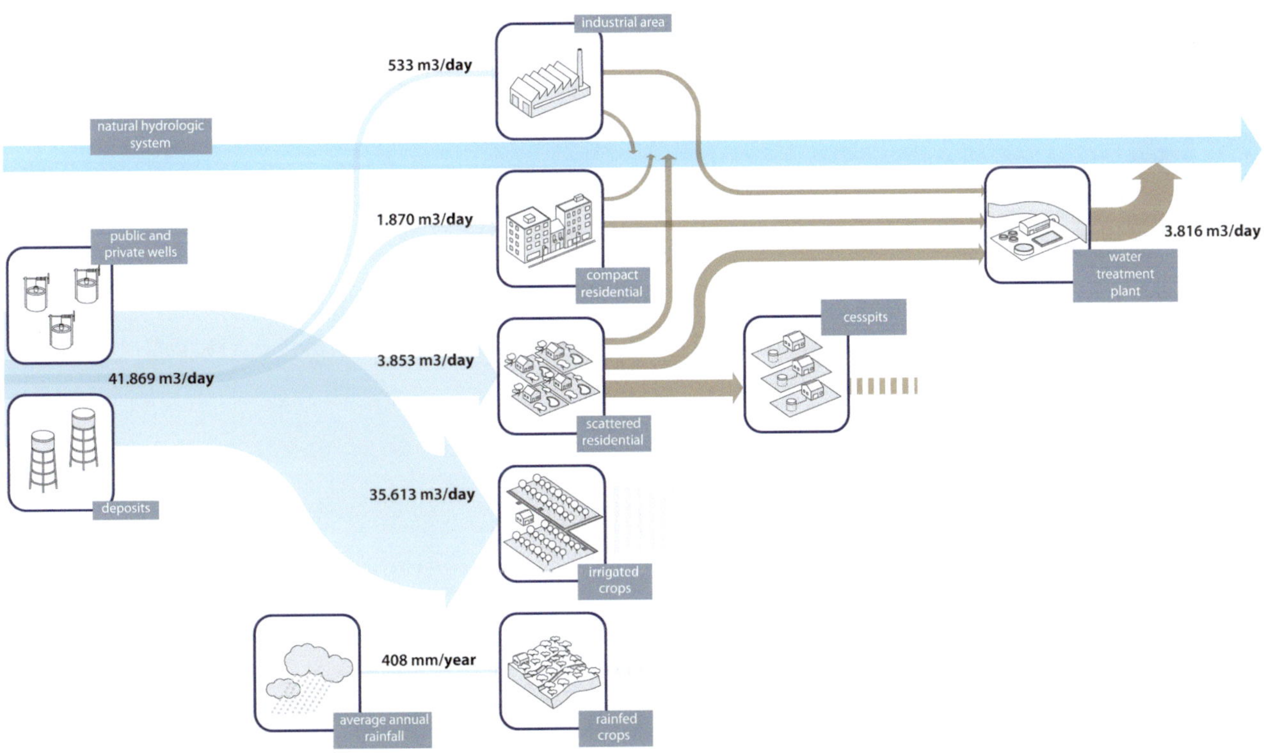

Fig. 3.85 Water. Estimated flows in high season. *Source* Own elaboration

Table 3.55 Water consumption: *Source* Own elaboration

	Compact residential (m^3/day)	Consumption per capita compact residential (litres/inhab/day)	Scattered residential (m^3/day)	Consumption per capita scattered residential (litres/inhab/day)	Total residential (m^3/day)	Irrigated crops (m^3/day)	Rainfed crops annual rainfall (mm/year)	Industry (m^3/day)	Total consumption (m^3/day)	Consumption per capita residential (litres/inhab/day)
Total strategic plan						35,613	408	533		
Permanent population	718	90	1,044	200	1,762				37,908	133.50
High season	1,870	90	3,853	200	5,723				41,869	142.92
NÁQUERA						15,623	326	356		
Permanent population	269	90	582	200	851				16,830	144.25
High season	807	90	2,328	200	3,135				19,114	152.13
SERRA						4,408	326	69		
Permanent population	212	90	200	200	411				4,888	122.78
High season	635	90	798	200	1,433				5,909	129.76
MARINES						5,926	463	83		
Permanent population	142	90	50	200	192				6,202	105.07
High season	185	90	126	200	310				6,320	115.73
OLOCAU						6,659	463	16		
Permanent population	54	90	184	200	238				6,912	156.57
High season	140	90	459	200	599				7,273	155.53
GÁTOVA						2,997	463	10		
Permanent population	41	90	28	200	70				3,076	116.08
High season	103	90	142	200	245				3,252	132.16

Table 3.56 Services in housing estates. *Source* Data provided by the municipalities, Diputación de Valencia and through direct observation

Housing estate	Garbage collection service	Public lighting	Water supply	Sewage	Power supply
NÁQUERA					
Urb. Monteamor A	3 times/week (summer)—1 time/week (winter)	no data	YES	YES	no data
Urb. El Peucal	3 times/week (summer)—1 time/week (winter)	YES	YES	NO	YES
Urb. La Carrasca	3 times/week (summer)—1 time/week (winter)	YES	YES	YES	YES
Urb. Els Trencalls	3 times/week (summer)—1 time/week (winter)	YES	en proyecto	NO	YES
Urb. La Forqueta	3 times/week (summer)—1 time/week (winter)	YES	YES	NO	YES
La Cava	3 times/week (summer)—1 time/week (winter)	NO	YES	no data	YES
Urb. Font de l'Oro	3 times/week (summer)—1 time/week (winter)	YES	YES	no data	YES
Fontanelles	3 times/week (summer)—1 time/week (winter)	NO	YES	YES	YES
Cañada de les Cerveres	3 times/week (summer)—1 time/week (winter)	NO	YES	no data	YES
Urb. El Collado dels Llops	3 times/week (summer)—1 time/week (winter)	YES	YES	YES	YES
Urb. L'Antara-Mellaos	3 times/week (summer)—1 time/week (winter)	YES	YES	no data	YES
Urb. El Paraíso	3 times/week (summer)—1 time/week (winter)	YES	YES	NO	YES
Barranc les Voltes	3 times/week (summer)—1 time/week (winter)	NO	YES	NO	YES
Urb. Corral Nou	3 times/week (summer)—1 time/week (winter)	YES	YES	YES	YES
Urb. La Lloma	3 times/week (summer)—1 time/week (winter)	YES	YES	YES	YES
Urb. Sant Miguel	3 times/week (summer)—1 time/week (winter)	YES	YES	YES	YES
Barranc del Sirer	3 times/week (summer)—1 time/week (winter)	NO	YES	NO	YES
Urb. Mont-Ros	3 times/week (summer)—1 time/week (winter)	YES	YES	YES	no data
Urb. L'Arenal	3 times/week (summer)—1 time/week (winter)	NO	YES	NO	NO
La Patá	3 times/week (summer)—1 time/week (winter)	YES	YES	NO	YES
Bassa Blanca	3 times/week (summer)—1 time/week (winter)	NO	no data	no data	YES
Urb. Cabeç Bord	3 times/week (summer)—1 time/week (winter)	NO	no data	no data	YES
Massapedro	3 times/week (summer)—1 time/week (winter)	NO	YES	no data	YES
Ampliación Las Lomas Urb. El Camp de Tir	3 times/week (summer)—1 time/week (winter)	YES	YES	YES	YES
Urb. El Corralet	3 times/week (summer)—1 time/week (winter)	NO	YES	no data	YES

(continued)

Table 3.56 (continued)

Housing estate	Garbage collection service	Public lighting	Water supply	Sewage	Power supply
Urb. Les Llomes de l'Algepsar	3 times/week (summer)—1 time/week (winter)	YES	YES	NO	YES
Urb. El Pla de l'Estepar I	3 times/week (summer)—1 time/week (winter)	NO	YES	NO	YES
Urb. El Pla de l'Estepar II	3 times/week (summer)—1 time/week (winter)	NO	YES	NO	YES
Urb. El Pla de l'Estepar III	3 times/week (summer)—1 time/week (winter)	NO	YES	NO	YES
Urb. Bonanza	3 times/week (summer)—1 time/week (winter)	YES	YES	NO	YES
Urb. Plans Bonanza	3 times/week (summer)—1 time/week (winter)	YES	YES	no data	YES
Els Plans/ Urb. Tres Rutes	3 times/week (summer)—1 time/week (winter)	YES	YES	NO	YES
Urb. Lloma Llarga	3 times/week (summer)—1 time/week (winter)	NO	YES	YES	YES
Mas de San Francisco	no data	no data	no data	no data	no data
El Plans	no data	no data	no data	no data	no data
SERRA					
Urb. Los Ríos	YES	NO	YES	no data	YES
Urb. Torre de Portacoeli	YES	YES	YES	YES	YES
MARINES					
Urb. El Real	YES	NO	NO	NO	YES
Urb. Pla d'Orcán	YES	no data	NO	NO	YES
Urb. La Contienda	YES	no data	NO	NO	YES
El Romeral	YES	YES	YES	YES	YES
OLOCAU					
Urb. La Pinadeta	no data	no data	YES	YES	no data
Urb. L'Arquet	no data	no data	YES	YES	no data
Urb. Rocha de Quito y el Colmenar	YES	no data	no data	no data	no data
Urb. Pla de Marco—Les Forquetes	YES	NO	no data	no data	YES
Urb. Els Puntals	YES	YES	YES	NO	YES
Urb. La Lloma—Los Picapiedra	YES	YES	YES	NO	YES
Urb. Pla de la Torre—La Seu	YES	NO	YES	NO	YES
Urb. Pla de Pedralvilla	YES	YES	YES	NO	YES
GÁTOVA					
Urb. Monte Roderro	YES	YES	YES	YES	YES
La Tejería	YES	YES	YES	YES	YES

3.9.3 Waste

The following aspects have been analysed:

Waste types: For the purposes of implementing this Plan, the production and emission of the following types of waste have been studied in each municipality:

- **Municipal Solid Waste (MSW)**: each municipality has an adequate collection system for MSW in towns, although with different degrees of selectivity (glass, plastics, paper). The frequency of collection may vary depending on the time of the year, increasing during the high season. At housing estates, it may occur less frequently. Olocau, Serra and Náquera have larger waste collection points. In the Public Participation Plan, complaints regarding garbage collection in the most popular recreational zones, in particular in Portaceli, were recorded (Table 3.56).
- **Construction and demolition waste**: The management of this kind of waste is the responsibility of the construction developer. Although three of the municipalities have waste collection points, illegal dumps have been detected.
- **Industrial waste**: As industrial activity has very little weight in Serra, Olocau and Gátova, there is no available information on industrial waste management. In Marines, the correct collection of this kind of waste by authorised waste managers is crucial, given the highly polluting nature of the paint factory. In Náquera, where the industrial activity is more important, the industrial waste is properly collected and treated by authorised managers, there being a treatment plant in the area for electrical and electronic equipment waste.
- **Livestock farming waste**: Intensive livestock farming is concentrated in Marines (pig farm of 2,080 heads) in particular and in Náquera, where 13 farms with a total of 1,350 heads of pigs, 340 cattle, 30 sheep and 11 horses are present. The extent of treatment of the waste and manure is unknown, but it is assumed it meets the existing regulations.
- **Gas emissions**: None of the five municipalities has a census of potentially atmospheric polluting activities nor performs controls on emission.
- **Forest, agricultural and gardening residues**: This organic waste can reach a significant volume in the area, and its volume has been estimated in order to consider its possible use for energy production.

Waste volume: In order to estimate the volume of each kind of waste (Figs. 3.86, 3.87 and Table 3.57), the following criteria have been used:

- **Municipal Solid Waste (MSW)**: The average ratio for the Valencian Region has been used, 1.28 kg/person/day.
- **Construction and demolition waste and waste from industrial activity**: This is very variable because it depends on the intensity of construction activity. An average production of 10.5 kg of CO^2 gas emissions/inhabitant/day, 7.2 kg of debris/inhabitant/day and 1.28 kg of industrial waste/inhabitant/day has been assumed (these ratios are given by the study of flows of energy, water, materials and information in the Region of Madrid). These ratios have been adjusted in accordance with the importance that the industrial or construction activity has in each municipality.
- **Livestock farming waste**: The volume has been estimated according to the importance of livestock farming in each municipality.
- **Forest, agricultural and gardening residues/waste**: An average production of 0.25 tons/hectare/year of forest residues is estimated and 1 ton/hectare/year of agricultural residues for natural and agricultural areas, respectively (ratio from the study "Biomass in Andalusia" developed in 2011 by the Andalusian Energy Agency, and assuming that 60% of agricultural land produces this kind of waste). It is also assumed that the production of garden waste is similar to the forest waste (0.25 Tons/hectare/year).

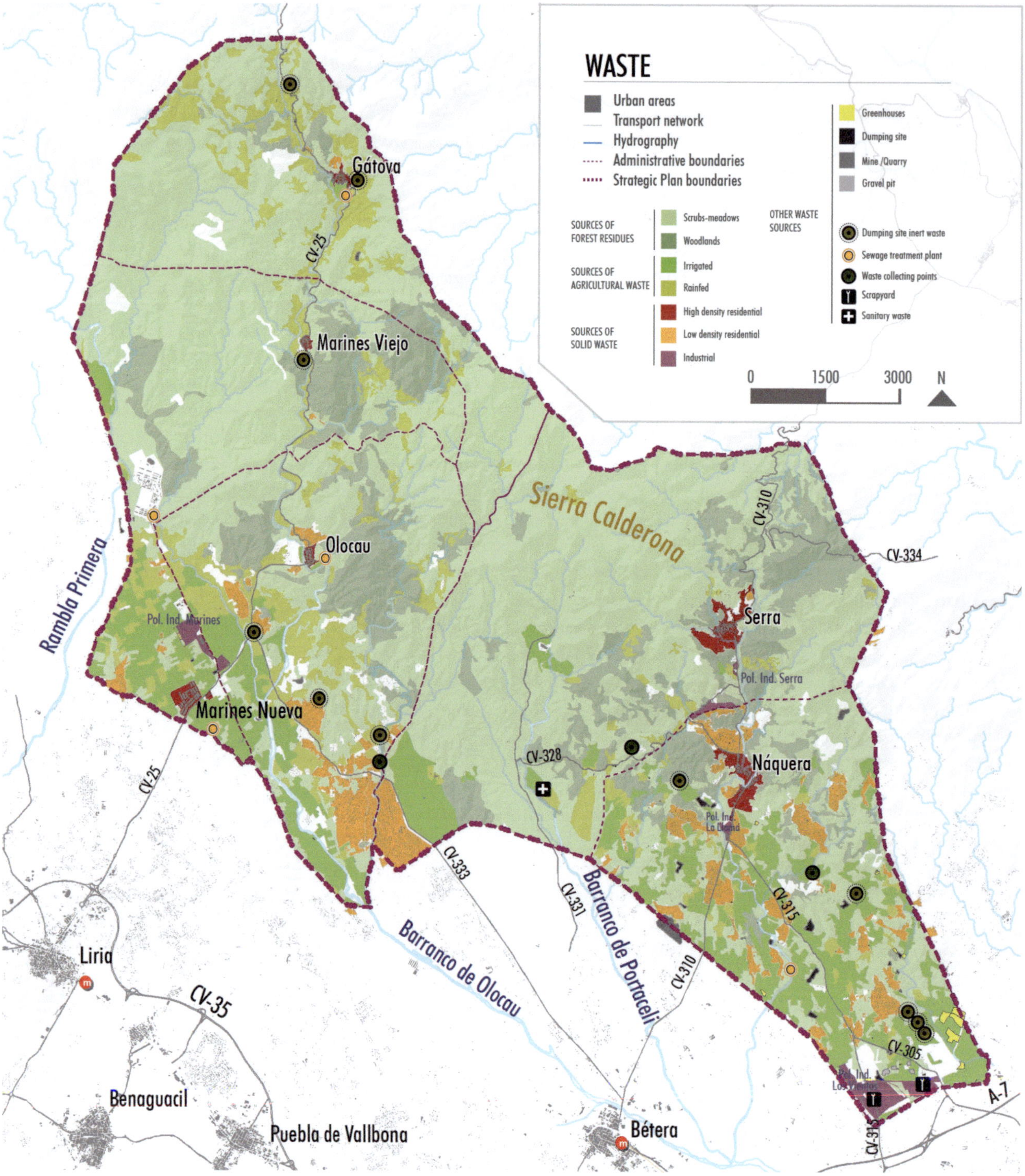

Fig. 3.86 Waste production and management. *Source* Own elaboration

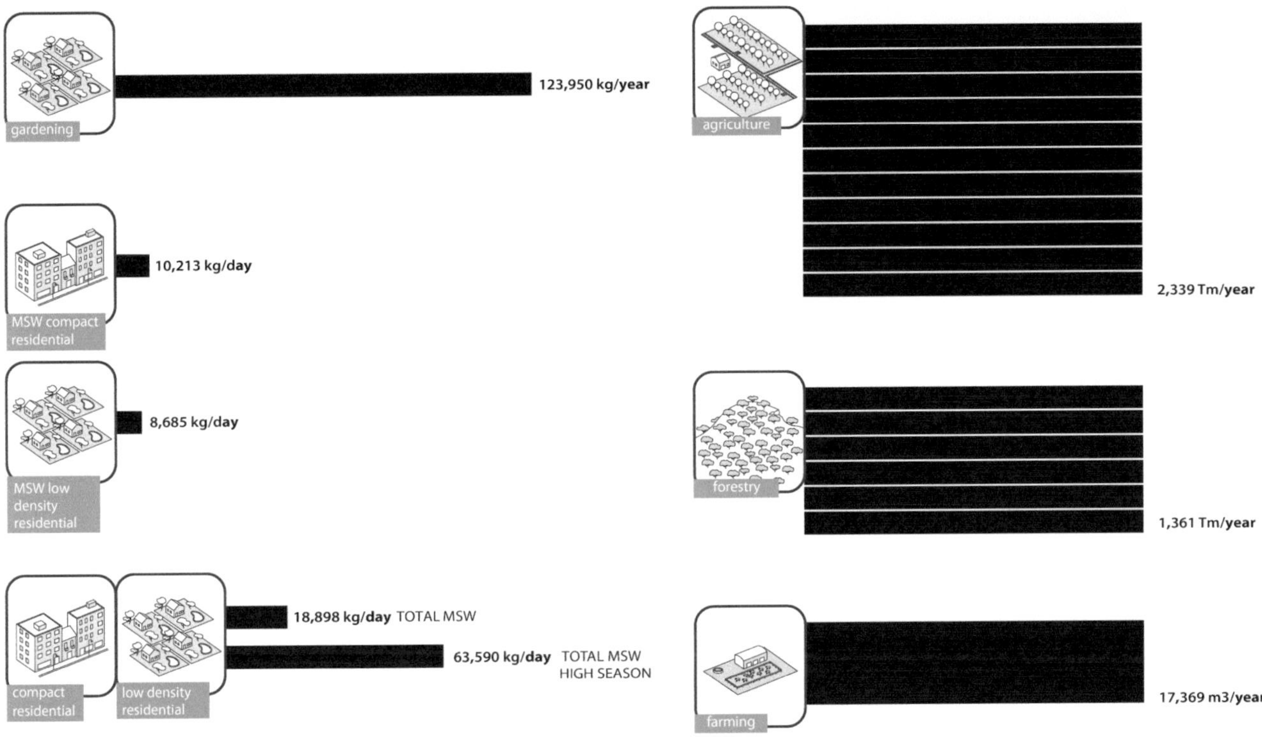

Fig. 3.87 Average production of waste. Annual estimate flows. *Source* Own elaboration

Table 3.57 Waste production *Source* Own elaboration

	MSW Compact Residential (kg/day)	MSW Scattered residential (kg/day)	TOTAL MSW (kg/day)	RSU PER CÁPITA (kg/hab/day)	Gardening (Tm/year)	Agriculture (Tm/year)	Forestry (Tm/year)	Livestock farming activity (m^3/year)
Total strategic plan					123.95	2,339	1,361	17,369
Permanent population	10,213	8,685	18,898	1.43				
High season	26,599	36,991	63,590	1.59				
NÁQUERA					64.40	762	218	9,165
Permanent population	3,827	4,842	8,669	1.47				
High season	11,482	22,349	33,830	1.64				
SERRA					3.02	248	530	150
Permanent population	3,009	1,661	4,670	1.39				
High season	9,028	7,665	16,692	1.51				
MARINES					9.48	413	203	7,574
Permanent population	2,024	418	2,441	1.33				
High season	2,631	1,205	3,836	1.43				
OLOCAU					45.71	610	346	450
Permanent population	767	1,528	2,294	1.51				
High season	1,993	4,406	6,400	1.66				
GÁTOVA					1.34	306	63	30
Permanent population	586	237	823	1.37				
High season	1,466	1,366	2,832	1.53				

3.9.4 Energy

The following aspects have been analysed:

Power Supply: No major issues have been identified related to the electricity distribution network. The main power lines follow the courses of the main ravines and roads. Only Náquera has a more complex network as a result of many illegal constructions, where some old network sections should be checked. Regarding the production of clean energy, only Náquera and Olocau, and to a lesser extent Serra, have installed production sources for this kind of energy. In these municipalities, solar farms are present and Náquera also possesses a Hydrothermal Carbonisation Plant of Biomass. Gátova's rabbit farm is self-sufficient in electricity supply. There is no evidence that any public facility makes use of renewable energy (Table 3.56).

Power Consumption: To determine energy consumption, total consumption data provided by the supply companies have been used. The distribution of such consumption by sector (industry, transport, tertiary, housing, etc.) has been estimated using standard ratios that have been adjusted according to the importance each sector has in each municipality (Fig. 3.88 and Table 3.58).

Renewable Energy Production: Given the constraints arising from the presence of the Natural Park in the area, the use of wind energy as renewable energy has been discarded. The use of biomass or the use photovoltaics to produce energy is the two possible alternatives that can be studied (Figs. 3.87 and 3.88) (Table 3.58).

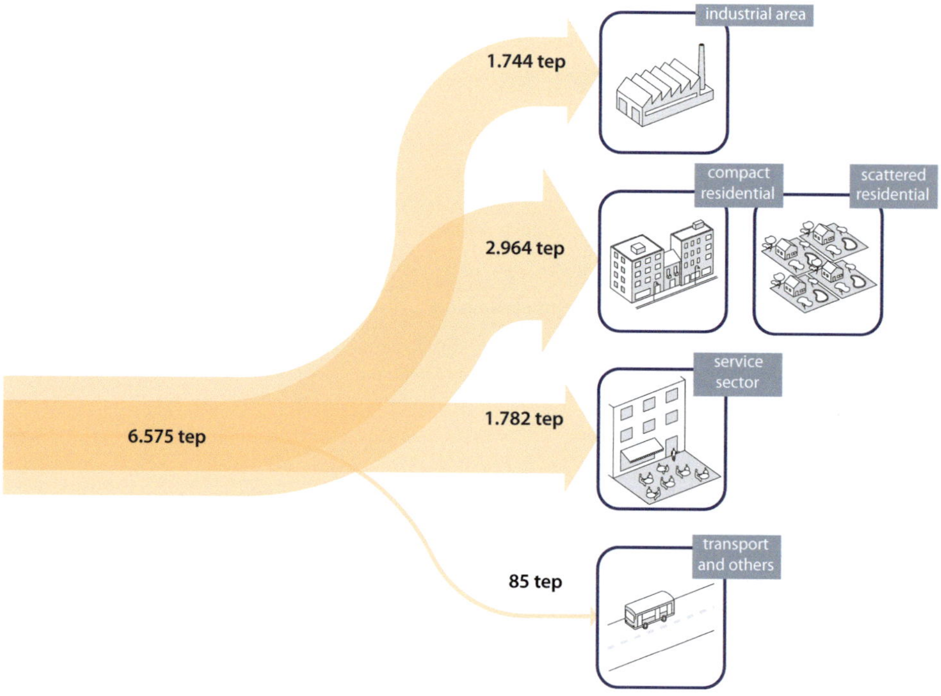

Fig. 3.88 Energy. Annual estimate flows. *Source* Own elaboration

Table 3.58 Energy comsumption. *Source* Own elaboration

	Total residential (tep/year)	Service sector (tep/year)	Industry (tep/year)	Transport and others (tep/year)	Total consumption (tep/year)	Consumption per capita (kWh/inhab/year)
Total strategic plan	2,964	1,782	1,744	85	6,575	5,794
NÁQUERA	1,061	900	1,298	18	3,277	6,460
SERRA	967	446	82	38	1,533	5,323
MARINES	461	212	293	19	985	6,250
OLOCAU	365	174	63	4	606	4,644
GÁTOVA	110	50	9	6	175	867

3.9.5 Ecological Footprints

Given the different behavioural profiles of permanent and seasonal residents in the area, and in order to investigate possible ways to create more sustainable spaces, infrastructures and ways of living, the ecological footprints of different types of persons have been coarsely estimated (Table 3.59).

Table 3.59 Ecological footprints. *Source* Own elaboration

	CO_2	Food	Accommodation	Goods and services	Total	Planets necessary with existing lifestyle
Average area strategic plan	11.68	17.19	4.79	9.75	43.40	2.76
Person living in urban centre working in an industrial estate nearby	4.90	16.50	3.60	7.50	32.62	2.08
Housewife living in urban centre	5.70	14.90	3.60	7.50	31.75	2.02
Person living in housing estate working in the city of Valencia	17.10	19.50	4.50	10.60	51.73	3.29
Part-time farmer working in the service sector in the area of the plan	5.70	16.50	6.10	12.70	40.92	2.60
Retired person living in a urban centre	4.90	12.40	3.20	6.60	27.03	1.72
Retired person living in a urban centre	15.20	17.00	6.50	9.10	47.73	3.04
Children living in a urban centre	5.70	16.50	2.40	11.60	36.22	2.31
Children living in a urban centre	17.40	19.50	7.40	12.00	56.24	3.58
Young person studying at university living in a urban centre or in a housing estate	9.00	16.50	3.60	7.50	36.59	2.33
Seasonal resident (summer)	22.80	19.50	6.50	12.00	60.82	3.87
Person serving in the military	20.10	20.30	5.30	10.10	55.78	3.55
National average	12.90	14.90	4.80	9.40	42.00	2.50

3.9.6 Synthesis and Diagnosis

WATER RESOURCES:

Drinking water supply: provision is effective throughout the whole area, although considering the age of the network, numerous leaks are assumed to exist.

Sewage and water treatment: the largest housing estates of Náquera and Olocau, some of which are located in areas with highly vulnerable aquifers, have a major deficit of this kind of infrastructure and therefore resolve the discharge of wastewater with cesspits. In villages, towns and housing estates with sewage systems, the wastewater is treated in treatment plants, except in Marines Viejo. There are not systems to reuse the water, either at urban level or within buildings/dwellings.

Water consumption: The average consumption per capita for domestic use slightly exceeds the Spanish average (the average consumption per capita in housing estates is estimated to be twice the average consumption per capita of compact towns and villages). The significant annual variation in water demand due to the seasonal use of the area, which multiplies up to eightfold in some areas, must also be noted, as it not only affects water consumption but also energy consumption and waste production.

WASTE:

Municipal Solid Waste (MSW): Each municipality has an adequate collection system for MSW in towns and villages although with different degrees of selectivity (glass, plastics, paper). The frequency of collection may vary depending on the time of the year, increasing during high season. In housing estates, it may occur less frequently. Olocau, Serra and Náquera have waste collection points. During the course of the development of the Public Participation Plans, complaints regarding the garbage collection in the most frequented recreational zones, in particular the one located in Portaceli, were recorded.

Forest, agricultural and gardening waste: Organic waste from agriculture, maintenance of gardens (which can be a significant amount given the large number of single-family houses) and the management of forests, can reach a significant volume, and this has been estimated in order to consider its possible use for energy production.

ENERGY:

Power and energy consumption: Power consumption is marked by a strong seasonality, as consequence of the numerous dwellings used as secondary residences and of the strong presence of low-density housing estates, where the use of private vehicles is necessary to cover many basic needs.

Renewable energy production: This is very limited in the area covered by the Strategic Plan. Given the impossibility of installing wind turbines in the elevated areas of the Natural Park, the main sources of renewable energy would be the use of biomass and photovoltaic energy.

Ecological footprint:

This shows how the demand for resources and the energy consumption per capita is directly linked to urban types and associated ways of living. The study of the ecological footprint highlights the poor sustainability level of monofunctional and low-density residential urban fabrics, as well as their associated mobility models.

3.10 Socio-demography

3.10.1 Generalities

The socio-demographic study included in the present sub-chapter examines the spatial and temporal distribution of the population, as well as population pyramids and levels of employment and unemployment by sectors.

3.10.2 Spatial and Temporal Distribution of the Population

Population density decreases towards the north-west of the studied area, Naquera having the highest density (175 inhabitants/km^2), Serra, Marines and Olocau intermediate densities (59 inhabitants/km^2) and Gátova the lowest one (20 inhabitants/km^2). Olocau and Náquera are the towns with the highest urban densities, as a consequence of their compact structure in the first case and of the combination of dense and not-so-dense urban fabrics in the second. Marines Nuevo is slightly less dense than the first two, but due to the subdivision of the original patio-houses, its density is now higher than the density planned at the moment of its foundation. The towns of Gátova and Serra have lower densities, given the strong seasonality of the urban area in the first case and the importance of housing estates in the second.

In consideration of the spatio-temporal distribution of the population living in towns/villages and housing estates, it must be noted that in Olocau, the number of permanent residents in housing estates exceeds that of those living in the compact town. In Naquera, these numbers are virtually the same; while in Marines, Gátova and Serra, the opposite situation occurs, although in Serra this is because of the many housing estates that are officially included as part of the central town.

During high season (summer, Easter holidays) and weekends, the population increases significantly both in the towns and, more remarkably, in the housing estates. However, these variations do not change the fact that in Marines and Gátova, both with few housing estates, and in Serra, the population in the urban centres is still larger than the overall population of their housing estates.

The total population in the area covered by the Strategic Plan is 13,198 inhabitants, 60.5% of which live in the five towns or villages and 39.5% in housing estates and scattered settlements. During the high season, the population increases to as much as 40,047 inhabitants, 51.8% of which are distributed among towns and villages and 48.2% in housing estates and scattered settlements.

According to these figures, the population triples during the high season (303%) in the whole area covered by the Strategic Plan, 260% in the compact towns and villages and 369% in the housing estates. The marked seasonality of the population and the importance of housing estates are very noticeable in Olocau, Náquera and Serra. This not only conditions the administrative operations of such municipalities, but also their range of services and their levels of sustainability. Marines and Gátova stand out as alternative and positive models (Fig. 3.89 and Tables 3.60, 3.61).

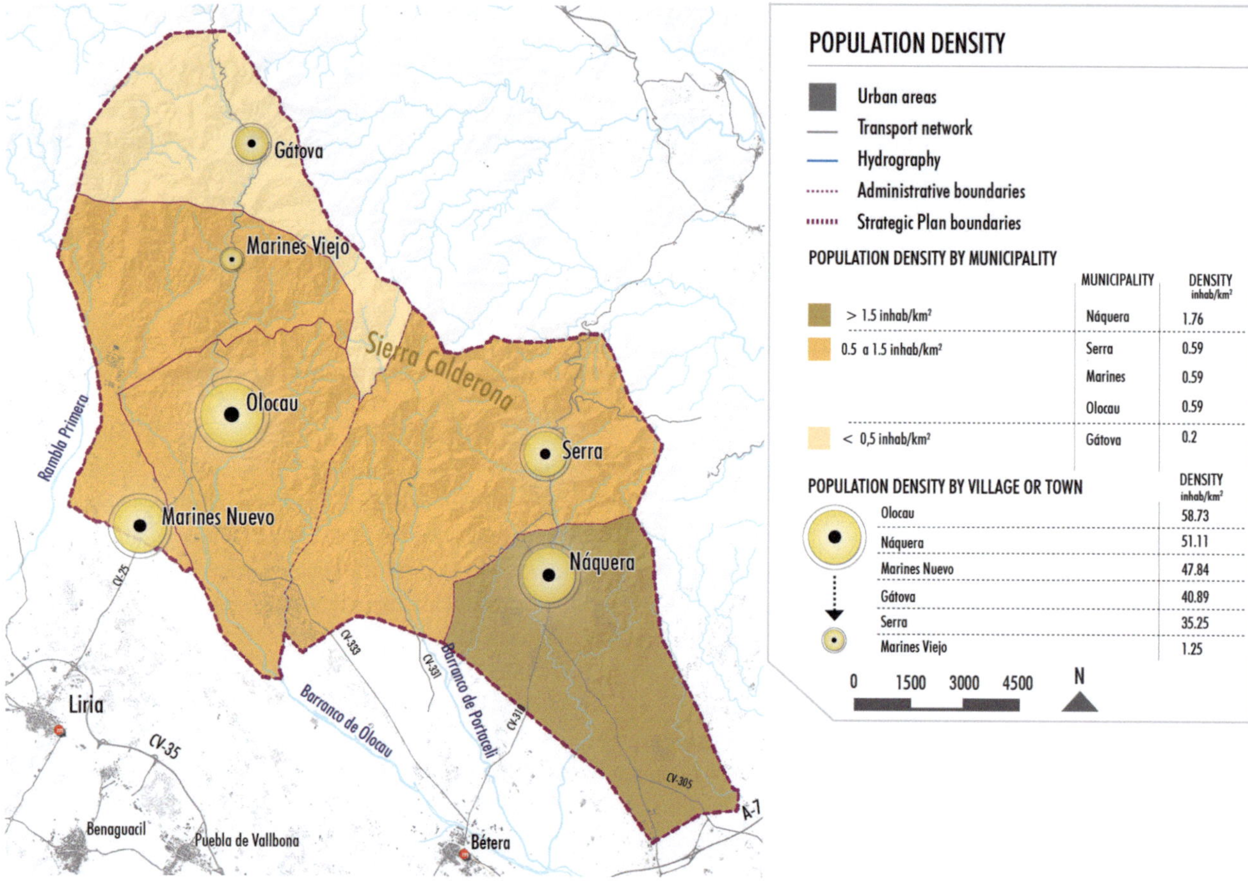

Fig. 3.89 Population densities by municipality and town. *Source* Own elaboration

Table 3.60 Population density. *Source* Own elaboration

Municipality	Area	Density (Inhab/ ha)
TOTAL PTE		0.66
NÁQUERA	Town	51.11
	Municipality	1.76
SERRA	Town	35.25
	Municipality	0.59
MARINES	NUEVO Town	47.84
	VIEJO Town	1.25
	Municipality	0.59
OLOCAU	Town	58.73
	Municipality	0.57
GÁTOVA	Town	40.89
	Municipality	0.20

Table 3.61 Population and seasonality. *Source* Own elaboration

Municipality		Permanent population	%	Non-permanent population	%	(%) Seasonality
NÁQUERA	Town	2,990	50.7	8,970	43.5	300.00
	Housing estates	2,910	49.3	11,640	56.5	400.00
	Total municipality	5,900	100.0	20,610	100.0	349.32
SERRA	Town	2,351	70.2	7,053	63.9	300.00
	Housing estates	998	29.8	3,992	36.1	400.00
	Total municipality	3,349	100.0	11,045	100.0	329.80
MARINES	Town	1,581	86.3	2,055	76.6	130.00
	Housing estates	251	13.7	628	23.4	250.00
	Total municipality	1,832	100.0	2,682.8	100.0	146.44
OLOCAU	Town	599	39.5	1,557	40.4	260.00
	Housing estates	918	60.5	2,295	59.6	250.00
	Total municipality	1,517	100.0	3,852.4	100.0	253.95
GÁTOVA	Town	458	76.3	1,145	61.7	250.00
	Housing estates	142	23.7	712	38.3	501.06
	Total municipality	600	100.0	1,856.5	100.0	309.42
TOTAL STRATEGIC PLAN	Town	7,979	60.5	20,781	51.9	260.44
	Housing estates	5,219	39.5	19,266	48.1	369.15
	Total municipality	13,198	100.0	40,047	100.0	303.43

3.10.3 Demographic Pyramids

The analysis of the demographic pyramids in the five studied municipalities shows a distribution by ages similar to the whole Valencian Region, with the exception of Gátova, where the ageing of the population is very clear (Figs. 3.90, 3.91, 3.92, 3.93 and 3.94).

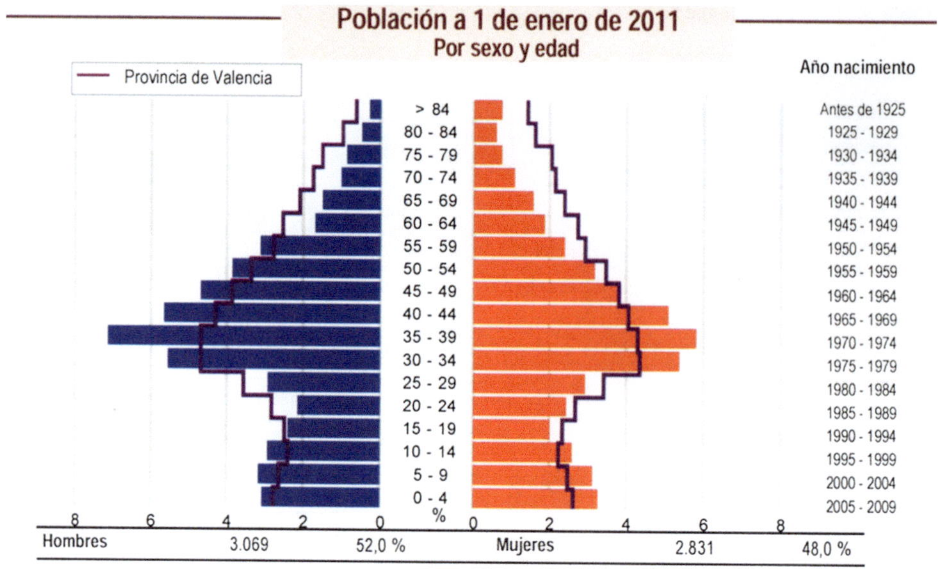

Fig. 3.90 Population pyramid. Náquera. *Source* I.V.E. 2011

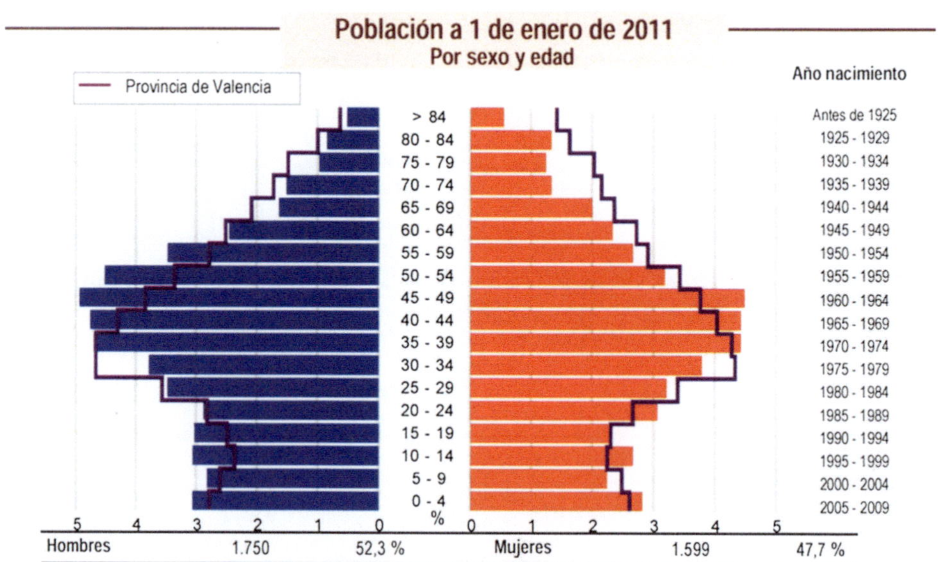

Fig. 3.91 Population pyramid. Serra. *Source* I.V.E. 2011

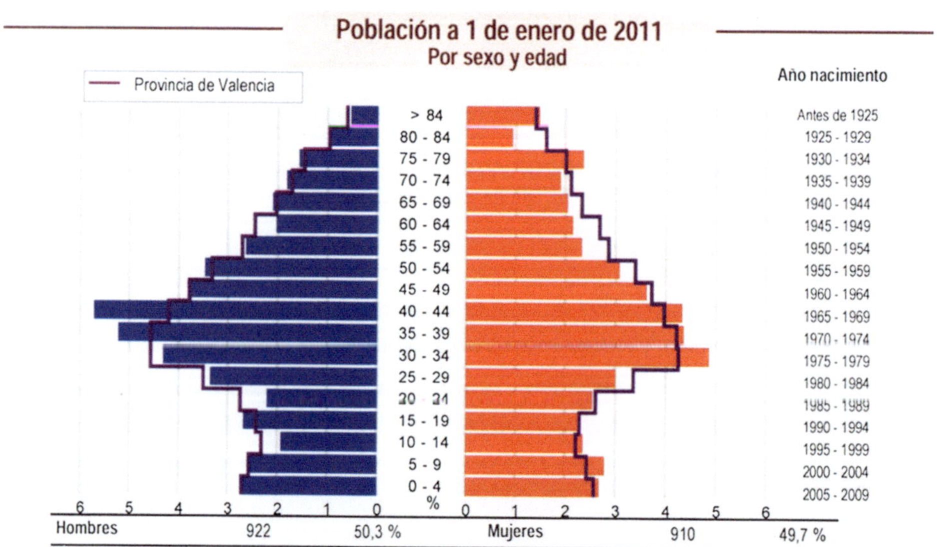

Fig. 3.92 Population pyramid. Marines. *Source* I.V.E. 2011

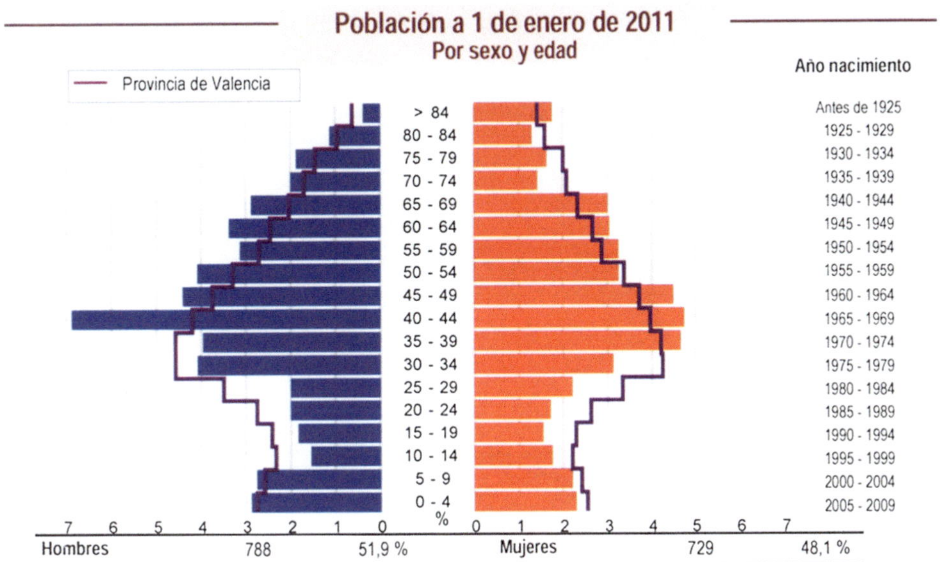

Fig. 3.93 Population pyramid. Olocau. *Source* I.V.E. 2011

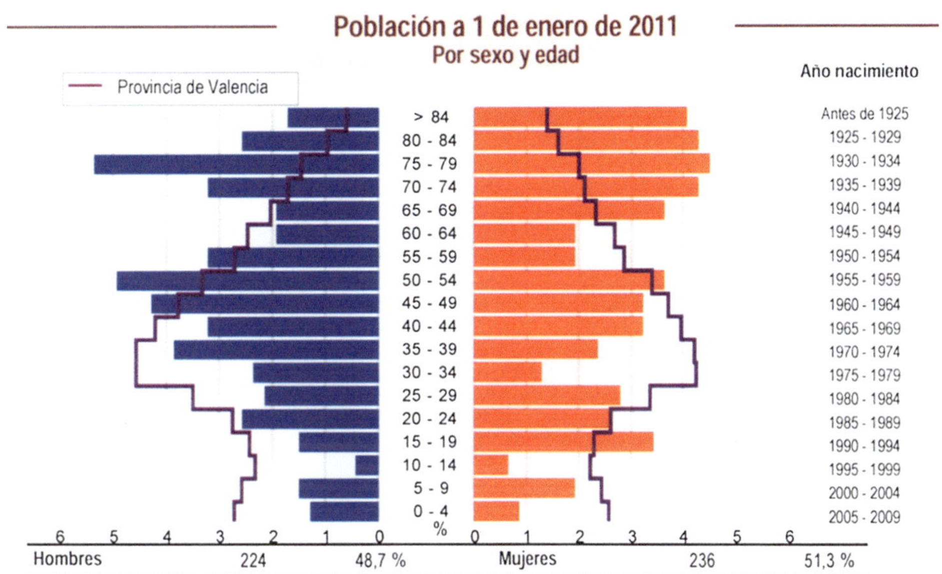

Fig. 3.94 Population pyramid. Gátova. *Source* I.V.E. 2011

3.10.4 Employment and Unemployment by Sectors

The distribution of employment by sectors in the area covered by the Strategic Plan shows significant differences depending on the municipality. Thus, Náquera and Olocau present a diversified distribution of their labour market; Serra and Gátova have a strong dependence on services and construction, while in Marines, the number of people working in the industrial sector and in construction stands out.

The unemployment levels vary significantly between municipalities. The highest rate occurs in Gátova (17.9%), probably due to its reduced economic activity, and, surprisingly, in Marines (14.4%) where the agricultural and industrial activities are relevant. These levels are slightly lower in Olocau, Serra and Náquera. It should be noted that a great part of the local population works outside the area of the Strategic Plan, in industrial estates and nodes of services located in the metropolitan area of Valencia.

The incidence of unemployment per sector (agriculture, construction, industry and services) with respect to the total unemployment shows that unemployment has not increased significantly in agriculture, except in Gátova that in the construction sector it has increased especially in Marines and Olocau and to a lesser extent in Gátova that in the industrial sector it remains stable in Marines and Náquera, while in Serra, it is growing at a fast pace. Finally, unemployment in the service sector, which generates around half of the jobs within the Strategic Plan's area, has increased particularly in Marines and Gátova (year 2012) (Tables 3.62, 3.63 and 3.64).

Table 3.62 Incidence of unemployment by activity sector [estimation]. *Source* Own elaboration based on existing municipal records of Caja España (2012)

	Unemployed agriculture	% unemployment agriculture	Unemployed construction	% unemployment construction	Unemployed industrial sector	% unemployment industrial sector	Unemployed service sector	% unemployment service sector
NÁQUERA	1	0.5	31	6.1	49	6.9	182	15.1
SERRA	1	3.8	5	3.6	11	29.7	40	13.8
MARINES	1	5.3	14	21.0	10	3.7	49	29.7
OLOCAU	1	4.9	5	16.9	6	12.3	18	12.4
GÁTOVA	0	29.0	3	11.3	1	10.5	9	21.9

Table 3.63 Unemployment by activity sector (2012). *Source* Own elaboration based on existing municipal records of Caja España (2012)

	% unemployment (2011)	Agriculture	Construction	Industry	Services	Without prior occupation
NÁQUERA	10.5	0.22	11.67	18.28	68.06	1.76
SERRA	11.9	1.05	9.12	18.95	68.42	2.46
MARINES	14.4	1.65	18.68	13.19	63.74	2.75
OLOCAU	12.8	2.24	17.16	20.9	57.4	2.24
GÁTOVA	17.9	2.13	19.15	8.51	65.96	4.26

Table 3.64 Unemployment by activity sector (2007). *Source* Own elaboration based on existing municipal records of Caja España (2012)

	NÁQUERA		SERRA		MARINES		OLOCAU		GÁTOVA	
	No.	%	No.	%	No.	%	No.	%	No.	%
Agriculture	116	4.6	16	3.3	24	4.5	14	5.9	1	1.3
Industry	712	28	37	7.6	279	51.9	52	21.8	11	14.5
Extractive industry	0		0		0		0		3	
Production industry	709		36		278		50		8	
production and distribution of energy, water and gas	3		1		1		2		0	
Construction	507	19.9	147	30.1	69	12.8	31	13	23	30.3
Services	1,207	47.5	288	59	166	30.8	142	59.3	41	53.9
Total	2,542	100	488	100	538	100	239	100	76	100

3.11 Local Economy

3.11.1 Generalities

This subchapter looks into those aspects considered more economically relevant for the development of the Sierra Calderona Strategic Plan. Occasionally, and with a prospective vocation, the possible new activities that could be implemented in the area are explored.

For the purpose of the Plan and in order to know the relative importance of the various productive sectors, it has been estimated roughly, as stated in the previous subchapters, the economic relevance of the most vulnerable sectors (agriculture) and of those showing a stronger potential, given the specific conditions in the area (tourism, forestry). In any case, it must be noted that a large part of the population works outside the area of the Strategic Plan and that an important part of the wealth generated in the area is exported off its limits.

Finally, the situation of the local administrations' accounts has been analysed, as well as possible ways to fund existing and potential projects.

In conclusion, this study aims to guide the definition of proposals that could activate the local economy and it has been structured as follows:

- Business activity (current situation)
- Forestry sector in the area covered by the Strategic Plan (estimation of the current situation and potentialities)
- Agricultural sector (estimation of the current situation)
- Tourism: visitors and seasonal residents (estimation of the current situation)
- Local public administrations (current situation)
- Rural development and funding

According to this, the economic magnitudes of the industrial and service sector are not specified, being partially treated when studying the tourism sector.

3.11.2 Business Activity

The business activity based in the area helps understand the strength of local economies and the level of entrepreneurship although it might be highly connected to external companies or people (suppliers or employees). It must be noted at this point that the companies based in the study area are linked to different sectors (i.e. agriculture, agro-industrial, industrial, construction and services) and that the local economic activity developed outside the business framework (freelancers, government agencies, etc.) is not included.

When determining the volume of business activity per capita, the relevance of Náquera can be seen as a regional economic hub. In Marines, due to the presence of an internationally important paint factory, the business activity per capita is high. Conversely, Olocau and especially Gátova and Serra have a very low volume of business per capita, which is especially remarkable in Serra, being the municipality with the second largest population in the region (Table 3.65).

Table 3.65 Volume of the business activity. *Source* www.informacion-empresas.com and SABI database

Business activity		
	Exploitation income	Business activity per capita
NÁQUERA	385,914,865.00 €	65,409.30 €
SERRA	9,719,914.00 €	2,902.33 €
MARINES	66,936,170.00 €	36,537.21 €
GÁTOVA	707,228.00 €	1,178.71 €
OLOCAU	11,690,880.00 €	7,706.58 €
Total	474,969,057.00 €	35,987.96 €

3.11.3 Forestry Sector

The economic activity in the forestry sector is practically non-existent, despite the great importance that this land use has in municipalities such as Serra and Gátova, though constraints arising from environmental protection considerably restrict the economic development associated with both the primary sector (forestry, and to a lesser extent agriculture) and the service sector.

For that reason, the business opportunities to implement forest exploitation have been explored considering the conditions described in the "Plan Rector de Usos y Gestión" or "PRUGSC" (in English: "Use and Management Governing Plan of the Sierra Calderona Natural Park") and the "Plan de Ordenación de los Recursos Naturales" or "PORNSC" (in English: "Natural Resources Management Plan").

For estimation purposes, the average yields and zones identified in the draft of PATFOR (Forestry Action Plan of the Valencian Region) for different kinds of exploitations (timber, biomass, cork, honey, etc.) have been used. In addition, restrictions imposed by the conditions of the Natural Park legislation have been applied, guided by the principle of prudence, which aims to avoid generating too ambitious expectations regarding the potential economic activity of this sector.

3.11.3.1 Timber Harvesting

According to the subchapter dedicated to the natural environment (Sect. 3.2), the estimated yields and revenues are shown in Table 3.66.

Table 3.66 Annual estimated revenue from timber harvesting. *Source* Own elaboration and PATFOR, Vaersa

Timber harvesting					
Production	Production (value used)	Area (ha)	Annual production (m^3/year)	Price timber (€/m^3)	Annual income
<0.5 m^3/ha/year	0.25	1,671.22	417.81	30.00 €	12,534.18 €
0.5–1 m^3/ha/year	0.75	2,656.24	1,992.18	30.00 €	59,765.50 €
1–1.5 m^3/ha/year	1.25	278.21	347.76	30.00 €	10,432.85 €
>1.5 m^3/ha/year	1.5	836.85	1,255.27	30.00 €	37,658.03 €
Total		5,442.52	4,013.02		120,390.56 €

3.11.3.2 Harvesting for Biomass

This has been differentiated between the production of pellets and chips for biomass. According to the subchapter dedicated to the natural environment (Sect. 3.2), the estimated yields and revenues are shown in Tables 3.67 and 3.68.

Table 3.67 Estimated annual revenues from the exploitation of biomass (chips). *Source* Own elaboration and PATFOR, Vaersa

Chips					
Production	Production (value used)	Area (ha)	Annual production (m^3/year)	Price chips (€/m^3)	Annual income
0.5–1 t/ha/year	0.75	286.00	214.50	40.00 €	8,580.00 €
0.2–0.5 t/ha/year	0.35	1,929.00	675.15	40.00 €	27,006.00 €
<0.2 t/ha/year	0.15	3,226.95	484.04	40.00 €	19,361.70 €

Table 3.68 Estimated annual revenues from the exploitation of biomass (pellets). *Source* Own elaboration and PATFOR, Vaersa

Pellets					
Production	Production (value used)	Area (ha)	Annual production (m^3/year)	Price pellets (€/m^3)	Annual income
0.5–1 t/ha/year	0.5	286.00	143.00	107.25 €	15,336.75 €
0.2–0.5 t/ha/year	0.2	1,929.00	385.80	135.03 €	52,094.57 €
<0.2 t/ha/year	0.1	3,226.95	322.70	55.36 €	17,863.06 €
Total		5,441.95	851.50		85,294.38 €

3.11.3.3 Cork Harvesting

According to the subchapter dedicated to the natural environment (Sect. 3.2), the estimated yields and revenues are shown in Table 3.69.

3.11.3.4 Truffle Harvesting

According to the subchapter dedicated to the natural environment (Sect. 3.2), the estimated yields and revenues are shown in Table 3.70.

In order not to overestimate the economic expectations regarding this use, the surface area potentially suitable for producing truffles has been significantly reduced in relation to the estimations included in the PATFOR (Forestry Action Plan of the Valencian Region). Low yields have been estimated in zones with an average estimated productivity, and the areas with a low-estimated productivity have not been considered at all.

3.11.3.5 Honey Harvesting

According to the subchapter dedicated to the natural environment (Sect. 3.2), the estimated yields and revenues are shown in Table 3.71.

3.11.3.6 Hunting

According to the subchapter dedicated to the natural environment (Sect. 3.2), the estimated yields and revenues are shown in the Table 3.72.

Table 3.69 Revenue per ten years shift and annual estimate from cork harvesting. *Source* Own elaboration and PATFOR, Vaersa

Cork harvesting							
Production	Production (value used)	Area (ha)	kg/10 years	Price cork (€/kg)	Income every 10 years	Annual income	Costs 10 years
Low < 1,000 kg/ha	300	224.00	67200.00	1.60 €	107,520.00 €	10,752.00 €	96,768.00 €
Average 1,000–2,000 kg/ha	1000	10.35	10350.00	1.60 €	16,560.00 €	1,656.00 €	14,904.00 €
High > 4,000 kg/ha	4,000	1.40	5600.00	1.60 €	8,960.00 €	896.00 €	8,064.00 €
Total		235.75	83150.00		133,040.00 €	13,304.00 €	119,736.00 €

Table 3.70 Estimated annual revenues from truffle harvesting. *Source* Own elaboration and PATFOR, Vaersa

Truffle					
Aptitude	Production (value used)	Area (h)	kg/year	€/kg	Annual income
Low: 0–10 kg/ha/year	0.5	719.87	359.94	500.00 €	179,967.70 €
Total		719.87	359.94	500.00 €	179,967.70 €

Table 3.71 Annual estimated revenue from honey harvesting. *Source* Own elaboration and PATFOR, Vaersa

Honey			
Area (h)	No. of hives (in 6 groups of 250 units)	€/hive/year	Annual income
2,600	1,500	20.70 €	31,050.00 €

Table 3.72 Estimated annual revenues from hunting. *Source* Own elaboration and PATFOR, Vaersa

CAZA													
	Income (€/ha)	Income (value used) (€/ha)	GÁTOVA		MARINES		NÁQUERA		OLOCAU		SERRA		Total strategic plan
			Area (ha)	Income (€)	Area (ha)	Income (€)	Area (ha)	Income (€)	Area (ha)	Income (€)	Area (ha)	Income (€)	Income (€)
Small game	16	10	1,836.00	67,082,318.84 €	1,722.60	62,938,999.15 €	3,447.00	125,943,765.28 €	2,781.90	101,642,866.44 €	4,828.50	176,419,921.86 €	534,027,871.57 €
Big game	0.76	0.76	183.60	216,411.77 €	172.26	203,045.16 €	344.70	406,302.49 €	278.19	327,906.26 €	482.85	569,141.73 €	1,722,807.41 €
Total			2,019.60	67,298,730.61 €	1,894.86	63,142,044.31 €	3,791.70	126,350,067.76 €	3,060.09	101,970,772.70 €	5,311.35	176,989,063.59 €	535,750,678.98 €

3.11.3.7 Summary

Adding all the forest exploitation already indicated, the following income could be obtained (Table 3.73).

3.11.4 Agricultural Sector

According to the surfaces devoted to each type of crop (land uses SIOSE 2011) and to the revenues and profits (€/hectare) estimated for each type of crop, the following economic data for the agricultural activity would be obtained (Table 3.74).

Table 3.73 Potential income from the forestry sector. *Source* Own elaboration and PATFOR, Vaersa

Total revenue forest harvesting (considering chips instead of pellets)	399,659.96 €
Total revenue forest harvesting (not considering truffle harvesting and considering chips instead of pellets)	219,692.26 €
Total revenue Hunting	147,270.82 €

Table 3.74 Estimated annual agricultural revenues and profits. *Source* Own elaboration

KIND OF CROP		Arable crops (not rice)	Citrus trees	Other fruit trees	Olive trees	Total
REVENUE (€/ha)		3,900.00	6,788.00	3,714.00	1,841.00	
INCOME (€/ha)		1,778.00	2,842.00	1,714.00	640.00	
GÁTOVA	Area (ha)	10		229	271	510
	revenue (€)	39,000.00 €	0.00 €	850,506.00 €	498,911.00 €	1,388,417.00 €
	income (€)	17,780.00 €	0.00 €	392,506.00 €	173,440.00 €	583,726.00 €
MARINES	Area (ha)	22	386	51	219	678
	revenue (€)	85,800.00 €	2,620,168.00 €	189,414.00 €	403,179.00 €	3,298,561.00 €
	income (€)	39,116.00 €	1,097,012.00 €	87,414.00 €	140,160.00 €	1,363,702.00 €
OLOCAU	Area (ha)	34	370	177	435	1,016
	revenue (€)	132,600.00 €	2,511,560.00 €	657,378.00 €	800,835.00 €	4,102,373.00 €
	income (€)	60,452.00 €	1,051,540.00 €	303.378.00 €	278,400.00 €	1,693,770.00 €
NÁQUERA	Area (ha)	59	1,061	127	22	1,269
	revenue (€)	230,100.00 €	7,202,068.00 €	471,678.00 €	40,502.00 €	7,944,348.00 €
	income (€)	104,902.00 €	3,015,362.00 €	217,678.00 €	14,080.00 €	3,352,022.00 €
SERRA	Area (ha)	15	235	85	79	414
	revenue (€)	58,500.00 €	1,595,180.00 €	315,690.00 €	145,439.00 €	2,114,809.00 €
	income (€)	26,670.00 €	667,870.00 €	145,690.00 €	50,560.00 €	890,790.00 €
Total strategic plan	Area (ha)	140	2,052	669	1,026	3,887
	revenue (€)	546,000.00 €	13,928,976.00 €	2,484,666.00 €	1,888,866.00 €	18,848,508.00 €
	income (€)	248,920.00 €	5,831,784.00 €	1,146,666.00 €	656,640.00 €	7,884,010.00 €

3.11.5 Tourist Sector

The impact of the tourist sector on the economy of the area covered by the Strategic Plan is especially difficult to quantify, as the number of visitors and their average spending must first be determined. In order to obtain an approximation of the economic magnitude, a distinction has been made between occasional visitors and seasonal visitors, the latter being those with secondary residences in the area. The occasional visitors have been subdivided into subgroups, depending on the activities they participate in. In a second stage, the number of visitors associated with each group or subgroup has been quantified, and the average expenditure per day and group has been estimated. For seasonal visitors, an average stay of 15 days/year has been assumed. After having set these parameters, the impacts of occasional and seasonal tourism in the area have been calculated. These figures should be considered as strictly indicative and are based on data collected for the annual reports of the Natural Park (occasional visitors and patterns of visit), in the population fluctuations registered during high season and in self-made estimations related to the average spending per day and visitor type (Table 3.75).

Table 3.75 Estimated revenue from the tourist activity. *Source* Own elaboration

Occasional visitors				
Profile	No. visitors	Duration stay (days)	Spending (€/day)	TotaL Spending
Group visiting recreative areas	17,503	1	1.5	26,254.76 €
hikers	19,448	1	3	58,343.92 €
group of sportsmen/women (mountain/road cyclist)	18,151	1	6	108,908.65 €
group visiting rural + urban areas	9,076	1	12	108,908.65 €
group staying overnight visiting the rural environment (*)	519	2.53	50	65,604.50 €
group staying overnight visiting the rural environment and town (*)	130	2.53	80	26,241.80 €
Total	64,827			394,262.29 €
Seasonal visitors				
In housing estates, towns and villages	25,000	20	15	7,500,000.00 €
Total	25,000			7,500,000.00 €

3.11.6 Economy of the Construction Sector

The economic crisis (2008–2014) had a strong impact on the construction sector in the area covered by the Sierra Calderona Strategic Plan. This impact becomes clear when comparing the small number of building permits accepted during 2011 with the total granted in previous years, when the construction activity was intense (Tables 3.76 and 3.77).

Table 3.76 Construction activity (year 2011). *Source* I.V.E.

Construction activity (year 2011)						
	GÁTOVA	MARINES	NÁQUERA	OLOCAU	SERRA	Total plan
TOTAL building licenses given (2011):	2	1	2	2	4	11
New construction	2	0	1	2	1	6
Rehabilitation	0	1	1	0	3	5
Demolition	0	0	0	0	0	0
New dwellings	1	0	4	2	4	11
New buildings	2	0	4	2	4	12

Table 3.77 Housing stock (year 2001). *Source* I.V.E. and population and housing census 2001

Housing stock (year 2001)	GÁTOVA	MARINES	NÁQUERA	OLOCAU	SERRA	Total plan
Buildings in bad conditions, ruin state	7	20	43	28	6	104
Main dwellings	204	515	1,124	400	745	2,988
With pending payments	12	144	305	99	175	735
Rental properties	4	12	69	11	37	133
Second residences	67	28	4,285	1,465	1,548	7,393
Empty second residences	472	332	115	314	44	1,277

3.11.7 Economy of the Public Local Administrations

The study of the 2009 municipal accounts (Table 3.78) showed a positive balance but, unlike the 2008 and previous years' balances, it indicated a marked increase in financial liabilities (loans contracted by the municipalities with other governmental agencies or financial entities) that in Náquera, Olocau and Serra reached 20, 20 and 26% of their respective budgets. In Gátova, the situation was slightly better at 16%). In Marines, an exceptionally low level of debt was retained at 2%.

The 2009 municipal accounts provide evidence of a clear fall in income, as result of the slowdown in the economy's pace and the subsequent economic crisis. In some municipalities, this had an especially intense effect due to the fall of activity in the construction sector and of the building permits.

Regarding public expenses or costs, it can be seen that the costs for staff, goods and services, and real investments account for up to 89% of the total municipal spending in Náquera and 96% in the municipalities of Gátova, Olocau, Marines and Serra.

The search for a more balanced budget without incurring new debts suggests the need for a selective reduction of expenses, which could initially be done in the "real investments" section. The reduction in that section would be especially effective in Gátova, where in 2009 this meant 41% of the total municipal spending, but also in Serra, Marines and Olocau, where it, respectively, meant 33, 32 and 25%, though it would have a lower effect in Náquera, where it reached only 14% of the total municipal spending. It must be understood that, in any case, the reduction of "real investments" would have a negative and direct impact on businesses and jobs linked to the construction sector and on the public works developed by the municipalities.

A second possible adjustment would be the reduction of costs for both staff and goods and services. These constitute 75% of the municipal budget in Náquera, 71% in Olocau, 63% in Marines and Serra and 56% in Gátova, but an indiscriminate cut in this section could have a direct and severe impact on the social services provided to the population. At this point, and in order to avoid as far as possible a reduction in the quality of the aforementioned services, the need arises to increase efficiency and reduce unnecessary expenses included in both kinds of costs.

The level of expenditure, fundraising and investment per capita is similar in the five municipalities, with the exception of Marines, which has lower levels for all three parameters.

The study of the situation of the municipal accounts between 2009 and 2013 shows a considerable increase of the municipal debt levels. This comprises previous debts, the fall of taxes and revenues, and especially, new debts owed to financial entities as a result of the implementation of "Real Decreto-Ley 7/2012, of March 9th", through which a fund to finance the payments to suppliers was created, and by which municipalities have to pay the pending payments to their suppliers even if the regional government has not yet paid its debts to local administrations. This situation has generated a difficult economic juncture that could be aggravated when the pre-amortisation period of 2 years finishes, and the municipalities should begin to return the capital received (at the moment they are only paying the interest of the newly incurred debts).

The five municipalities included in the area covered by the Sierra Calderona Strategic Plan have very diverse characteristics concerning their public accounts. However, two main and common lines of action are proposed:

- **Increase municipal revenues**: This is a difficult goal due on one hand to the fall in the construction sector, with the subsequent reduction in building permit taxes and in building property taxes (IBI), and on the other hand, to the current difficulties to obtain payment of pending sums. Some possibilities could be:
  - To regulate the tax payment of non-regularised dwellings. (This measure would be more effective in those municipalities with many non-regularised units).
  - To obtain the pending payments from the regional government.
  - To apply for external subsidies and funding with other public administrations. In order to do that, viable and solid programmes and projects must be defined.
  - To strengthen the local economic activity in order to be able to increase the collection of direct and indirect taxes.
  - To promote a regional compensation for the environmental services provided and for the positive externalities generated in the Valencian Region and the metropolitan area of Valencia.

- **Reduce municipal expenses**: Assuming that the reduction should not affect the services provided to the population by local administrations, costs could be dropped by:
  - Pooling, sharing or coordinating certain services, in order to generate economies of scale.
  - Reducing unnecessary costs of representation and management.
  - Increasing the efficiency of all administrative processes and delivery of services.

Table 3.78 Municipal Budget 2009. *Source* Ministerio de Hacienda y Administraciones Públicas

	GÁTOVA		MARINES		NÁQUERA		OLOCAU		SERRA	
	Total (€/year)	%	Total (€/year)	%	Total (€/year)	%	Total (€/year)	%	Total (€/year)	%
Total municipal income Liquidation 2009	1,017,442.00 €	100	1,810,136.40 €	100	12,911,533.30 €	100	2,494,227.9€	100	4,616,375.0 €	100
Direct taxes	147,159.70 €	14.46	387,815.00	21.42	2,975,096	23.04	681,168	27.31	1,288,288.2	27.91
Indirect taxes	30,348.20 €	2.98	13,360.50	0.74	184,345.4	1.43	20,674.7	0.83	190,942.1	4.14
Fees and other charges	168,335.90 €	16.55	191,486.40	10.58	2,813,917.3	21.79	611,116.5	24.50	641,232.3	13.89
Current transfers	245,208.20 €	24.10	426,052.30	23.54	1,704,508.3	13.20	409,323.5	16.41	812,427.7	17.60
Patrimonial revenues	5,566.60 €	0.55	20,809.70	1.15	2,015,403.5	15.61	315.2	0.01	22,792.8	0.49
Disposal real investments	3,000.00 €	0.29	0.00	0.00	238,153.8	1.84	0	0.00	0	0.00
Capital transfers	256,165.40 €	25.18	743,349.00	41.07	974,609	7.55	321,630	12.89	476,166.8	10.31
Financial assets	0.00 €	0.00	0.00	0.00	5500	0.04	0	0.00	0	0.00
Financial liabilities	161,658.00 €	15.89	27,263.50	1.51	2,000,000	15.49	450,000	18.04	1,184,525.1	25.66
Total municipal spending Liquidation 2009	790,767.90 €	100	1,586,546.20	100	8,285,498.40 €	100	2,332,286.80 €	100	4,638,710.30 €	100
Staff	244,516.2	30.92	562,139.40	35.43	3,205,831.9	38.69	685,685.3	29.40	1,449,380.7	31.25
Common goods and services	200,112.6	25.31	449,152.50	28.31	2,980,789.2	35.98	963,509.9	41.31	1,477,667.3	31.86
Financial	7,700.6	0.97	15,631.90	0.99	124,572.9	1.50	25,774.6	1.11	60,659.7	1.31
Current transfers	8,319.9	1.05	43,481.90	2.74	156,898.8	1.89	20,118.6	0.86	75,977.9	1.64
Real investments	324,738.6	41.07	515,500.80	32.49	1,195,697.3	14.43	581,166	24.92	1,532,531.8	33.04
Capital transfers	0	0.00	0.00	0.00	0	0.00	0	0.00	0	0.00
Financial assets	0	0.00	0.00	0.00	5,500	0.07	0	0.00	5,865.8	0.13
Financial liabilities	5,380	0.68	639.70	0.04	616,208.3	7.44	56,032.4	2.40	36,627.1	0.79
Spending per capita	992.48		571.03		1,089.19		1,115.07		952.2	
Collection per capita	396.22		226.52		556.24		474.54		481.2	
Investment per capita	724.86		291.08		210.51		392.95		498.55	

- Renegotiating and adjusting payments to external service suppliers. (In the area covered by the Strategic Plan, there are already examples of reductions of up to 30–40% in external services bills costs without reducing the quality of the services received).
- Diminishing real investments, keeping those that maintain a line of continuous improvement in the quality of services, and when contracting new building works, following strict principles of budgetary austerity.
- Reviewing the relationship between the "statutory" and "non-statutory" services provided by the municipalities to the population, in order to reduce non-compulsory municipal expenditures (those in which Local Councils assume the provision of public services that should be provided by regional or national administrations). At this point, it must be recalled that, since most of the municipalities located in the area of the Sierra Calderona Strategic Plan have less than 5000 inhabitants, the distribution of competences would be:
  - *Services or competences that correspond to the municipalities "statutory municipal services"*: public lighting, cemetery, waste collection, road cleaning, drinking water supply, sewage, access to urban settlements and paving of public streets.
  - *Services or competences that do not correspond to the municipalities ("non-statutory municipal services")* (i.e. those that are the responsibility of other governmental agencies but are assumed exceptionally and voluntarily by the municipalities): local police, libraries, adult schools, sports complex, day-care centre for the elderly, nursery, training workshops, etc.

Concurrently, it should be assumed that a very significant portion of public investments aimed to maintain, manage and improve the Sierra Calderona Natural Park should remain in the area and in its local population by creating jobs, supporting local companies providing services to the Natural Park, etc. This is especially critical considering that 49% of the area covered by the Sierra Calderona Strategic Plan is included within the limits of the Natural Park, 80% is regulated by the Park's "Plan de Ordenación de Recursos Naturales (PORNSC)", and its entirety is included in the Park's socio-economic influence area.

Thereby, considering that the five municipalities included in the Strategic Plan account for 56% of the surface of the Natural Park, as well as that their 2010 investment budgets amounted to €3,348,859 (Management Report 2010), it might have been expected that a similar percentage of the investments would have reverted in benefits for these municipalities, generating jobs within the local population linked to the management of natural spaces.

As displayed in the last rows of Table 3.78, the expenditure levels, tax collection and investment per capita are similar in all five municipalities, with the exception of Marines, which presents lower levels for the three parameters and of Gátova, which shows a higher level of investment per capita.

3.11.8 Rural Development and Funding

Socio-economic development in the limits between predominantly urban and rural areas is particularly complex, as they are characterised by dynamics linked to both systems. This becomes particularly evident the area covered by this Strategic Plan. Furthermore, there are constraints and potentials derived from the Natural Park status, whose "Plan Rector de Uso y Gestión (PRUGSC)" and "Plan de Ordenación de Recursos Naturales (PORNSC)" propose remarkable goals, which assume the necessity to support local communities in order to manage the complex forest–agricultural–urban mosaic that characterises most rural Mediterranean areas. The implementation of those generic goals requires clear strategies and specific mechanisms, some of which are already listed in the two statutory Plans mentioned above, but whose elaboration is currently still pending l (i.e. Sierra Calderona Forestry Plan, Hunting Plan, Livestock farming Plan).

In addition, the harmonisation between rural development and the preservation of environmental, cultural and landscape values of rural areas is a priority objective for the European Union that has also been assumed by the national government, the Valencian Region and the provincial council, who already defined and allocated funds for this purpose.

The analysis of the purposes intended for these funds shows a clear main intention: to help public and private initiatives that, being respectful of the character of the rural environment, would facilitate the economic activation and improvement of it. Based on the highly positive nature of this approach, these are the issues that would require a more detailed consideration:

(1) Real benefits generated from the received funds and investments.
(2) Degree of knowledge amongst public and private agents of the necessary procedures to apply for external funding.
(3) Potential to create consortiums or groups (like local action groups) to become eligible for external funding.

Finally, and given the difficult economic situation in the year 2013 of all public administrations (national, regional and local), a reduction on the funds assigned for rural development must be considered. At this point, it is possible to justify with solid arguments the need of maintaining those funds, as the investments in this concept are relatively low in relation to the high benefits obtained by the whole society. Assuming, nevertheless, that external funds might be reduced, or that municipal organisations should take a more active role in fundraising to start up their own projects, it would be necessary to study mechanisms to facilitate and promote:

(1) The mobilisation of private economic resources of the local population in programmes and projects that promote rural sustainable development.
(2) The mobilisation of the knowledge capital (trained young people and professionals) in local projects and programmes.
(3) The attraction of external private investment to carry out programs and projects that promote rural sustainable development.

It is crucial to emphasise that local population should always participate, promote and benefit from these initiatives and the necessity of generating efficient and effective structures that could help local and external investors to start up projects in the area.

The south-central area of Sierra Calderona needs to urgently develop all the above listed aspects, given the strong link that in recent decades the area has had with the construction sector, now in recession, as well as the need to improve and diversify its range of services, as the classical model, based on secondary residences, becomes obsolete and the competition with other regional, national or foreign areas increases (Fig. 3.95).

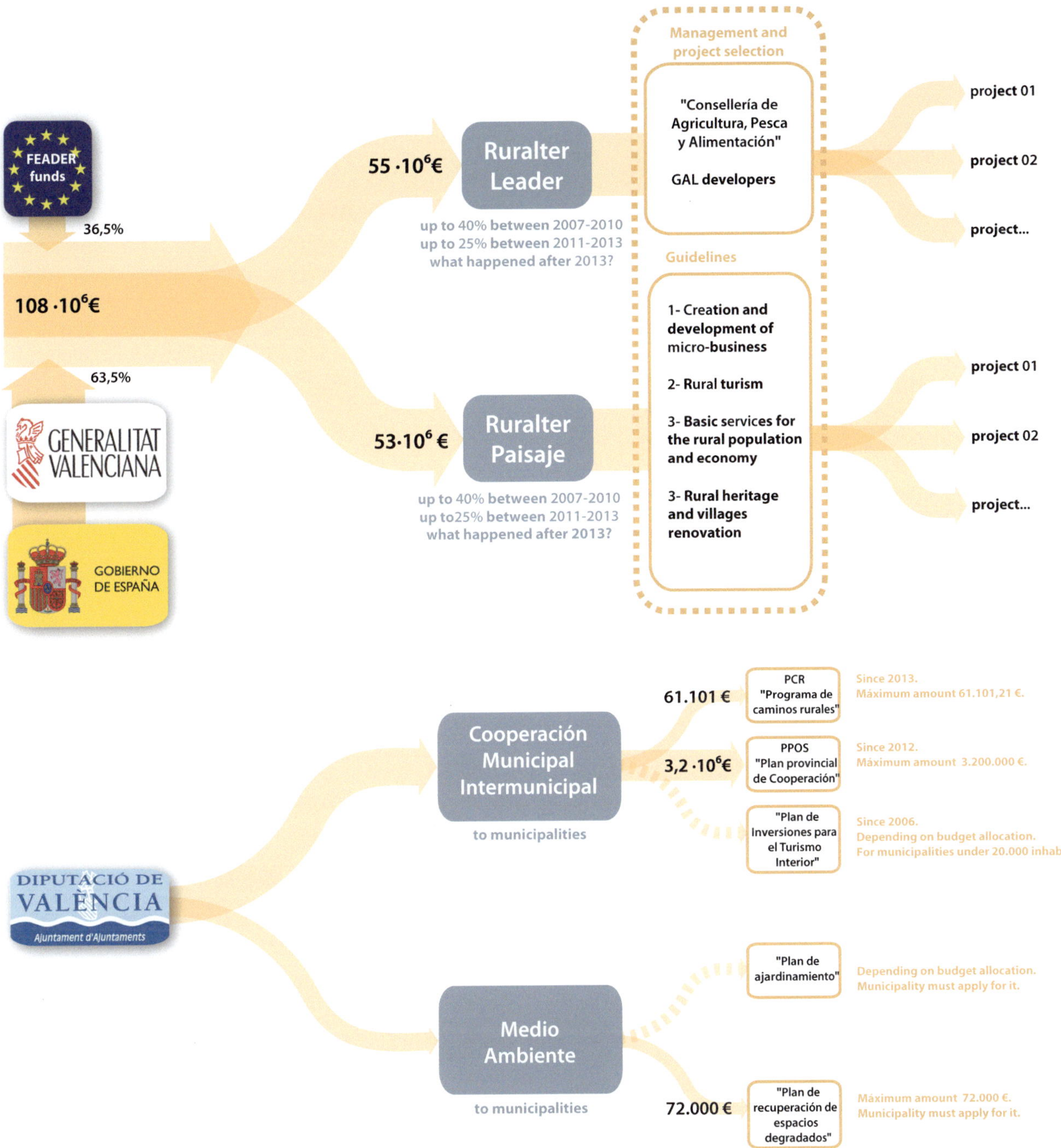

Fig. 3.95 Rural Development Funds (RURALTER programmes (EU, Spanish Government and Valencian Region Government) and funds from Diputación de Valencia (provincial council)).

3.11.9 Synthesis and Diagnosis

Business activity: The calculation of the volume of business activity per capita shows the relevance of Náquera as a regional economic hub. In Marines, due to the presence of an internationally important paint factory, the business activity per capita is remarkable. Olocau and especially Gátova and Serra present a very low volume of activity business per capita, which is especially remarkable in the case of Serra that, despite being the second largest municipality, lacks a strong business infrastructure.

Forestry sector: The economic activity in the forestry sector is practically non-existent despite the great importance that this land-use has in municipalities such as Serra and Gátova, in which the constraints arising from environmental protection considerably restrict the possibilities of economic development associated with the primary sector (forestry and, to a lesser extent, agriculture) or the service sector. A first estimation of the potential income derived from forest activities would be €546,930/year, €147.270 euros/year of which would be related to hunting and €179,967/year to potential truffle harvesting.

Agricultural sector: The study of agricultural activity shows that this sector has an appreciable contribution to the local economy (€18,848.500/year) with a level of profitability highly dependent with the type of crop. The high profitability of some new crops or some traditional crops in the process of replantation (cherry trees) also needs to be considered. It should also be noted that most of the agricultural activity is practised on a part-time basis, there is frequent cultivation of horticultural crops for self-consumption, and there is an increasing predominance of citrus tree plantations in the agricultural mosaic.

Tourism sector: The estimate of the volume of economic activity associated with occasional and seasonal tourism (e.g. residents in secondary homes) is based on an approximate quantification of the number of people in each group, their average expenditure per day and the duration and frequency of their visits. The figures obtained (€394,000/year for occasional visitors and €7,500,000/year for seasonal visitors) highlight the importance that seasonal visitors have in the local economy and the low incidence that occasional visitors have. An increase in the provision of high quality and specialised services could raise both the number of occasional visitors and their average expenditure per visit.

Construction sector: The current economic crisis has had a strong impact in the construction sector in the area covered by this Strategic Plan. This impact becomes clear when comparing the small number of building permits granted during 2011 with the number granted in previous years, when the construction activity was intense.

Municipal accounts: The study of the situation of the municipal accounts between 2009 and 2013 shows a considerable increase in the municipal debt levels, derived from previous debts, the fall in revenues and, especially, new debts owed to financial entities as a result of the implementation of the "Real Decreto-Ley 7/2012, of March 9th" through which a fund to finance the payments to suppliers is created, and through which municipalities have to pay the pending payments to their suppliers, even if the regional government has not yet paid its debts to local administrations.

The five municipalities included in the area covered by the Sierra Calderona Strategic Plan have very diverse characteristics, but two main lines of action are proposed to improve the situation of the municipal accounts:

- **Increase municipal revenues**: This is a difficult goal due on the one hand to the decline in the construction sector with the subsequent reduction in building permit taxes and in building property taxes (IBI), and on the other hand, to the current difficulties to obtain payment of debts. Some possibilities could be:
    - To regulate the tax payment of not-regularised dwellings. (This measure would be more effective in those municipalities with many non-regularised units.)
    - To obtain the pending payments from the regional government.
    - To apply for external subsidies and funding, together with other public administrations. In order to do that, viable and solid programmes and projects must be defined.
    - To strengthen the local economic activity, in order to be able to increase the collection of direct and indirect taxes.
    - To promote regional compensation for the environmental services provided and for the positive externalities generated on the Valencian Region and the metropolitan area of Valencia.
- **Reduce some municipal expenses**: Assuming that the reduction should not affect the services

provided to the population by local administrations, costs could be decreased by:

- Pooling, sharing or coordinating certain services, in order to generate economies of scale.
- Reducing unnecessary costs of representation and management.
- Increasing the efficiency of all administrative processes and delivery of services.
- Renegotiating and adjusting payments to external service suppliers. (In the area covered by the Strategic Plan, there are already examples of reductions of up to 30–40% in external services bills without decreasing the quality of the services received.
- Reducing real investments, keeping those that maintain a strategy of continuous improvement in the quality of services, and when contracting new building works, following strict principles of budgetary austerity.
- Reviewing the relationship between "statutory" and "non-statutory" services provided by the municipalities to the population, in order to reduce non-compulsory essential municipal expenditure.

Rural development and funding: Finally, and given the difficult economic situation in the year 2013 of all public administrations (national, regional and local), it is necessary to consider its possible impact on the funds assigned for rural development. Those funds need to be maintained due to their high social profitability and be supplemented with:

The mobilisation of private economic resources of the local population in programmes and projects that promote rural sustainable development.

The mobilisation of the knowledge capital (trained young people and professionals) in local projects and programmes.

The attraction of external private investment to carry out programs and projects that promote rural sustainable development.

It is crucial to highlight that local populations should always participate in, promote and benefit from these initiatives, and there is a necessity for generating efficient and effective structures that could help local and external investors to initiate projects in the area.

3.12 Main Consulted Sources

The following list includes the main sources consulted in the drafting of the this chapter in order of use

- Draft of "Plan de Acción Territorial Forestal de la Comunidad Valenciana" (PATFOR)
- Cartography from "Instituto Cartográfico Valenciano"
- "Plan Rector de Uso y Gestión del Parque Natural de la Sierra Calderona" (PRUGSC)
- "Plan de Ordenación de Recursos Naturales del Parque Natural de la Sierra Calderona" (PORNSC)
- "Plan de Uso Público del Parque Natural de la Sierra Calderona"
- Management reports of "Parque Natural de la Sierra Calderona" (years 2009, 2010 y 2011)
- Thematic cartography from "Consellería de Infraestructuras, Territorio y Medio Ambiente"
- Website "Parques Naturales" managed by "Consellería de Infraestructuras, Territorio y Medio Ambiente"
- Municipality of Naquera: www.naquera.com
- Municipality of Olocau: www.olocau.es
- Thematic cartography from the cross-municipal coordination area of "Diputación Provincial de Valencia"
- Report of the agrarian sector (year 2006 et seq.) from "Generalitat Valenciana, Consellería de Agricultura, Pesca y Alimentación"
- Study "DIRA_Huerta de Alboraya y Almàssera_2012"
- Study "Cullera_PROYECTO DE REPARCELACIÓN EN EL SECTOR NPR-5 "VEGA-PUERTO" DEL PLAN GENERAL DE CULLERA".
- Report "Córdoba_2010"
- Report "2011_Informe del Sector Agrari Valencià"
- Weekly price report "CAPA_2013"
- "Precio de la Cereza" by Gregorio López Ortega and Diego Frutos Tomás
- "Estudio técnico-económico de los procesos de producción agrícola y de transformación de las principales orientaciones hortofrutícolas de la región de Murcia" (AMOPA)
- Reports from "Instituto Murciano de Investigación y Desarrollo Agrario y Alimentario" (IMIDA)
- "Datos Algarrobo" by Hermosilla, 1988
- "Libro de aforos" (2012) from "Departamento de carreteras, Diputación Provincial de Valencia"
- Map of traffic (year 2011) from "Ministerio de Fomento"
- Road Report (years 2008–2012) from "Consellería de Infraestructuras, Territorio y Medio Ambiente"

- "Agencia Valenciana de Movilidad" (AVM). Metrobus Valencia
- "Generalitat Valenciana: Portal Argo_Conselleria de Educación"
- "Generalitat Valenciana: Portal Argo_Instalaciones Deportivas"
- Municipal planning cartography from "Colegio Oficial de Arquitectos de la Comunidad Valenciana"
- Reports on cultural heritage from "Generalitat Valenciana: Conselleria de Educación, Cultura y Deporte"
- Municipality of Serra: www.serra.es
- Municipality of Marines: www.marines.es
- "AVT—Agencia Valenciana de Turisme (2010). Turismo familiar en la Comunitat Valenciana" from February 2010.
- "AVT—Agencia Valenciana de Turisme (2011). Perfil del turista alojado en oferta reglada 2011".
- "AVT—Agencia Valenciana de Turisme (2011a). Turismo rural en la Comunidad Valenciana" from August 2011.
- "AVT—Agencia Valenciana de Turisme (2011a). Turismo rural en la Comunidad Valenciana" from August 2011.
- "AVT—Agencia Valenciana de Turisme (2012a). Turismo activo en la Comunidad Valenciana" from November 2012.
- "AVT—Agencia Valenciana de Turisme (2012b). Turismo cultural en la Comunidad Valenciana" from September 2012.
- "AVT—Agencia Valenciana de Turisme (2012c). Viajes turísticos de los españoles. Datos provinciales. Tercer trimestre 2012".
- "AVT—Agencia Valenciana de Turisme (2013). Encuesta de ocupación hotelera por zonas turísticas de la Comunidad Valenciana" from February 2013.
- "AVT—Agencia Valenciana de Turisme. Turismo rural en la Comunidad Valenciana" from April 3rd 2013.
- "Caja España (2011). Fichas municipales de Gátova, Marines, Náquera, Olocau y Serra".
- "Estrategia Territorial de la Comunidad Valenciana. Conselleria de Medio Ambiente".
- EC-OE-European Confederation of Outdoor Employers (2010). Defining the outdoors.
- ETC-European Travel Commission (2005). City Tourism and Culture: The European Experience.
- "IET—Instituto Español de Turismo (2011). Familitur 2011—Comunitat Valenciana"
- "Plan de Espacios Turísticos de la Comunidad Valenciana (2006)"
- "Ayuntamiento de Marines: Agenda 21 Local"
- "Boletín Informativo de Cifras INE_2008_Agua"
- Report "Residuos Ganaderos" from "Dirección General de Investigación Agraria y Ganadera".
- "Conselleria de Agricultura, Pesca y Alimentación".
- "IVE (Instituto Valenciano de Estadística). Fichas municipales".
- "Documento de Síntesis Ley para el desarrollo rural sostenible" by "Conselleria de Agricultura, Pesca y Alimentación"
- Database SABI (sistema de Análisis de Balances Ibéricos)
- Website www.informacion-empresas.com

Regional Objectives and Strategies 4

Abstract

Based on the results of the "Analysis and Diagnosis" and the Public Participation Plan, developed during the second stage of the Sierra Calderona Strategic Plan, this chapter proposes a set of fifty-two "Regional Objectives and Strategies" that were informed by a Public Participation Plan and that were organised in the following topics: Natural Environment, Agriculture and Livestock Farming, Urban Environment and Well-being, Infrastructures and Mobility, Cultural Heritage, Tourism and Public Use, Landscape, Sustainability, Socio-Demography, Economic activity and Governance. The strategies for the Sierra Calderona are presented in the form of brief texts and schematic drawings. The information in this chapter guided the drafting of ten Sectoral Plans and eighteen Pilot Projects, which defined the main frameworks and tools for achieving the defined objectives through the proposed strategies.

4.1 Introduction and Methodology

Based on the results of the "Analysis and Diagnosis" and the Public Participation Plan, developed during the first phase of the Sierra Calderona Strategic Plan, this chapter presents the subsequent "Regional Objectives Strategies" organised in the following sections:

- NATURAL AND FOREST ENVIRONMENT
- AGRICULTURE AND LIVESTOCK FARMING
- URBAN ENVIRONMENT AND WELL-BEING
- INFRASTRUCTURE AND MOBILITY
- CULTURAL HERITAGE
- TOURISM AND PUBLIC USE
- LANDSCAPE
- SUSTAINABILITY
- SOCIO-DEMOGRAPHY
- ECONOMIC ACTIVITY
- GOVERNANCE AND REGIONAL PLANNING & MANAGEMENT

The fifty-two objectives and strategies for the Sierra Calderona are presented in the form of brief texts and schematic drawings. They are deducted not only from the conclusions of the first stage of the Sierra Calderona Strategic Plan but also from the guidelines of the Territorial Strategy of the Valencian Region, as well as from the "Plan Rector de Usos y Gestión" or "PRUGSC" (in English: "Use and Management Plan of the Sierra Calderona Natural Park") and the "Plan de Ordenación de los Recursos Naturales" or "PORNSC" (in English: "Natural Resources Management Plan of the Sierra Calderona Natural Park"), as these two last documents affect a very significant portion of the area covered by the Strategic Plan.

The information in this chapter guided the drafting of the "Sectoral Plans and Pilot Projects", which defined the basic conditions and tools for achieving the proposed objectives through the suggested strategies. Those Sectoral Plans and Pilot Projects are included in the second volume or book of the Sierra Calderona Strategic Plan.

4.2 Regional Objectives and Strategies

4.2.1 Natural Environment

1. Forest Planning and Management: **TO DEVELOP PLANNING AND MANAGEMENT INSTRUMENTS AS FORESEEN IN THE CURRENT FORESTRY LEGISLATION** (Fig. 4.1).

- By elaborating a forest management Plan as an instrument for regional planning of forestry in the whole area included in the Natural Park.

J. Galan, *Analysis and Strategies for Sustainable Regional Planning*,
https://doi.org/10.1007/978-3-319-96508-6_4

- Alternatively, by drafting technical instruments for forest management in woodlands of either public domain or public interest, as defined in the Ley Forestal de la Comunidad Valenciana (Valencian Forestry Act).
- By developing a feasibility study of the use of biomass for energy production. This study should cover the production and management of biomass, as well as its energetic assessment and final commercialisation.

2. Economic uses: **TO INTEGRATE ENVIRONMENTAL PROTECTION IN THE ECONOMIC EXPLOITATION OF THE FOREST** (Fig. 4.2).

- By promoting economic and sustainable productive uses in forested areas, as referred to in the Natural Park's regulations for the park and its buffer zone, as a means of retaining functional links between the local population and natural environment, as well as promoting adequate forest management.
- By promoting the participation of the local population in the maintenance and management of the forest through communal initiatives.
- By promoting protective, ecological and landscape values in those forested areas utilised for economic purposes and by applying appropriate silvicultural treatments in order to generate forested areas with a balanced density and composition.
- By encouraging sustainable private productive uses (i.e. honey) in privately owned forestlands, primarily in those areas that are on abandoned agricultural land.

3. Public use: **TO OPTIMISE PUBLIC USE OF NATURAL SITES ACCORDING TO THEIR POTENTIALS AND LIMITATIONS** (Fig. 4.3).

- By defining areas with different intensities of public use and by establishing routes of high scenic and educational value, properly signposted, designed, integrated and connected with the main urban settlements.

Fig. 4.1 Strategy 1

Fig. 4.2 Strategy 2

4. Fire prevention: **TO COORDINATE FIRE PREVENTION WITH PRODUCTIVE ACTIVITIES IN THE RURAL ENVIRONMENT IN ORDER TO OPTIMISE THE USE OF INFRASTRUCTURES AND REDUCE THEIR VISUAL IMPACT** (Fig. 4.4).

- By integrating economic uses of natural areas, such as extensive livestock farming and rainfed agriculture, into the fire prevention strategies.
- By combining the infrastructures that give access to emergency vehicles, firebreaks, agricultural roads and public accesses, in order to minimise their impact on the environment.

5. Ecology: **TO FACILITATE EVOLUTION TOWARDS DEVELOPED AND BIODIVERSE WOODLANDS IN THE GREEN INFRASTRUCTURE INCLUDING THE ABANDONED AGRICULTURAL AREAS LOCATED BETWEEN URBANISED AREAS** (Fig. 4.4).

- By promoting the ecological development of natural areas towards advanced successional stages, thereby conditioning their management and the techniques for their possible economic exploitation.
- By promoting the natural regeneration of abandoned agricultural plots located in-between settlements, as well as of the blue infrastructure (ravines, brooks) and their adjacent areas.
- By promoting the diversification of shrub species, preventing the implantation of monospecific thickets and generating diverse types of vegetation cover, regarding both their layer-structure and composition.

6. Green infrastructure and spatial planning: **TO PROMOTE THE CREATION OF A GREEN INFRASTRUCTURE OF NATURAL CHARACTER, CONNECTING URBAN AREAS WITH NATURAL SPACES, MAKING USE OF TOPOGRAPHIC FEATURES AND WATER SPACES AND CONTROLLING URBAN SPRAWL** (Fig. 4.5).

7. Integration of uses: **TO COORDINATE PRODUCTIVE AND RECREATIONAL USES IN ORDER TO**

Fig. 4.3 Strategy 3

Fig. 4.4 Strategies 4 and 5

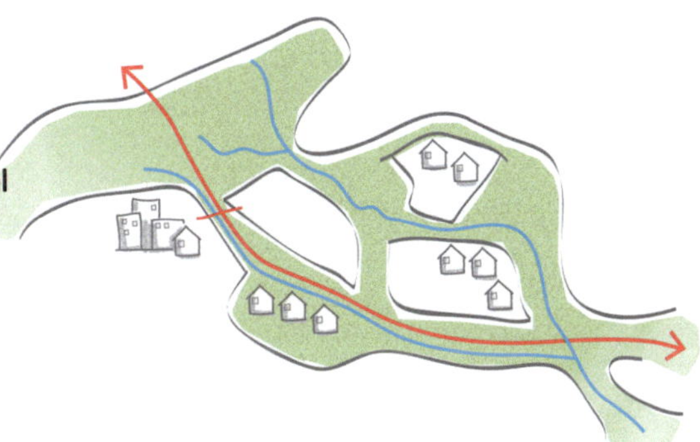

Fig. 4.5 Strategy 6

PROMOTE A PROPER MANAGEMENT, ENJOYMENT, ECONOMIC USE AND IMPROVEMENT OF NATURAL SPACES (Fig. 4.6).

- By promoting the regeneration of rainfed agricultural corridors along rural roads and forest tracks where such agriculture was practised in the past.
- By promoting extensive livestock farming in the aforementioned corridors and in those forest areas considered suitable for such use.
- By generating routes for standard vehicles, 4WD, bicycles and hikers properly connected with towns and villages, and designed according to the following criteria:
  - Continuity (incorporation of circular routes).
  - Scenic and ecological value.
  - Provision of services and car parks (in urban settlements, these could also function as hubs and points of departure/arrival for visitors, in picnic areas and water springs, in accommodation facilities, such as lodges and cottages, strategically located in the isolated areas of Sierra Calderona).
  - Controlled access to areas of high environmental value (i.e. only for hikers and emergency vehicles).
  - A range of activities offered for different kinds of visitors ("Transcalderona" trail, daily excursions, short excursions, etc.).

4.2.2 Agriculture and Livestock Farming

8. Mountain agriculture: **TO REGENERATE A MONTANE AGRICULTURAL SYSTEM SUPPORTED BY RURAL PATHS AND FACILITATED BY PLACING ADDED VALUE ON ITS PRODUCTS AND BY USING ABANDONED FIELDS** (Fig. 4.7).

- By generating a montane agricultural system located along the main rural tracks and defining an integrated network of firebreaks.
- By facilitating the integration and use of abandoned fields as part of the agricultural mountain system (i.e. by including them in an open land bank).

Fig. 4.6 Strategy 7

- By improving the production, processing and marketing processes, as well as by added value to products (i.e. by creating quality brands, a seal of "Denomination of origin" or by recovering traditional crops and promising varieties).
- By allowing the introduction of soft and well-integrated irrigation systems.
- By identifying terraced areas with cultural and landscape value, in order to prioritise their recovery and protection.
- By protecting historical and traditional irrigation systems around historical settlements.

9. Rainfed agriculture in plains: **TO COORDINATE RAINFED AGRICULTURE ON THE PLAINS WITH MONTANE AGRICULTURE IN ORDER TO INCREASE THEIR MUTUAL VIABILITY** (Fig. 4.7).

- By generating sufficient yields to make production, processing and marketing viable.
- By introducing irrigation systems in some rainfed plots.
- By promoting the spontaneous transformation of abandoned rainfed plots into forested areas, in order to create a forest-based green infrastructure that will contribute to preventing the merging of housing estates and scattered settlements, and that will facilitate their visual integration into the landscape.

10. Extensive livestock farming: **TO PROMOTE EXTENSIVE LIVESTOCK FARMING, THE USE OF CATTLE TRAILS AND GRAZING IN RAINFED AGRICULTURAL AREAS (AGRICULTURAL CORRIDORS IN MOUNTAIN AREAS AND PLAINS), AS WELL AS IN SUITABLE WOODLANDS** (Fig. 4.7).

- By integrating extensive livestock farming into fire prevention Plans by using grazing as a means of managing the understory in mountainous and agricultural areas.
- By promoting the consumption of local meat under a quality seal or specific identifier.

11. Irrigated agriculture in the plains: **TO DIVERSIFY AND IMPROVE THE PRODUCTIVE AND COMMERCIALISATION PROCESSES** (Fig. 4.8).

- By promoting more competitive and profitable irrigated agriculture by reducing its costs (full-time or professional farmers, plot groupings, leasing, land banks, cooperatives, etc.) and by adding value to the products (through quality seals, organic agriculture, etc.).
- By diversifying crops according to the soil and climatic conditions.
- By recognising and placing value on the existing hydrological infrastructure in order to improve it.

12. Suburban orchards and vegetable gardens: **TO PRESERVE AND ENHANCE HISTORIC ORCHARDS AND VEGETABLE GARDENS AROUND HISTORICAL VILLAGES AND TO PROMOTE THEIR ESTABLISHMENT AROUND HOUSING ESTATES** (Fig. 4.9).

- By promoting consumption of local products among the local population (i.e. by creating a network of commercial outlets).
- By facilitating the cultivation of abandoned plots through the creation of a land bank.
- By promoting the creation of community gardens in existing irrigated urban plots, particularly where historical irrigation systems exist, in the margins of rivers passing through urban areas, and around housing estates.

13. Green infrastructure of agricultural character: **TO PROMOTE THE CREATION OF A GREEN INFRASTRUCTURE OF AGRICULTURAL CHARACTER, INCLUDING AREAS WITH SOILS OF HIGH AGROLOGICAL CAPACITY, WITH A STRONG LANDSCAPE/CULTURAL INTEREST, WITH A POTENTIAL ROLE TO FACILITATE A BALANCED RELATIONSHIP BETWEEN URBAN AREAS OR INFRASTRUCTURES AND NATURAL AREAS, AS WELL AS WITH THE POTENTIAL TO CONTRIBUTE TO FIRE PREVENTION** (Fig. 4.10).

14. Planning and management of agriculture and livestock farming: **TO FACILITATE THE AGRICULTURAL UTILISATION OF THE MOST FERTILE SOILS AND IMPROVE THE PRODUCTION, PROCESSING AND MARKETING PROCESSES** (Fig. 4.11).

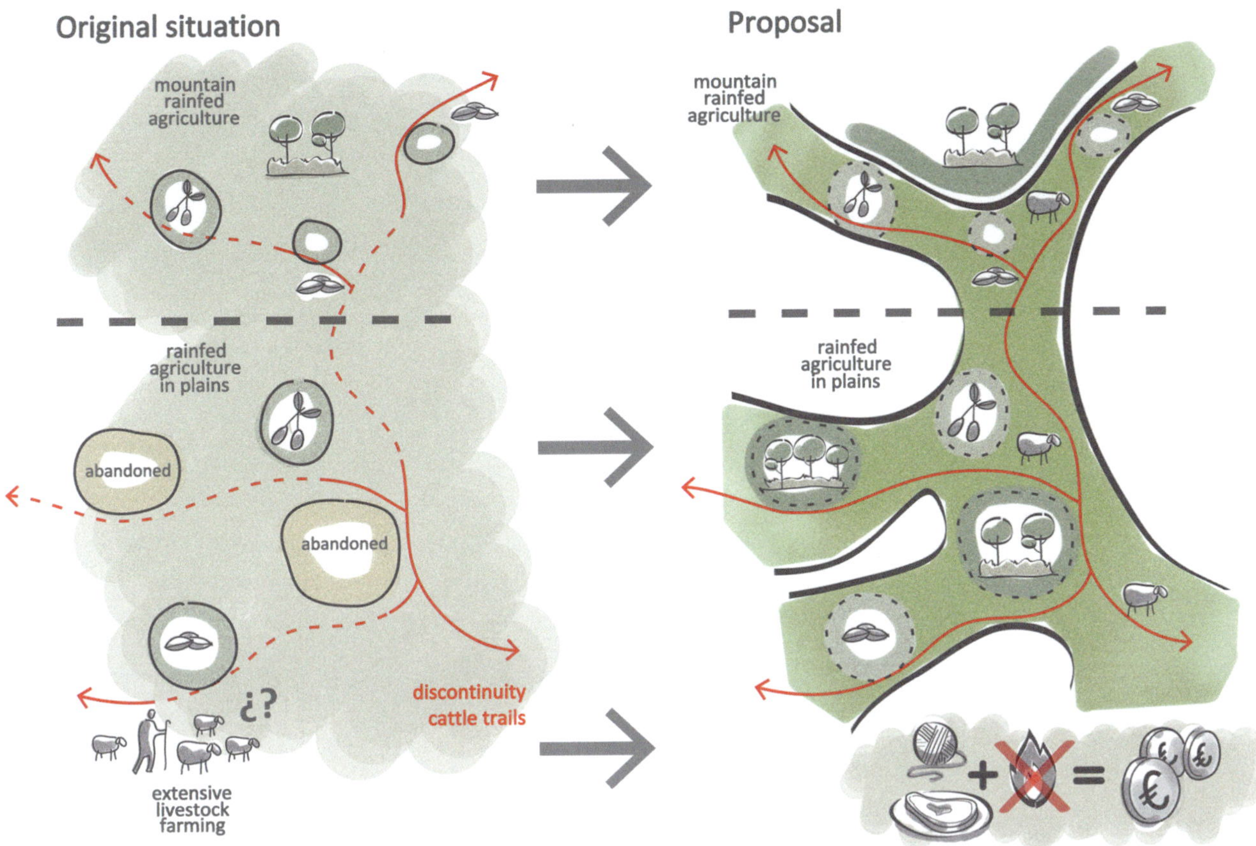

Fig. 4.7 Strategies 8, 9 and 10

Fig. 4.8 Strategy 11

- By ensuring that the most fertile soils remain in agricultural use and are not urbanised.
- By improving production, processing and marketing processes using a diversified range of agricultural crops.
- By establishing procedures to facilitate the cultivation of abandoned agricultural land through agreements that benefit both landowners and potential farmers.
- By drafting an agriculture and livestock farming plan.

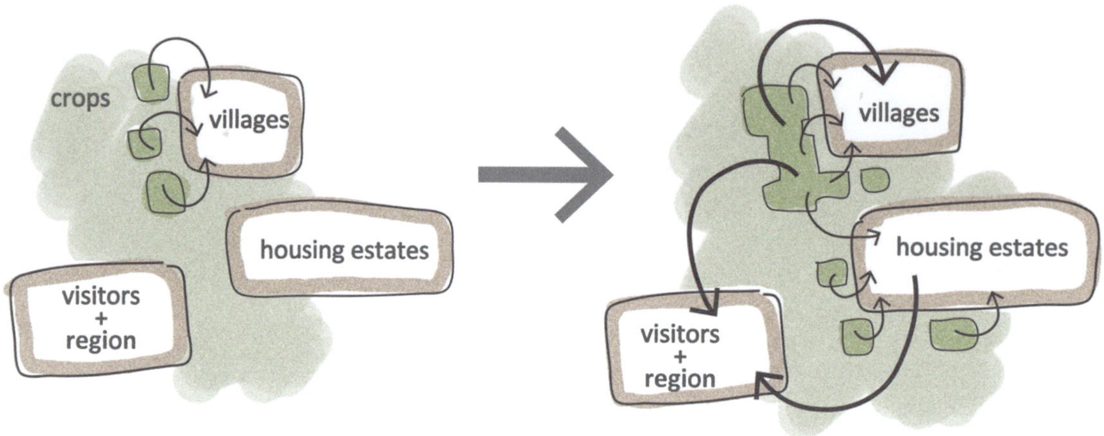

Fig. 4.9 Strategy 12

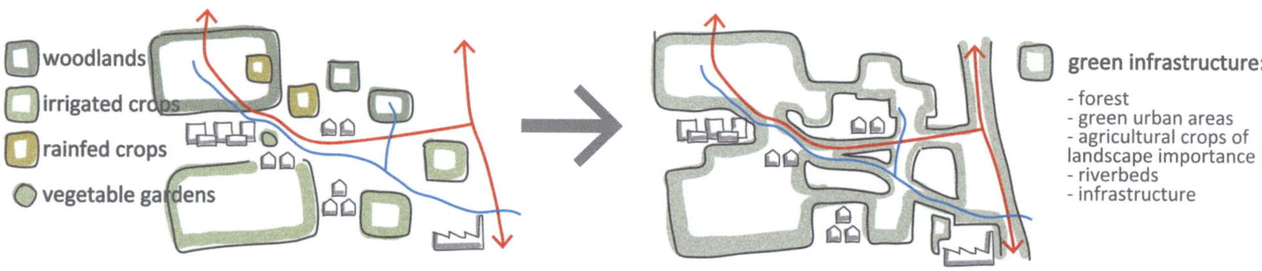

Fig. 4.10 Strategy 13

Fig. 4.11 Strategy 14

4.2.3 Urban Environment and Well-Being

15. Towns and villages: **TO PROMOTE A COMPACT URBAN MODEL OF HIGH-MEDIUM DENSITY, MULTIFUNCTIONAL AND WITH HIGH-QUALITY PUBLIC SPACES** (Fig. 4.12).

- By promoting towns and villages as service hubs for both residents and visitors.
- By promoting the utilisation of uninhabited dwellings, restoration of buildings and construction in empty urban lots.
- By planning compact high-medium density multifunctional and well-integrated new urban fabrics when those are needed.
- By promoting green urban infrastructures (i.e. parks, squares, boulevards, pedestrian streets) with easy access to the regional green infrastructure and highly related to

riverbeds, topographic features and transport infrastructure.
- By improving the quality of public space and facilitating tits use by locals and visitors (i.e. signposting, car parking).

16. Urban territorial model: **TO IDENTIFY THE MAIN POTENTIALS OF EACH TOWN IN ORDER TO SUPPLEMENT THEIR SERVICES AND FUNCTIONS, AVOID UNNECESSARY COMPETITION AND OPTIMISE THE USE OF THE AVAILABLE RESOURCES** (Fig. 4.13).

- RURAL CENTRE (particularly intense relationship with mountainous and forest areas, access via mountain roads): SERRA, MARINES VIEJO AND GÁTOVA.
  - Offering rural lodgings (main supplier).
  - Offering activities in the natural environment.
  - Offering tertiary services at a local scale.
  - Offering retail outlets for handicrafts.
  - Centralising the management of the natural environment.
- RURAL NODE (strong relationship with mountainous and forest areas, functions as supramunicipal services and facilities' hub): OLOCAU.
  - Offering rural lodgings (main supplier).
  - Offering activities in the natural environment.
  - Centralising the management of the natural environment.
  - Offering supramunicipal facilities and services (education, sport, health, culture).
  - Offering retail and trading venues.
- URBAN NODE (links to the agricultural plains of Camp del Turia, functions as a supramunicipal services and facilities' hub also serving the biggest housing estates, linked to industrial and agricultural activities): MARINES NUEVO.
  - Offering supramunicipal facilities and services (education, sport, health, culture).
  - Node for industrial activities.
  - Node for agricultural activities (production, storage and processing).
  - Offering retail and trading venues.
- MIXED NODE (links to both plains and mountainous areas, functions as a supramunicipal service hub also serving the biggest housing estates, linked to industrial and agricultural activities): NÁQUERA.
  - Offering rural lodgings and activities in the natural environment (secondary supplier).
  - Offering supramunicipal facilities and services (education, sport, health, culture).
  - Node for industrial activities.
  - Node for agricultural activities (production, storage and processing).
  - Offering retail and trading venues.
- REGIONAL NODES: LÍRIA ("metropolitan node", according to the Territorial Strategy of the Valencian Region) and BÉTERA (due to the services it offers and its metropolitan connections through different forms of public transport).

17. Housing estates: **TO DISCOURAGE THIS MONO-FUNCTIONAL LOW-MEDIUM DENSITY URBAN MODEL AND UPGRADE EXISTING CASES BY**

Fig. 4.12 Strategy 15

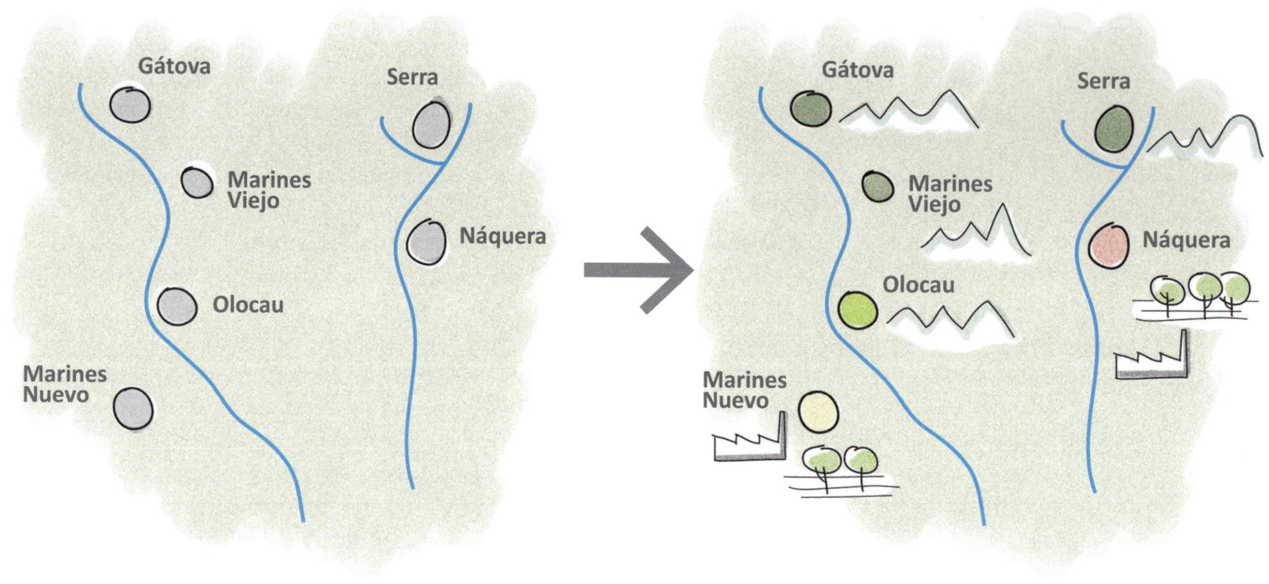

Fig. 4.13 Strategy 16

IMPROVING THEIR FUNCTIONALITY, QUALITY AND LEVELS OF SUSTAINABILITY (Fig. 4.14).

- By upgrading the existing housing estates in order to improve their levels of sustainability and their functional relationships with the nearby town. These objectives could be mainly achieved by densification of the estates, diminishing seasonal fluctuations of their population, promoting multifunctional tissues and increasing their functional relationship with the nearest towns and housing estates.
- By discouraging the expansion or creation of new housing estates.
- By improving the visual quality of their edges and the quality of their inner public spaces through the adjustment of their street sections and the creation of internal green infrastructures.
- By promoting the utilisation of empty dwellings and construction of vacant plots.

18. Scattered Settlements: **TO AVOID THE EXPANSION OF THIS URBAN TYPE AND DEFINE PROCEDURES TO FACILITATE THEIR INTEGRATION INTO THE LANDSCAPE, REGULATING, IF APPROPRIATE, ILLEGAL CONSTRUCTIONS** (Fig. 4.15).

- By upgrading scattered settlements according to the following criteria:
  - Isolated buildings: Improving the enclosure elements (i.e. walls, fences and hedges) to facilitate their integration into the landscape.

Fig. 4.14 Strategy 17

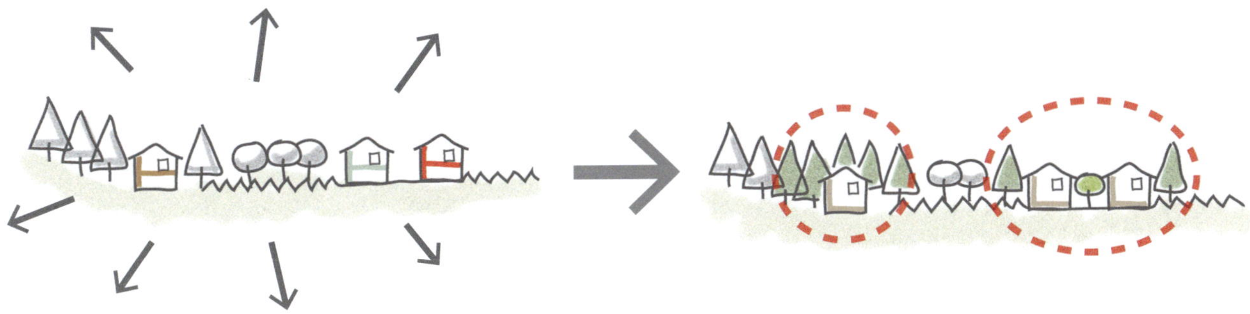

Fig. 4.15 Strategy 18

- Compact group of buildings: Establishing an urbanisation perimeter that should not exceed the current limits, allowing internal improvements based on the existing constructions and building new housing in available inner vacant plots.
- Scattered groups of buildings: Improving the enclosure elements (walls, fences and hedges) to facilitate their integration into the landscape and ensuring that interstitial spaces maintain an agricultural or forest land use.

- By promoting continuity of the agricultural activity around them or regression into forest.

19. Areas of economic activity: **LOCATE THE MAIN AREAS OF ECONOMIC ACTIVITY (INDUSTRY, TERTIARY, ETC.) ACCORDING TO A SUPRAMUNICIPAL STRATEGY** (Fig. 4.16).

- In rural centres and rural nodes (Serra, Olocau, Gátova and Marines Viejo) by offering services to visitors and by managing the forest/agricultural environment and the processing of their products.
- In urban nodes or mixed nodes (Marines Nuevo and Náquera) by expanding the industrial and the tertiary sectors on an inter-municipal scale, encouraging the hiring of local workers and facilitating commuting through soft transport systems and public transport.
- By re-using vacant industrial lots before generating new industrial land.

20. Available housing stock and vacant lots: **TO PROMOTE THE USE OF THE EXISTING HOUSING STOCK, THE CONSTRUCTION OF VACANT LOTS AND THE DIMINUTION OF SEASONALITY** (Fig. 4.17).

- By generating a network or structure that will facilitate the rental or acquisition of empty dwellings (banks of dwellings, websites to facilitate the rental or purchase of dwellings, etc.).
- By promoting the continued use of the existing housing stock instead of its seasonal use (favouring their use as primary residences and incorporating those still used as secondary residences to networks of dwelling exchange).
- By encouraging, selectively, the construction of strategic vacant plots (those inside the towns, in interstices between separated urban fabrics, in highly visible urban fronts, etc.).

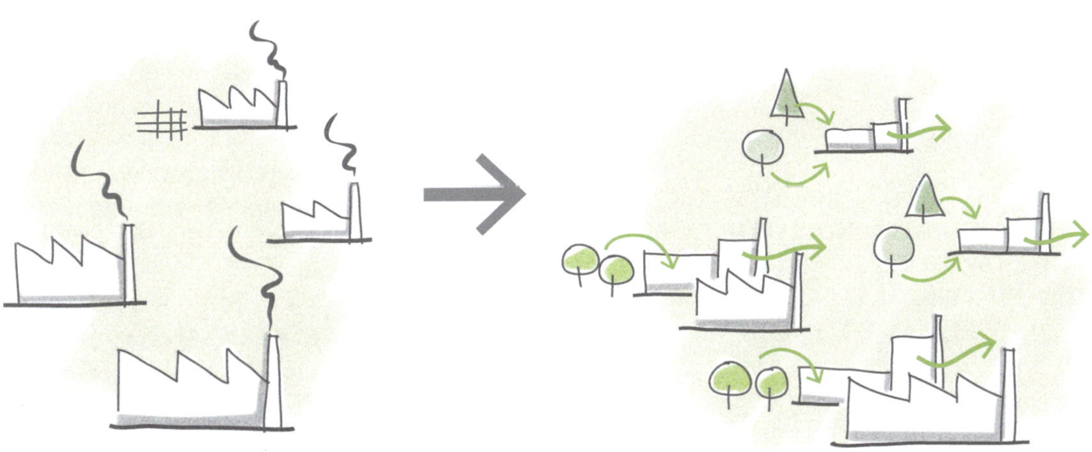

Fig. 4.16 Strategy 19

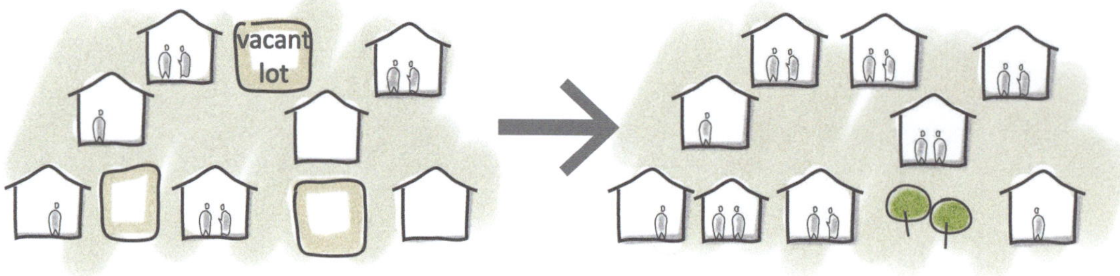

Fig. 4.17 Strategy 20

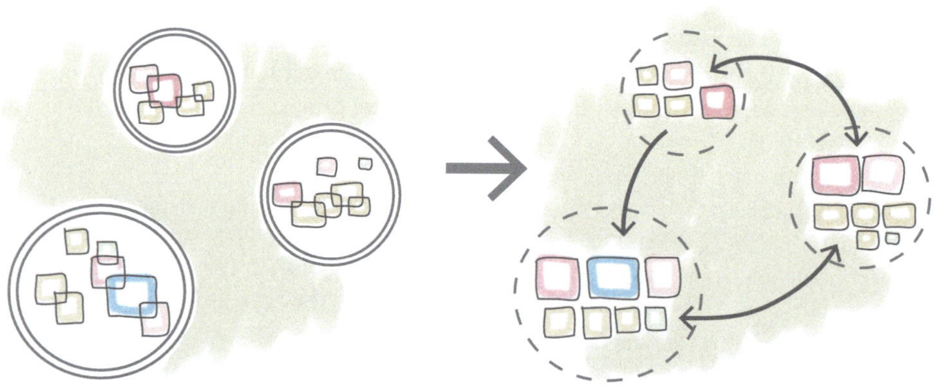

Fig. 4.18 Strategy 21

21. Services and facilities: **TO GENERATE COMMON SERVICES AND FACILITIES** (Fig. 4.18).

- By promoting inter-municipal services and facilities as tools to generate economies of scale, increasing the amount and type of services and the number of users.
- By coordinating the management of inter-municipal services and facilities (cultural centres, libraries, sports complexes, etc.) with effective and frequent public transport connections.
- By avoiding unnecessary duplications with metropolitan urban nodes (Liria, Moncada, Sagunto).

4.2.4 Infrastructure and Mobility

22. Public transport: **TO PROMOTE, THROUGH ATTRACTIVE FARES AND SHORT TRAVEL TIMES, THE USE OF PUBLIC TRANSPORT WITHIN THE AREA (BETWEEN TOWNS AND HOUSING ESTATES), AS WELL AS WITH THE METROPOLITAN NODES** (Fig. 4.19).

- By promoting intermodal means of transport outside the area included in the Sierra Calderona Strategic Plan through a competitive offer of both inexpensive fares and short travel times (travel cards for the local population, improvements in the train and bus services, etc.).
- By promoting public transportation and soft mobility within the area covered by the Strategic Plan and around it.

23. Soft mobility: **TO PROMOTE SOFT MOBILITY (BY FOOT, BICYCLE) WHEN VIABLE** (Fig. 4.20).

- By strengthening the functional relationships between towns, and between towns and housing estates, and by improving soft mobility infrastructures (i.e. cycle lanes, adequate sidewalks for pedestrians).
- By promoting pedestrian mobility and the use of bicycles within towns and between them and the nearest housing estates and by encouraging cycling between housing estates and public transport stops (bus or, with greater difficulty, train).

24. Infrastructures: **TO IMPROVE THE QUALITY OF THE VISUAL ENVIRONMENT ALONG THE ROADS, ESPECIALLY IN ROADS THAT ARE NOT LOCATED IN FOREST AREAS, AS WELL AS THE CONTINUITY, ACCESS AND SIGNPOSTING OF FOREST TRACKS, RURAL ROADS AND TRAILS** (Fig. 4.21).

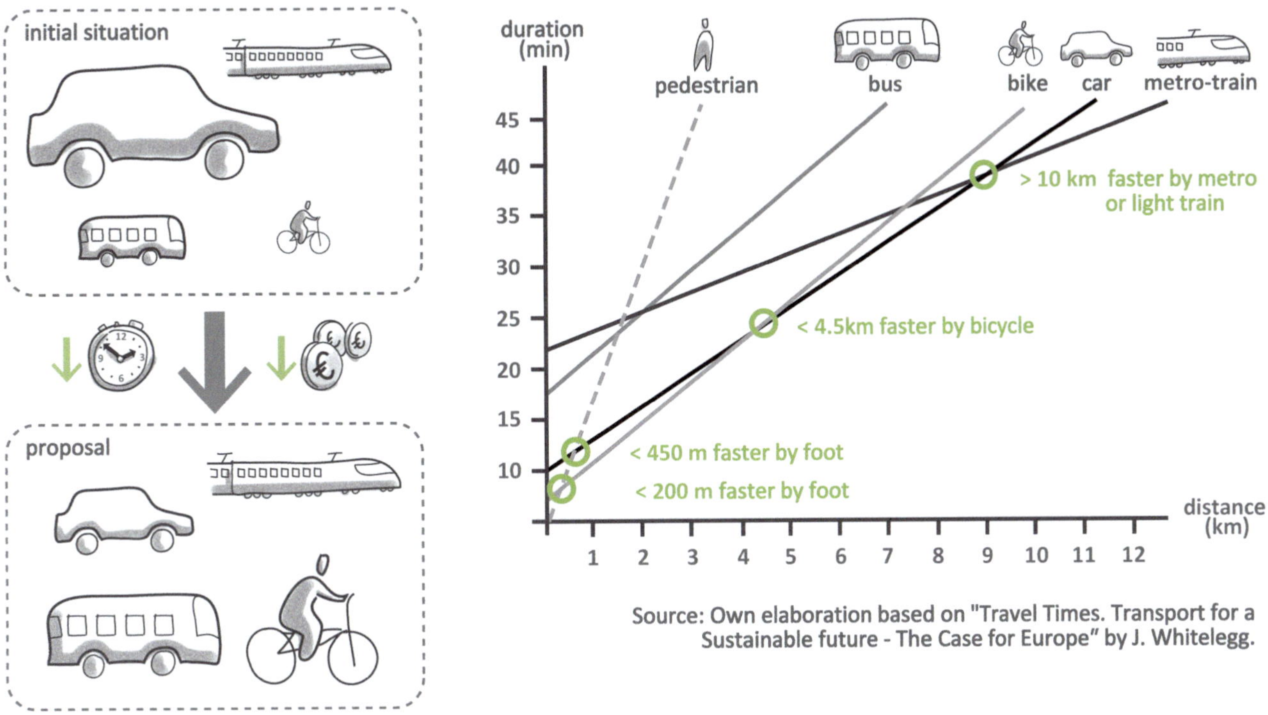

Fig. 4.19 Strategy 22

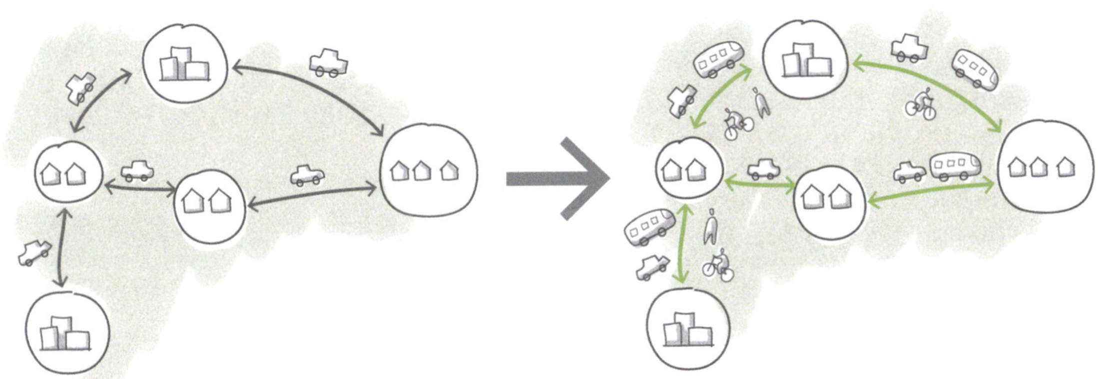

Fig. 4.20 Strategy 23

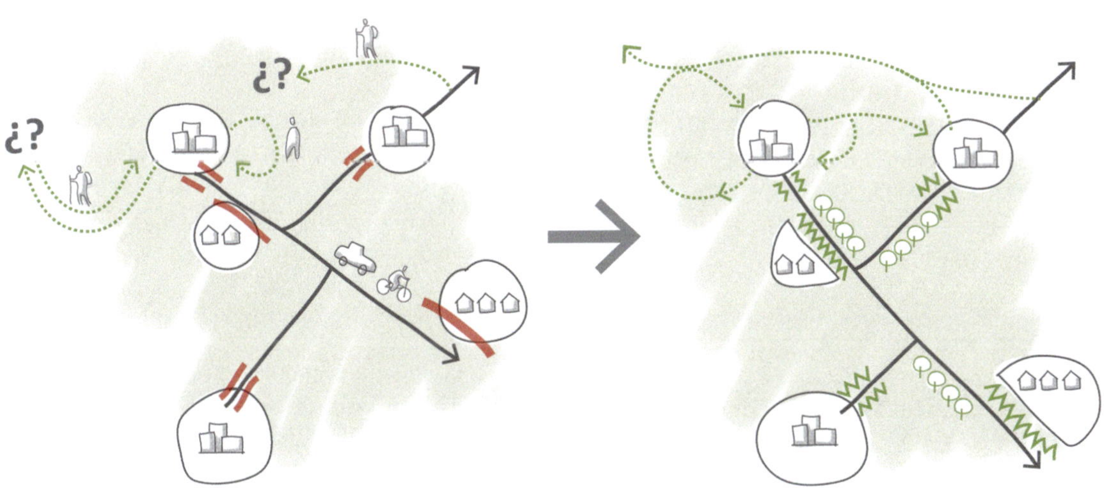

Fig. 4.21 Strategy 24

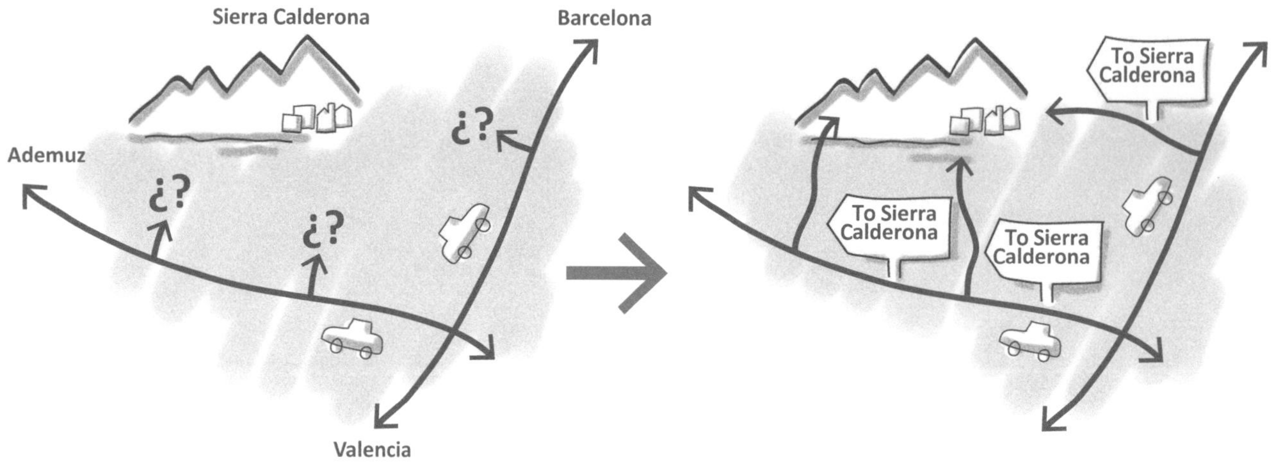

Fig. 4.22 Strategy 25

25. Location in the territory: **TO INDICATE THE EXITS LEADING TO THE SIERRA CALDERONA NATURAL PARK FROM THE MAIN NATIONAL AND REGIONAL TRANSPORT INFRASTRUCTURES (Roads and highways AP7, V21, CV35)** (Fig. 4.22).

4.2.5 Cultural Heritage

26. Analysis and diagnosis. **TO IDENTIFY THOSE ELEMENTS WHOSE RESTORATION HAS PRIORITY (DEPENDING ON THEIR CULTURAL VALUE, DEGREE OF DETERIORATION AND VIABILITY FOR PUBLIC OR PRIVATE USE)** (Fig. 4.23).

27. Heritage network: **TO INTEGRATE HERITAGE SITES IN A SET OF SPATIAL NETWORKS OR SYSTEMS (Ravines of Gátova-Olocau, L'Assut-Náquera and Portaceli)** (Fig. 4.23).

- By generating and unifying signposting and by developing informative materials for both locals and visitors.

28. Restoration and reactivation of heritage sites: **TO GENERATE FUNDRAISING INITIATIVES TO FINANCE THE RESTORATION OF HERITAGE SITES AND TO FORESEE THE ALLOCATION OF NEW USES (PUBLIC, MIXED OR PRIVATE) THAT WILL ENABLE THEIR FEASIBLE USE AND MAINTENANCE** (Fig. 4.23).

4.2.6 Tourism and Public Use

29. Marketing: **TO GENERATE A REGIONAL BRAND AND CARRY OUT JOINT MARKETING CAMPAIGNS AT REGIONAL, NATIONAL AND INTERNATIONAL LEVELS** (Fig. 4.24).

- By creating a common website to inform about accommodation, dining and other services offered.
- By creating common promotional materials (for culture, nature, services).

30. Natural and cultural heritage: **TO IMPROVE THE NETWORK OF ITINERARIES AND TO ADAPT THEM TO DIFFERENT KINDS OF USERS** (Fig. 4.25).

- By generating well connected and signposted itineraries (with continuity, with access from urban areas, well indicated from roads, etc.) and service points for different types of visitors interested in the natural environment and/or cultural heritage:
  - Family tourism
  - Cyclists
  - Hikers, etc.
- By organising guided tours for specific types of experiences (visits to caves, bird-watching, ecology and forestry information, self-guided heritage tours, food tastings, etc.).
- By integrating cultural heritage into inter-municipal information networks, in information panels and guided visits.
- By improving and unifying tourist information (maps, brochures, points of information, information panels, signposting along roads, marking of trails, etc.).

31. Tourism and economic development: **TO PROMOTE A KIND OF TOURISM THAT SUPPORTS LOCAL ECONOMIES AND THE MANAGEMENT**

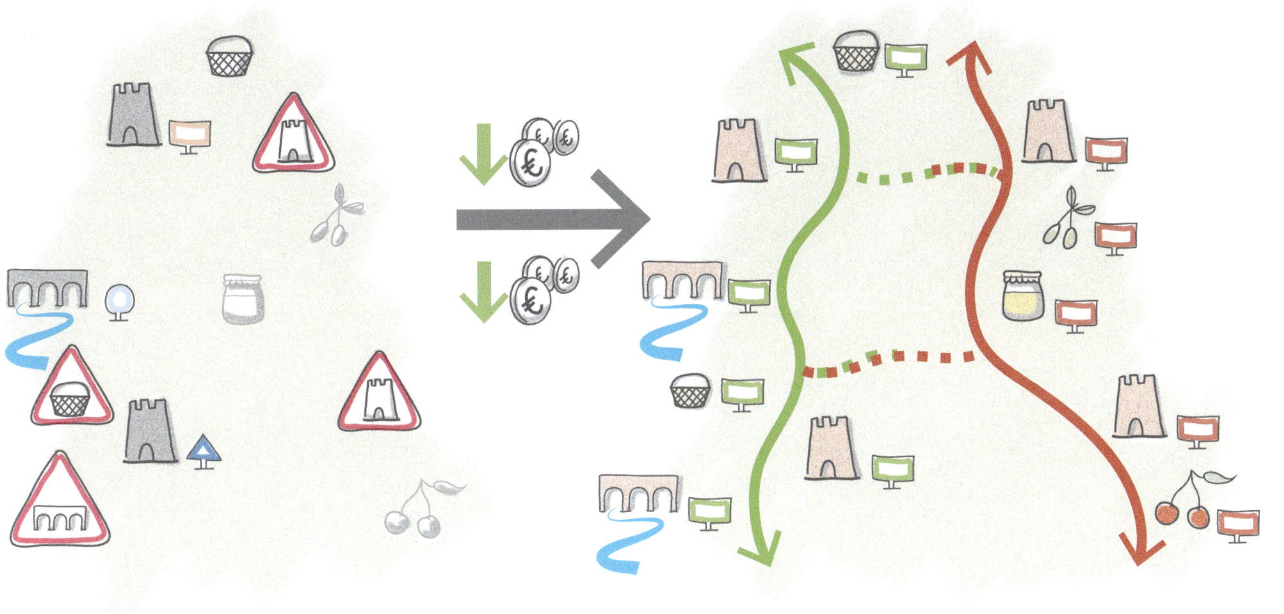

Fig. 4.23 Strategies 26, 27 and 28

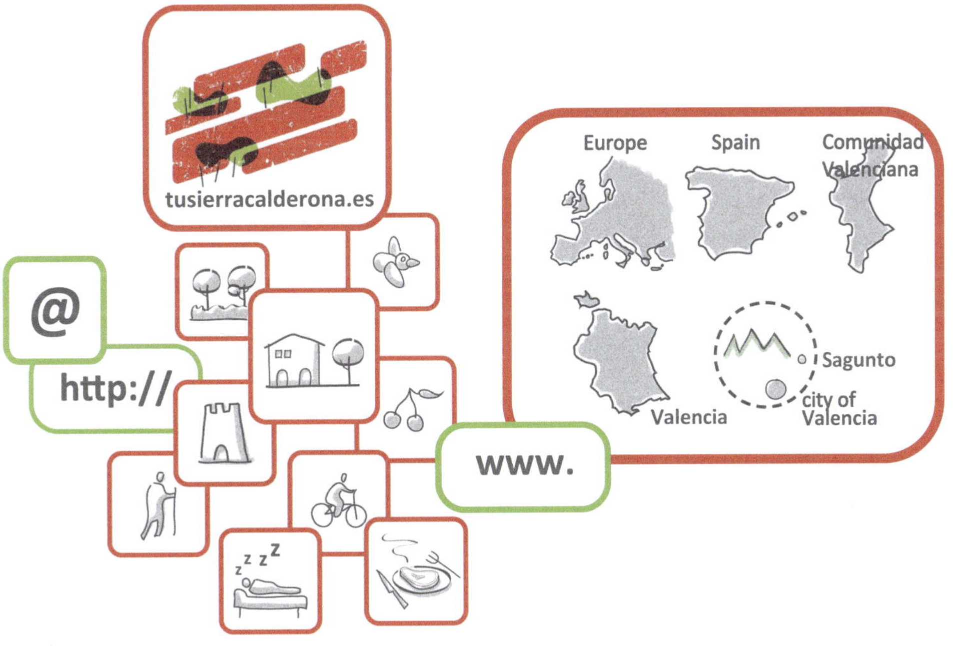

Fig. 4.24 Strategy 29

OF THE NATURAL ENVIRONMENT AND HERITAGE (Fig. 4.26).

- By favouring positive interactions of locals, visitors and the natural environment.
- By facilitating and promoting the guided knowledge of the natural and cultural environment (guided group visits to areas with restricted access or to singular sites, tastings of local products, etc.).
- By offering local gastronomic products and crafts and promoting their sale.
- By improving the provision of accommodation, dining and specialised services and by making sure that these actions have a positive impact on local employment and the professional training of local people.
- By providing the skills and knowledge necessary to create and manage a wider touristic provision (educational courses for the creation and management of tourist

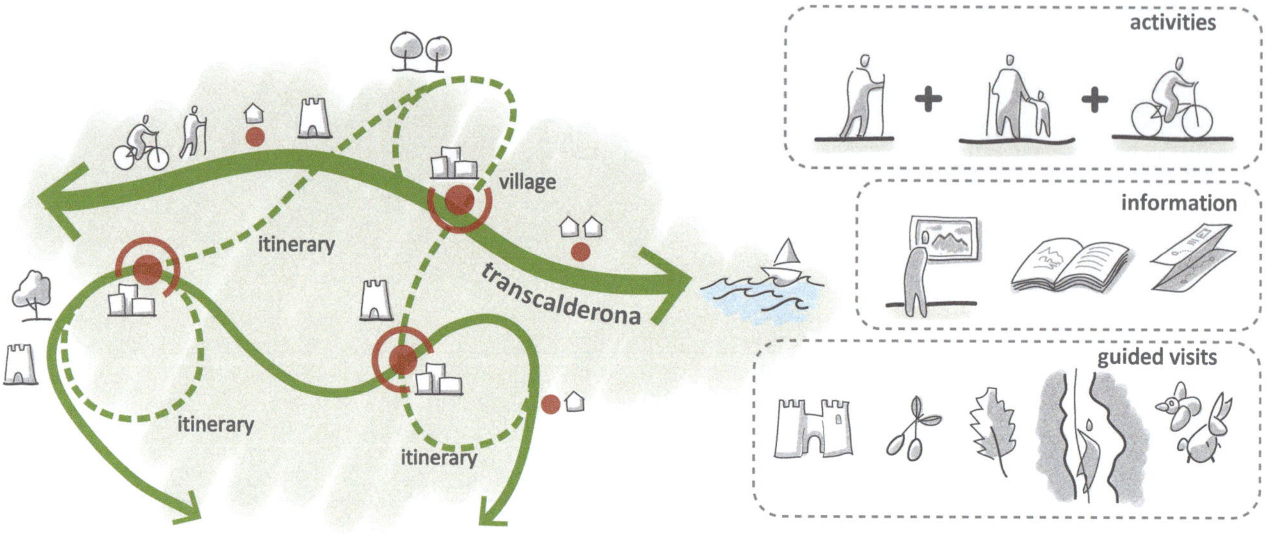

Fig. 4.25 Strategy 30

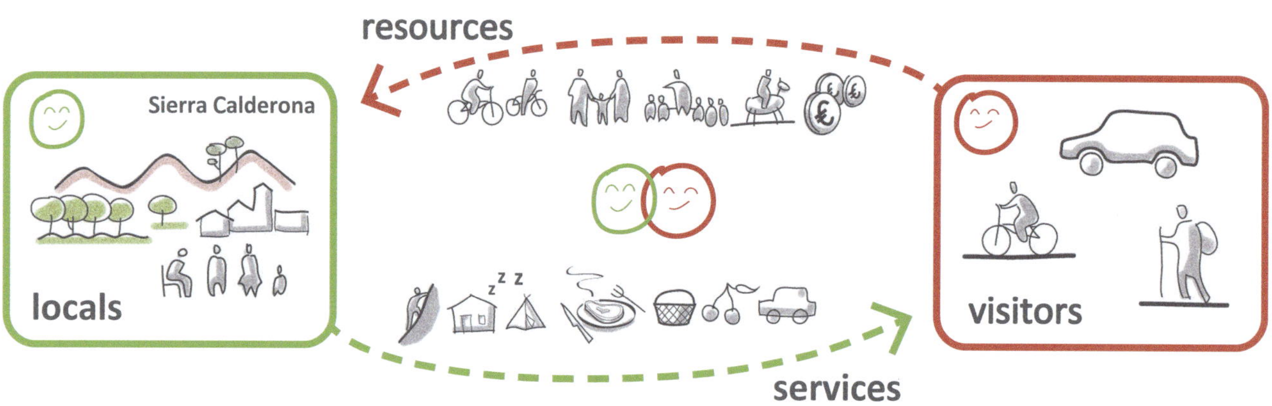

Fig. 4.26 Strategy 31

services, tourist guides, production and commercialisation of agricultural products, crafts, organic agriculture, sustainable construction, stonework, hostelry, catering, etc.).

- By promoting joint and complementary services and by training and certifying the professionals involved in the tourism sector.

4.2.7 Landscape

32. Agricultural landscape: **TO SUPPORT FEASIBLE AND COST-EFFECTIVE AGRICULTURE CONTRIBUTING INDIRECTLY TO THE QUALITY OF THE LANDSCAPE.**

- By defining regulations or recommendations for the design, construction and management of trails, fences and rural constructions.

33. Natural landscape: **TO PROMOTE AN ECOLOGICALLY DEVELOPED, SAFE AND PRODUCTIVE NATURAL ENVIRONMENT** (Fig. 4.27).

- By promoting the successional development of forests towards mature stages and by coordinating that goal with economic activities that could be beneficial for their protection.
- By protecting and promoting the agricultural system in the mountainous areas as a buffer system between rural tracks and forests.
- By promoting the presence of all the layers that characterise Mediterranean mature forests (herbs, shrubs and tree species), managing them when needed for fire protection.
- By matching up firebreaks, forest trails, rural tracks and roads, in order to minimise their aggregated impact on the landscape.

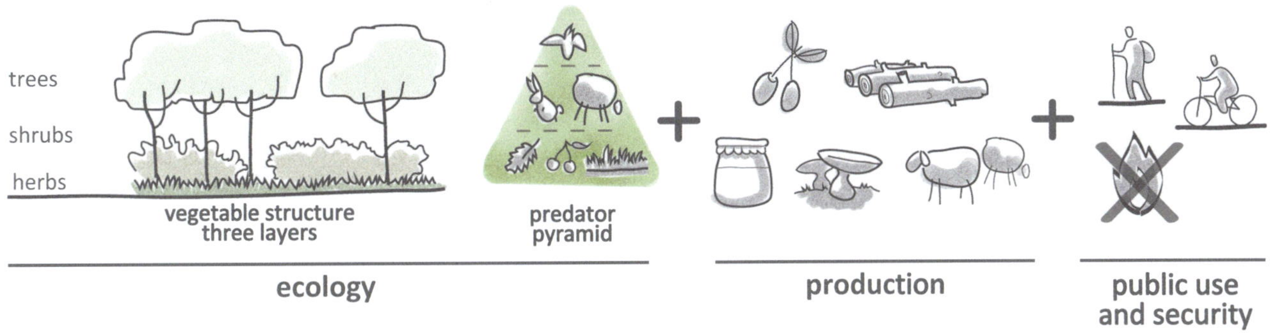

Fig. 4.27 Strategy 33

34. Urban landscape: **TO IMPROVE THE QUALITY OF URBAN LANDSCAPES BOTH INTERNALLY AND EXTERNALLY (in towns, villages, housing estates, scattered settlements and industrial areas)** (Fig. 4.28).

- By regulating the position, height, volume and character of new constructions in towns and villages, in order to preserve their character and improve the appearance of their margins and skylines.
- By defining a good system of open spaces and streets, following existing physiographic elements (ravines, topographic elevations). By improving the areas for pedestrians (sidewalks, pedestrian streets). By minimising the visual intrusion of vehicles and by promoting the presence of adequate urban vegetation and urban furniture, as well as the design of both façades and enclosures (walls and fences).
- By regulating the finishing of façades and enclosures.
- By providing recommendations for street sections in new housing areas and urban developments.
- By defining transitional spaces between towns, villages and transport infrastructures, ensuring their adequate design and maintenance.

35. Landscape from infrastructure: **TO IMPROVE THE PERCEPTION OF THE LANDSCAPE WHEN VIEWED FROM TRANSPORT INFRASTRUCTURES** (Fig. 4.28).

- By improving the adjacent visual environment (hard shoulders and 10 m margins on both sides).
- By generating a high-quality and comfortable environment with tree shade along bicycle lanes and rural roads.

36. Green infrastructure: **TO DEFINE A SYSTEM OF OPEN SPACES THAT WILL PHYSICALLY AND**

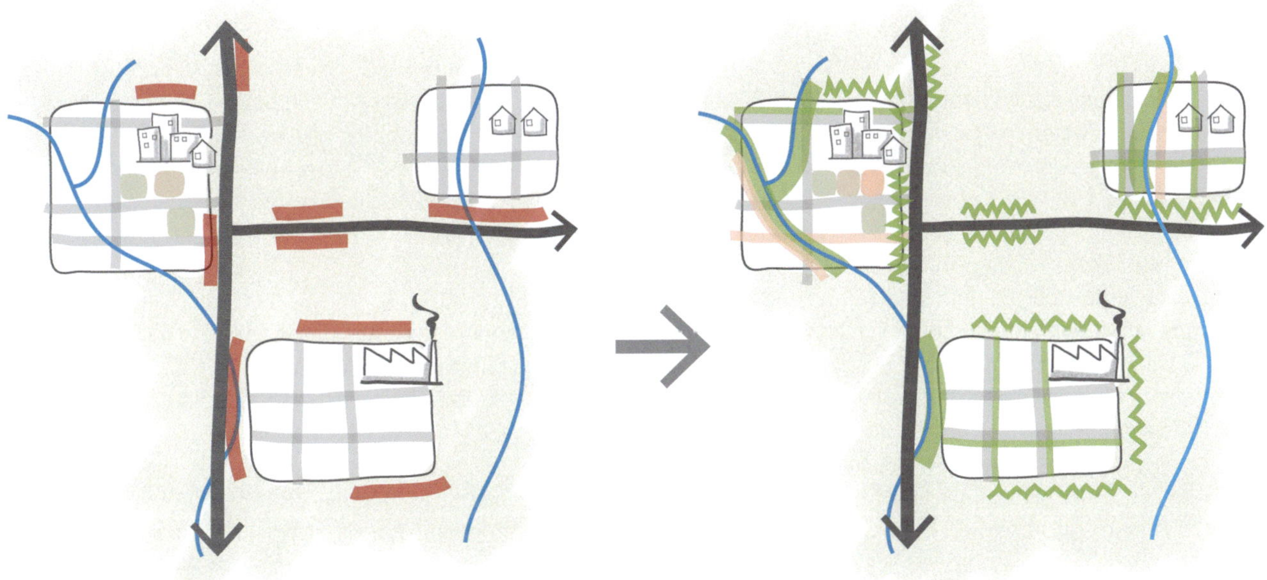

Fig. 4.28 Strategies 34 and 35

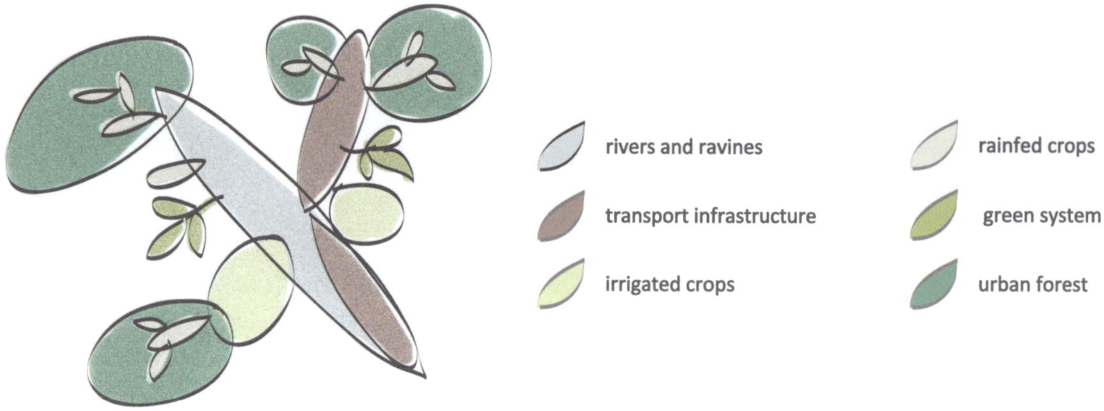

Fig. 4.29 Strategy 36

ENVIRONMENTALLY CONNECT NATURAL AND AGRICULTURAL SPACES, RIVERBEDS AND ADJACENT FLOOD AREAS, AS WELL AS MAJOR URBAN AND SUBURBAN PUBLIC SPACES (Fig. 4.29).

- By reinforcing the functions of ravines and topographic elevations as both open spaces and structural elements, in both the natural and the urban environment.
- By including in the green infrastructure a network of agricultural and forested spaces, avoiding the fusion of housing estates and branching towards the south of the Calderona Mountain Range (Sierra Calderona).

4.2.8 Sustainability

37. Water resources: **TO OPTIMISE WATER CONSUMPTION AND WASTEWATER MANAGEMENT** (Fig. 4.30).

- By encouraging agricultural practices that will minimise and optimise water consumption.
- By promoting domestic practices that will minimise and optimise drinking water consumption.
- By ensuring proper treatment of wastewater, especially in those settlements lacking sewage networks.
- By evaluating cesspits and by studying their location, current status and possible improvements.
- By installing meters in the water distribution and sewage networks, in the entrances and exits from towns, villages and housing estates.
- By promoting the "Water" recommendations proposed in "Guías de Sostenibilidad en la Edificación Residencial" by "Instituto Valenciano de la Edificación" or "IVE".

38. Waste: **TO DECREASE WASTE PRODUCTION AND TO PROMOTE RECYCLING AND REUSE** (Fig. 4.30).

- By enhancing the internal use of waste produced in garden maintenance and in forest management.
- By promoting the consumption of eco-friendly products.
- By promoting flea markets and transfer/sharing of goods or equipment not in use.
- By promoting the "Waste" recommendations proposed in "Guías de Sostenibilidad en la Edificación Residencial" by "Instituto Valenciano de la Edificación" or "IVE".

39. Energy: **TO REDUCE ENERGY CONSUMPTION, INCREASE ENERGY EFFICIENCY AND PROMOTE THE USE OF RENEWABLE ENERGY** (Fig. 4.30).

- By promoting the consumption of local products, in order to reduce transport footprints.
- By promoting production and consumption of clean energy, making effective use of public and private spaces and resources.
- By establishing and promoting sustainability requirements in the construction and urban planning sectors, as well as in the mobility systems.
- By promoting the "Energy" recommendations proposed in "Guías de Sostenibilidad en la Edificación Residencial" by "Instituto Valenciano de la Edificación" or "IVE".

40. Efficient and multifunctional territory: **TO PROMOTE A MULTIFUNCTIONAL AND EFFICIENT TERRITORY THAT FAVOURS INTERNAL FLOWS OF ENERGY AND MATERIALS** (Figs. 4.30, 4.31).

- By promoting internal metabolisms that increase domestic production of energy and reuse of waste.
- By establishing and promoting sustainability requirements in the construction and urban planning sectors, as well as in transport systems.

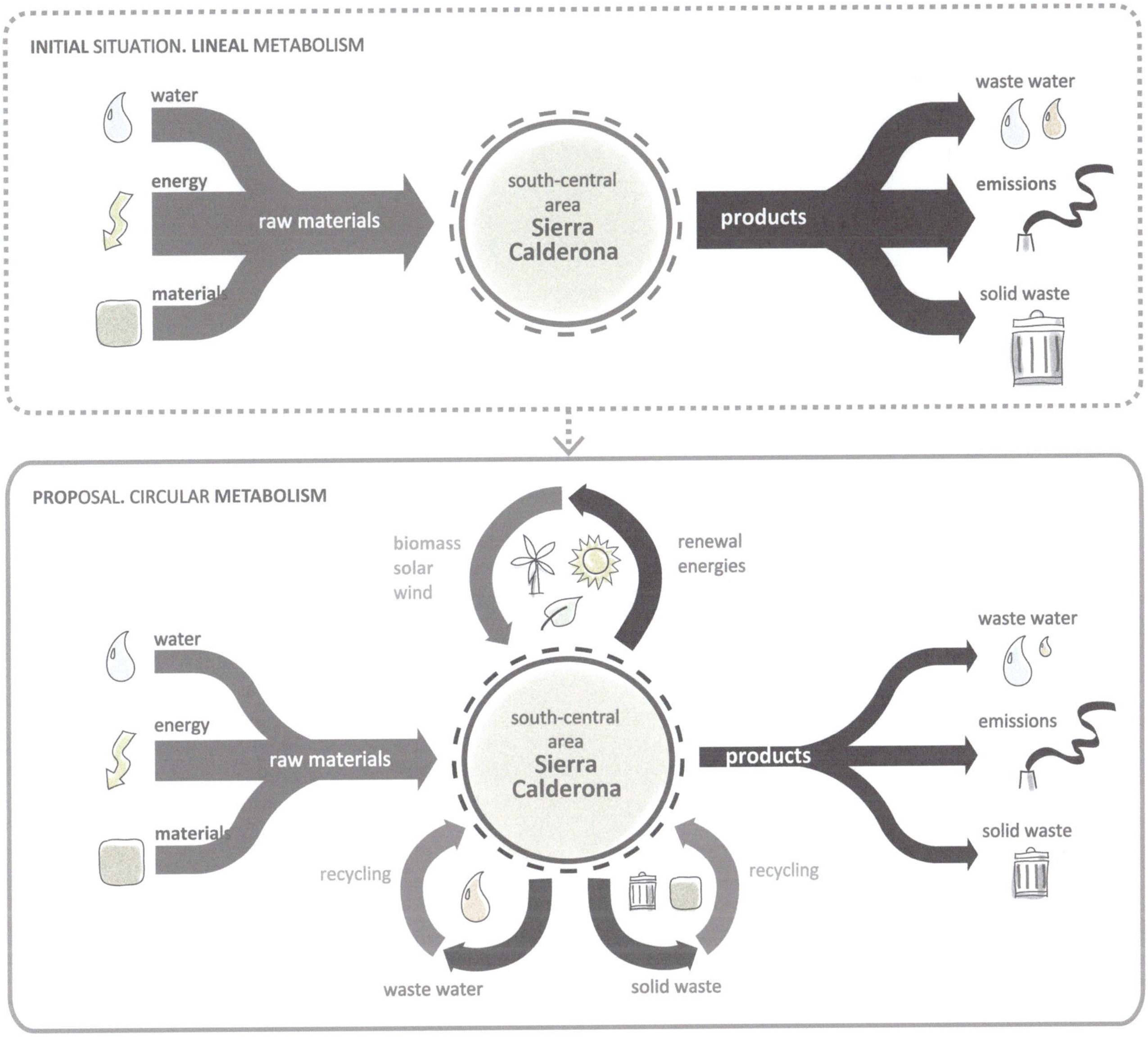

Source: own elaboration based on "Cities for a small planet" by Richard Rogers and Philip Gumuchdjian.

Fig. 4.30 Strategies 37, 38, 39 and 40

- By promoting circular economies and the cyclical local use of materials and energy.
- By promoting urban multifunctionality at different planning scales: places to live, work and rest.

41. Ecological footprint: **TO MONITOR LEVELS OF SUSTAINABILITY WITH THE ACTIVE PARTICIPATION OF THE POPULATION** (Fig. 4.32).

- By developing a regional observatory of sustainability, enabling and promoting the participation of citizens in the monitoring of the regional levels of sustainability (i.e. through a website calculating individual ecological footprints, by collecting ideas and opinions).
- By promoting the south-central area of Sierra Calderona as a sustainability model.
- By publicly recognising and rewarding individuals or groups developing exemplary sustainable initiatives.

4.2.9 Socio-Demography

42. Dynamic and diverse population: **TO PROMOTE A POPULATION DIVERSE IN AGE, DYNAMIC IN ATTITUDES, PROFESSIONALLY TRAINED AND PROVIDED WITH QUALITY SERVICES** (Fig. 4.33).

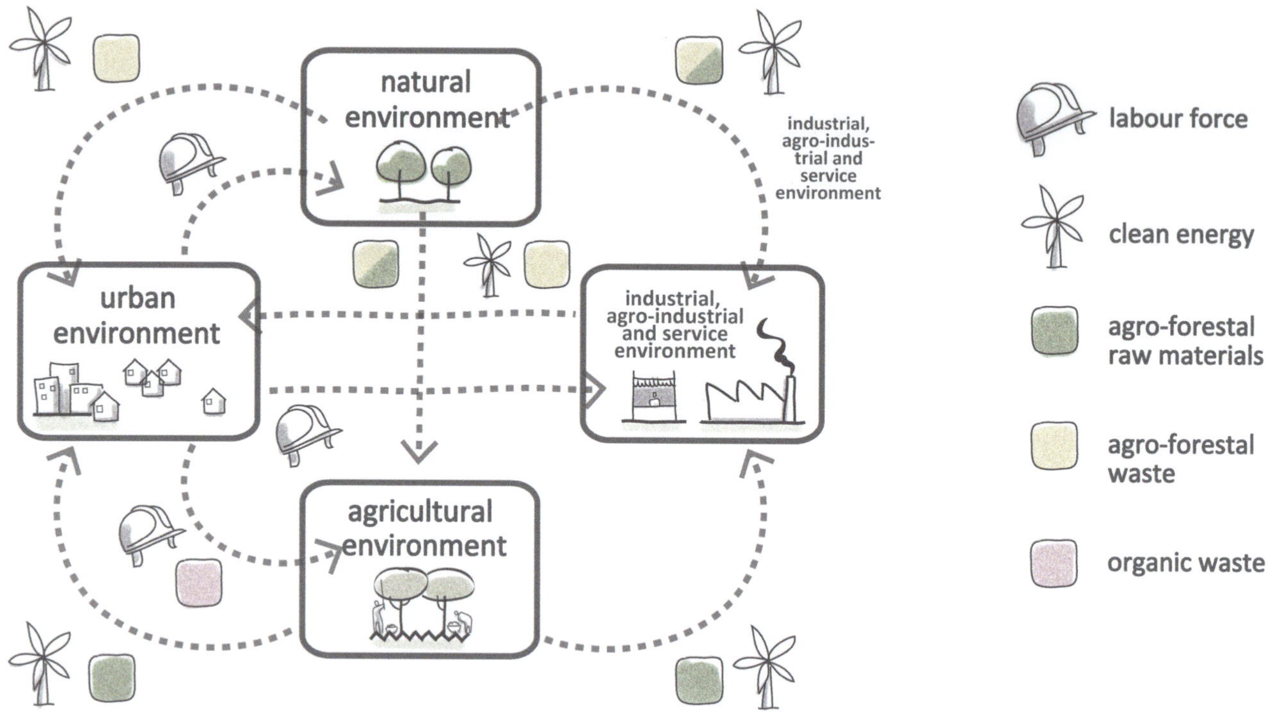

Fig. 4.31 Strategy 40

Fig. 4.32 Strategy 41

- By fostering the arrival of new residents through easy access to housing and support to new businesses, especially those connected with either services or agro-forestry.
- By offering high-quality and accessible public services (health, education, sport, culture).
- By improving quality of life and services for seniors, both locals and newcomers attracted by the benign conditions of the Calderona Mountain Range (Sierra Calderona).
- By providing high-quality public services for families with children (education, childcare, cultural activities, health, etc.), while the employment of women.
- By providing practical training in order to improve the population's professional qualifications (construction, catering, activities in the natural environment, tour guides, foreign languages, high-quality agrarian production, crafts, care for the elderly, entrepreneurship, creation and management of business, etc.).

4.2.10 Economic Activity

43. Business activity: **TO PROMOTE BUSINESS, ESPECIALLY THOSE RELATED TO THE AGRO-FORESTRY SECTOR, SERVICES AND TOURISM** (Fig. 4.34).

- By promoting local initiatives with local capital and character.
- By supporting the creation of new businesses and companies through existing or new municipal tools (advisory services, easy licencing, access and use or public facilities, etc.).
- By facilitating the dissemination and promotion of business through local websites and media.
- By training local people in entrepreneurship, fundraising, marketing, creation and management of businesses, etc.

Fig. 4.33 Strategy 42

- By making feasibility studies for the use of untapped resources (mineral water, outdoor sports, local food products, new types of culinary services and accommodation, etc.).

44. Primary sector: **TO PROMOTE SPECIALISED AGRICULTURAL, LIVESTOCK FARMING AND THE FORESTRY SECTOR** (Fig. 4.35).

- By promoting and training for the diversification and optimisation of the agricultural, livestock farming and forest processes, as well as for the revaluation of raw products in order to generate added values.
- By promoting local and regional markets for agricultural and livestock farming products.
- By promoting and re-establishing food fairs.
- By generating quality brands and labels associated with the agricultural, livestock farming and forestry local products (both raw and processed). Consider at this point the brand: "Calderona Natural Park".
- By creating a market in the area (fixed, itinerant, permanent, temporary or seasonal) accessible to locals and visitors where the local products should be prioritised.
- By promoting cooperatives for production, processing, distribution and marketing, as well as their integration into major networks related to organic and suburban agriculture.
- By facilitating the sale of local fruits, vegetables, firewood, etc., in shops and public hubs.
- By generating a land bank in order to mobilise unused agricultural land.
- By promoting the consumption of products derived from the management and maintenance of mountainous areas (biomass, firewood, etc.).
- By offering general educational and cultural services related to sustainable agricultural and forestry (i.e. farm-schools, agricultural parks, training centres).
- By promoting processed and artisanal agro-alimentary products.

45. Tourist and service sector: **TO DIVERSIFY AND EXPAND THE PROVISION OF SERVICES AND ACTIVITIES** (Fig. 4.36).

- By promoting the development of services with local capital and staff.
- By diversifying the supply and quality of services with special attention to activities in the natural environment and specialised tourism (i.e. health and well-being, seniors, families, gourmets, lodging).
- By training local people in the creation, management and promotion of the above-mentioned businesses.
- By advertising the provided services through websites, regional radio and television, as well as in national and international forums and media.
- By promoting the integration of local services in international networks for independent tourism (i.e. Tripadvisor or similar).

Fig. 4.34 Strategy 43

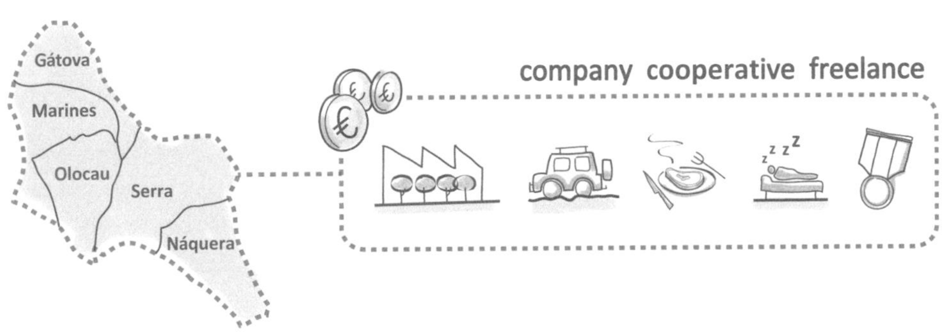

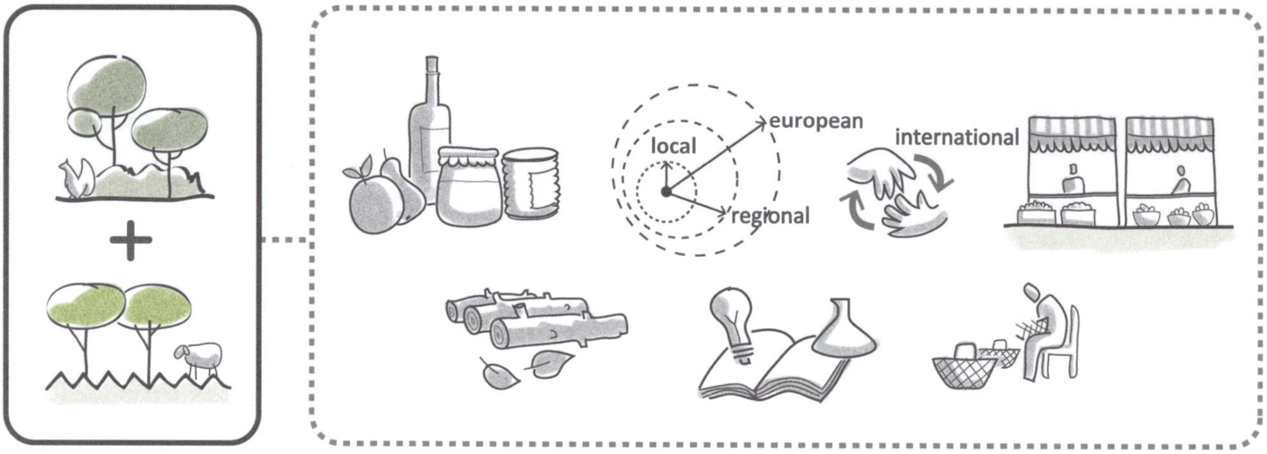

Fig. 4.35 Strategy 44

- By generating quality brands associated with local services.
- By advertising local services in shops and points of public use (town halls, sports complexes, restored heritage sites, etc.).
- By attracting high-quality initiatives that might become activators and references in the service and cultural sector: restaurants in travel guides and websites, a foundation displaying their art collection, a spa or hydrothermal centre (the water of Gátova and Serra have a recognised prestige), a singular hotel in a heritage building, an equestrian centre, a high-performance sports centre, businesses offering a wide selection of outdoor activities, guided tours in the natural environment (with special activities or visits) and to the cultural heritage (tastings of local products, interpretative activities in monuments, etc.), a quality campsite, a network of lodges in historic centres, nursing homes, a specialised hospital or rehabilitation centre, etc.

46. Industrial and construction sector: **TO PROMOTE LOW-IMPACT INDUSTRIAL ACTIVITY AND A QUALITY CONSTRUCTION SECTOR** (Fig. 4.37).

- By promoting the use and occupation of existing industrial areas near transport infrastructures (mainly in the industrial estate "Los Vientos"), the employment of local population and access by public transport.
- By promoting light and low-impact industry linked to the agro-forestry sector or to new technologies, mainly in the small industrial estates located in inner areas.
- By promoting high-quality and sustainable architecture, both in new and restored buildings, and by ensuring the aesthetic, and architectural integration of

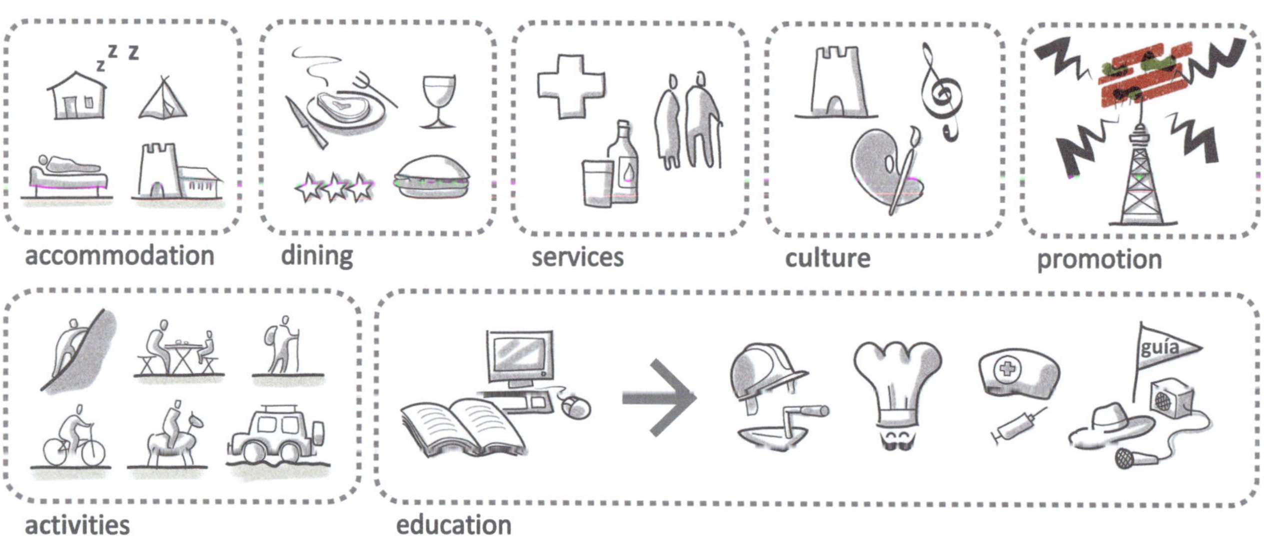

Fig. 4.36 Strategy 45

Fig. 4.37 Strategy 46

new buildings (without necessarily following traditional models).

47. Retail: **TO PROMOTE A DIVERSIFIED RETAIL PROVISION FOR BOTH RESIDENTS AND VISITORS** (Fig. 4.38).

- By favouring small shops as a high-quality alternative to big supermarkets and shopping malls, both for residents, in towns and in housing estates, and for visitors.
- By incorporating some retail activity in the housing estates.
- By studying the possible creation of medium-sized malls or clusters of local shops in strategic sites and with flexible timetables, in order to compete with larger and distant shopping malls, that have been used increasingly by local residents and visitors to the area.
- By promoting the production of crafts and processed agricultural, livestock farming or forestry products to be purchased by visitors.

48. Governmental agencies and public services: **TO JOIN, AND OPTIMISE GOVERNMENTAL AGENCIES AND PUBLIC SERVICES IN ORDER TO MAINTAIN OR UPGRADE THEIR QUALITY** (Fig. 4.39).

- By increasing municipal incomes:
  - Updating the tax payment of not-regularised dwellings (this measure would be more effective in those municipalities with a significant number of illegal houses).
  - Receiving pending payments from the regional government.
  - Developing a fundraising Plan with local agents and public administrations. In order to achieve this goal, the applications for funded programmes and projects must be well defined and justified.
  - Strengthening local economies in order to increase the collection of direct and indirect taxes.
  - Claiming compensation for the environmental services and positive externalities generated by the Sierra Calderona for the whole metropolitan area of Valencia.
- Decreasing expenses:
  - Joining, sharing or coordinating services in order to generate economies of scale.
  - Reducing unnecessary costs of representation and management.
  - Increasing the efficiency of all administrative processes and provision of public services.
  - Renegotiating and adjusting payments to external service suppliers (in the area covered by the Plan, there are examples where reductions of up to 30–40% have been achieved without decreasing the quality of the services provided).
  - Decreasing real investments and concentrating on those that are necessary to increase the quality of public services. When contracting new building

Fig. 4.38 Strategy 47

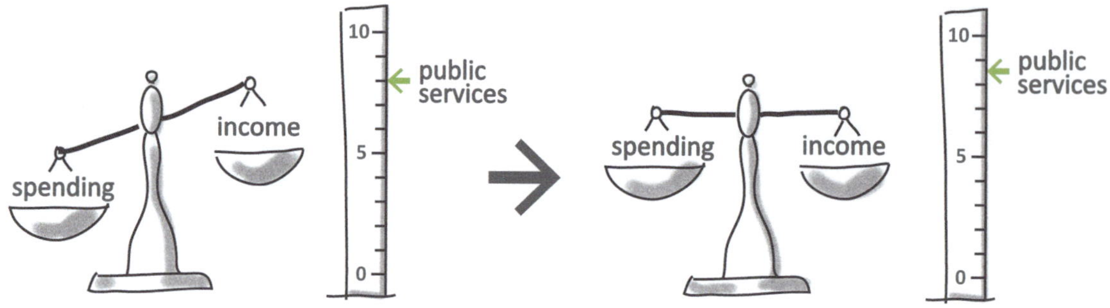

Fig. 4.39 Strategy 48

works, to follow strict principles of budgetary austerity.

– Reviewing the relationship between "statutory" and "non-statutory" services provided by the municipalities to the population, in order to reduce non-compulsory public expenditures.

49. Funding and rural development: **TO MOBILISE AND RAISE PUBLIC AND PRIVATE RESOURCES AT A LOCAL, REGIONAL, NATIONAL AND INTERNATIONAL SCALES** (Fig. 4.40).

- By generating projects and Plans that are duly documented and justified to obtain funds from the European Union.
- By generating fundraising projects for rural development at the national, regional and provincial levels (RURALTER programme and funds from the provincial council). Consider the possibility of creating an inter-municipal consortium to develop these initiatives and the effectiveness of existing local action groups (LAGS).
- By mobilising private economic resources of the local population and by administratively and technically supporting those initiatives aligned with this Strategic Plan and with rural sustainable development.
- By mobilising the knowledge capital and trained labour in local projects and programmes.
- By attracting and supporting external private investment.

4.2.11 Governance and Regional Planning & Management

50. A framework for inter-municipal governance: **TO EXPLORE ALTERNATIVES TO CONSTITUTE AN EFFECTIVE SUPRAMUNICIPAL UNIT** (Fig. 4.41).

- By pooling services in order to avoid unnecessary duplications and achieve economies of scale.
- By studying possible structures that could facilitate the joint administration of some inter-municipal issues (i.e. Community of municipalities).

51. Joint sectoral planning: **TO DRAFT SECTORAL PLANS FOR JOINT ACTIVITIES AND STRATEGIC TOPICS** (Fig. 4.41).

- Drafting an Agricultural and Livestock farming Plan (to be coordinated with the Livestock farming Plan foreseen in the Plan de Ordenación de los Recursos Naturales of the Sierra Calderona Natural Park).
- Drafting a Forestry Plan (to be coordinated with the Livestock farming Plan foreseen in the Plan de

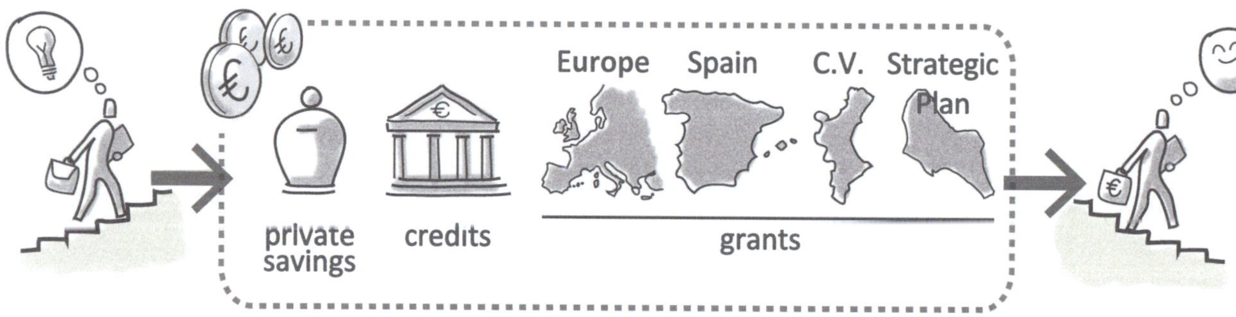

Fig. 4.40 Strategy 49

Ordenación de los Recursos Naturales of the Sierra Calderona Natural Park).
- Drafting a Tourism Plan.
- Drafting an Urban Development Plan that would contain objectives, principles, guidelines and projects to improve towns, villages, housing estates, scattered settlements and industrial areas.

52. Tools for regional management: **TO GENERATE TOOLS THAT WILL FACILITATE AND OPTIMISE REGIONAL MANAGEMENT** (Fig. 4.41).

- By creating a Land Bank: to facilitate the use of those agricultural plots whose landowners are not interested in or are not capable of cultivating them.
- By creating a Housing Bank: to facilitate the utilisation of empty dwellings.
- By creating a Job Bank: to streamline job supply and job search.
- By creating an Office for Administrative and Technical Advice: to advise individuals and groups, local or external, interested in investing and developing appropriate activities or projects in the area covered by the Strategic Plan.
- By creating a Credit Union: to help fund projects set up by local entrepreneurs.
- By pooling some Public Services: to share services and reduce costs.
- By creating a heritage network: to effectively rehabilitate, use and promote the existing heritage (guided tours, explanatory material with a unified image, etc.).
- By creating a tourism network: to effectively promote services for visitors (accommodation, dining, activities in the natural environment, culture, etc.).
- By creating a vocational/professional training network: to facilitate professional specialisation.
- By launching a website and promotional materials of the south-central area of Sierra Calderona.

4.3 Public Participation Plan—Phase 2

This subchapter includes the results of the second phase of the Public Participation Plan. This Public Participation Plan collected the opinions of a representative sample of the local population and visitors of the area, as well as of the opinions of those who responded to the survey at the website: www.tusierracalderona.es.

The proposed question was: Which goals and strategies would you define for the following sectors or geographic areas?

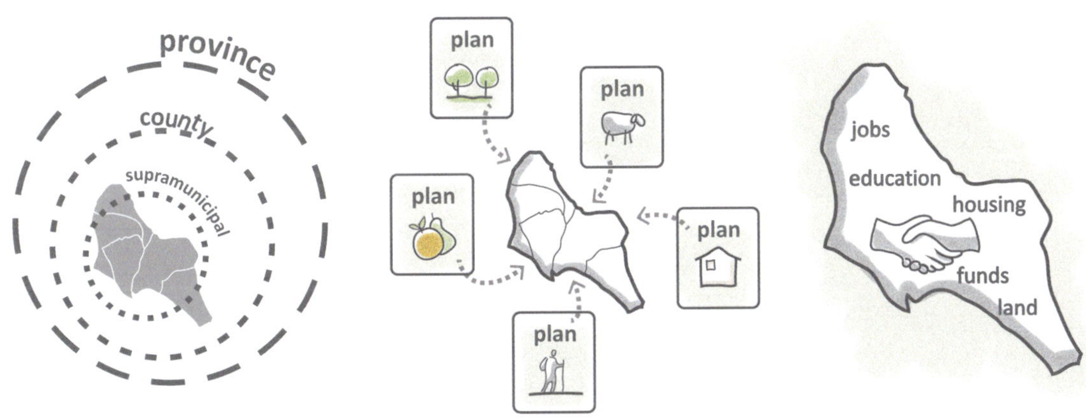

Fig. 4.41 Strategies 50, 51, 52

4.3.1 Evolution of the Natural Environment

See Table 4.1.

Table 4.1 Evolution of the natural environment (results of the Public Participation Plan_phase 2)

		TOTAL		GÁTOVA		MARINES		SERRA		NÁQUERA		OLOCAU	
		No. votes	%	No. votes	%	No. votes	%	No. votes	%	No. votes	%	No. votes	%
1	To involve the population in the cleaning and maintenance of the mountains	75	64.7	11	84.6	10	40.0	17	68.0	29	85.3	8	42.1
2	To support forest squads (more resources. Annual instead of seasonal …)	49	42.2	6	46.2	8	32.0	15	60.0	13	38.2	7	36.8
3	To improve fire prevention	28	24.1	0	0.0	13	52.0	4	16.0	4	11.8	7	36.8
4	To improve and strength firebreaks	27	23.3	3	23.1	6	24.0	7	28.0	6	17.6	5	26.3
5	To restore paths and trails	22	19.0	4	30.8	4	16.0	7	28.0	5	14.7	1	5.3
6	To allow economic activities in the natural environment (e.g. agriculture)	16	13.8	2	15.4	3	12.0	5	20.0	3	8.8	3	15.8
7	To increase the flexibility of restrictions/prohibitions in general	16	13.8	5	38.5	1	4.0	3	12.0	4	11.8	2	10.5
8	To increase the flexibility of restrictions/prohibitions: collection of plants. Flowers. Wood …	15	12.9	4	30.8	2	8.0	5	20.0	2	5.9	2	10.5
9	To protect the mountains	13	11.2	0	0.0	5	20.0	5	20.0	1	2.9	1	5.3
10	To improve the state of conservation of paths. Making them suitable for fire extinction vehicles	12	10.3	0	0.0	1	4.0	6	24.0	2	5.9	3	15.8

4.3.2 Evolution of the Agricultural Environment and Activity

See Table 4.2.

Table 4.2 Evolution of the agricultural environment and activity (results of the Public Participation Plan_phase 2)

		TOTAL		GÁTOVA		MARINES		SERRA		NÁQUERA		OLOCAU	
		No. votes	%	No. votes	%	No. votes	%	No. votes	%	No. votes	%	No. votes	%
1	Do not know/no answer	33	28.4	2	15.4	6	24.0	9	36.0	10	29.4	5	26.3
2	To allow the cultivation of plots located in the Natural Park (more permissiveness. Removing bureaucratic obstacles …)	27	23.3	2	15.4	8	32.0	5	20.0	7	20.6	5	26.3
3	To restore crops such as olive trees, vines, cereals (wheat), almond trees …	22	19.0	4	30.8	1	4.0	6	24.0	7	20.6	4	21.1
4	To restore grazing as a way to clean the mountains: goats and sheep	16	13.8	4	30.8	2	8.0	3	12.0	7	20.6	0	0.0
5	To restore grants and subsidies	15	12.9	2	15.4	6	24.0	1	4.0	1	2.9	5	26.3
6	To create an agriculture cooperative	10	8.6	2	15.4	1	4.0	1	4.0	3	8.8	3	15.8
7	To place value on agriculture: fair prices to improve profitability	10	8.6	3	23.1	3	12.0	2	8.0	2	5.9	0	0.0
8	To subsidy agricultural machines, operations and resources (water. Fertilisers …)	10	8.6	2	15.4	2	8.0	1	4.0	4	11.8	1	5.3
9	Others	8	6.9	1	7.7	3	12.0	0	0.0	3	8.8	0	0.0
10	To restore the production of wine	7	6.0	1	7.7	0	0.0	3	12.0	2	5.9	1	5.3

4.3.3 Evolution of the Urban Environment in Towns and Villages

See Table 4.3.

Table 4.3 Evolution of the urban environment in towns and villages (results of the Public Participation Plan_phase 2)

		TOTAL		GÁTOVA		MARINES		SERRA		NÁQUERA		OLOCAU	
		No. votes	%	No. votes	%	No. votes	%	No. votes	%	No. votes	%	No. votes	%
1	None. The urban centres are fine	28	24.1	4	30.8	10	40.0	5	20.0	6	17.6	3	15.8
2	Others	21	18.1	6	46.2	3	12.0	2	8.0	7	20.6	2	10.5
3	To improve the quality of the public space (trees …)	16	13.8	1	7.7	4	16.0	2	8.0	7	20.6	2	10.5
4	To implement regulations in order to control building heights, promote the use of adequate building materials …	16	13.8	0	0.0	3	12.0	6	24.0	5	14.7	2	10.5
5	To improve pedestrian accessibility	15	12.9	2	15.4	5	20.0	3	12.0	5	14.7	0	0.0
6	To improve the condition of ravines and brooks	14	12.1	0	0.0	1	4.0	2	8.0	4	11.8	7	36.8
7	Do not know/no answer	12	10.3	0	0.0	3	12.0	4	16.0	4	11.8	0	0.0
8	To implement regulations to unify the image of the façades	11	9.5	0	0.0	1	4.0	3	12.0	4	11.8	3	15.8
9	To improve the accessibility to some areas (churches. Buses to the centres. Sports centre in Marines …)	10	8.6	0	0.0	4	16.0	2	8.0	2	5.9	2	10.5
10	To restore the traditional building typology	10	8.6	2	15.4	2	8.0	2	8.0	1	2.9	3	15.8

4.3.4 Evolution of the Urban Environment in Housing Estates

See Table 4.4.

Table 4.4 Evolution of the urban environment in housing estates (results of the Public Participation Plan_phase 2)

		TOTALES		GÁTOVA		MARINES		SERRA		NÁQUERA		OLOCAU	
		No. votes	%	No. votes	%	No. votes	%	No. votes	%	No. votes	%	No. votes	%
1	Do not know/no answer	30	25.9	3	23.1	8	32.0	9	36.0	5	14.7	4	21.1
2	None. Housing estates are already fine	23	19.8	8	61.5	4	16.0	3	12.0	5	14.7	3	15.8
3	To improve accesses	15	12.9	0	0.0	6	24.0	1	4.0	6	17.6	2	10.5
4	To prevent their sprawl	15	12.9	0	0.0	2	8.0	5	20.0	7	20.6	1	5.3
5	To improve the surface of their streets. Access roads	13	11.2	0	0.0	5	20.0	1	4.0	4	11.8	3	15.8
6	To improve public services: waste collection	13	11.2	0	0.0	2	8.0	2	8.0	5	14.7	4	21.1
7	Others	13	11.2	0	0.0	4	16.0	2	8.0	6	17.6	1	5.3
8	To improve public services: cleaning and garbage collection	11	9.5	0	0.0	2	8.0	0	0.0	7	20.6	2	10.5
9	To improve their relation with urban centres (towns and villages). Integrating them into their activities	10	8.6	0	0.0	0	0.0	3	12.0	4	11.8	3	15.8
10	To improve public services: sewage network	8	6.9	0	0.0	1	4.0	0	0.0	5	14.7	2	10.5

4.3.5 Evolution of the Urban Environment in Scattered Settlements

See Table 4.5.

Table 4.5 Evolution of the urban environment in scattered settlements (results of the Public Participation Plan_phase 2)

		TOTALES		GÁTOVA		MARINES		SERRA		NÁQUERA		OLOCAU	
		No. votes	%	No. votes	%	No. votes	%	No. votes	%	No. votes	%	No. votes	%
1	Do not know/no answer	37	31.9	2	15.4	8	32.0	8	32.0	11	32.4	7	36.8
2	To legalise them	33	28.4	6	46.2	5	20.0	7	28.0	12	35.3	3	15.8
3	To prevent their sprawl	25	21.6	2	15.4	5	20.0	8	32.0	4	11.8	5	26.3
4	To preserve the existing ones	12	10.3	2	15.4	2	8.0	2	8.0	4	11.8	2	10.5
5	To urbanise the existing ones (sewage network. Access roads …)	8	6.9	0	0.0	2	8.0	1	4.0	5	14.7	0	0.0
6	None. The scattered settlements are already fine	7	6.0	0	0.0	3	12.0	0	0.0	3	8.8	1	5.3
7	Others	6	5.2	1	7.7	4	16.0	0	0.0	0	0.0	1	5.3
8	To make them responsible of their cleaning and maintenance	5	4.3	0	0.0	2	8.0	1	4.0	1	2.9	1	5.3
9	To control cesspits (their leaks pollute the soil)	5	4.3	0	0.0	0	0.0	1	4.0	3	8.8	1	5.3
10	To control their sprawl towards the mountains	5	4.3	0	0.0	0	0.0	1	4.0	4	11.8	0	0.0

4.3.6 Evolution of Industrial Land

See Table 4.6.

Table 4.6 Evolution of industrial land (results of the Public Participation Plan_phase 2)

		TOTALES		GÁTOVA		MARINES		SERRA		NÁQUERA		OLOCAU	
		No. votes	%	No. votes	%	No. votes	%	No. votes	%	No. votes	%	No. votes	%
1	Do not know/no answer	38	32.8	3	23.1	6	24.0	9	36.0	11	32.4	8	42.1
2	To develop more industrial land to create jobs	18	15.5	4	30.8	8	32.0	1	4.0	2	5.9	3	15.8
3	There is already too much industrial land. There is no need for more industry	18	15.5	0	0.0	1	4.0	11	44.0	4	11.8	2	10.5
4	None. The industry is fine as it is	8	6.9	0	0.0	2	8.0	0	0.0	4	11.8	2	10.5
5	To implement industry linked to agriculture and forestry: agriculture cooperative, oil presses, wineries, rabbit farms …	8	6.9	1	7.7	0	0.0	2	8.0	2	5.9	3	15.8
6	To improve the location of the industry: close to the main roads. Away from the urban centres (Náquera)	8	6.9	0	0.0	0	0.0	1	4.0	7	20.6	0	0.0
7	To coordinate industrial uses appropriate with the Natural Park	6	5.2	2	15.4	0	0.0	3	12.0	0	0.0	0	0.0
8	To facilitate the employment of locals	5	4.3	1	7.7	3	12.0	1	4.0	0	0.0	0	0.0
9	To facilitate the creation of new industry by reducing the taxes and by helping entrepreneurs …	5	4.3	0	0.0	4	16.0	0	0.0	1	2.9	0	0.0
10	To restore traditional manufactures: rodeno stone	4	3.4	2	15.4	1	4.0	1	4.0	0	0.0	0	0.0

4.3.7 Evolution of Public Services

See Table 4.7.

Table 4.7 Evolution of public services (results of the Public Participation Plan_phase 2)

		TOTAL		GÁTOVA		MARINES		SERRA		NÁQUERA		OLOCAU	
		No. votes	%	No. votes	%	No. votes	%	No. votes	%	No. votes	%	No. votes	%
1	To extend health services: visiting hours, emergency service, ambulances …	35	30.2	4	30.8	4	16.0	11	44.0	12	35.3	4	21.1
2	None. Public services are already fine	25	21.6	1	7.7	10	40.0	4	16.0	6	17.6	4	21.1
3	To extend the offer of cultural and sport facilities (sports centre in Gátova …)	19	16.4	3	23.1	3	12.0	4	16.0	7	20.6	2	10.5
4	To extend the offer of activities during the summer: outdoor cinema, theatre, summer schools, activities for the youth …	19	16.4	1	7.7	2	8.0	7	28.0	2	5.9	7	36.8
5	To improve education facilities building proper schools and eliminating barracks. Create more day-care facilities for babies and toddlers	17	14.7	1	7.7	3	12.0	1	4.0	10	29.4	2	10.5
6	Do not know/no answer	15	12.9	2	15.4	2	8.0	3	12.0	5	14.7	2	10.5
7	Inadequate waste collections service: insufficient during high season. High taxes. Insufficient selective collecting points …	13	11.2	1	7.7	1	4.0	2	8.0	6	17.6	2	10.5
8	To extend the opening hours of cultural-sport facilities: library, sports centres …	11	9.5	1	7.7	2	8.0	6	24.0	1	2.9	1	5.3
9	To create more schools so the children do not need to travel long distances	9	7.8	1	7.7	3	12.0	1	4.0	3	8.8	1	5.3
10	To improve the quality of libraries: book stock, number of computers …	9	7.8	0	0.0	1	4.0	3	12.0	4	11.8	1	5.3

4.3.8 Evolution of Transport and Mobility Infrastructures

See Table 4.8.

Table 4.8 Evolution of transport and mobility infrastructures (results of the Public Participation Plan_phase 2)

		TOTAL		GÁTOVA		MARINES		SERRA		NÁQUERA		OLOCAU	
		No. votes	%	No. votes	%	No. votes	%	No. votes	%	No. votes	%	No. votes	%
1	To increase the frequency and schedule of public transport	53	45.7	3	23.1	9	36.0	17	68.0	13	38.2	10	52.6
2	To extend the public transport network	40	34.5	1	7.7	8	32.0	9	36.0	10	29.4	11	57.9
3	To design a new road section between urban centres that will allow lorries, cars, cyclist and pedestrians to circulate separately	23	19.8	5	38.5	5	20.0	2	8.0	10	29.4	1	5.3
4	To improve the roads: surface and signposting. Secure areas with risk of landslides …	21	18.1	5	38.5	8	32.0	1	4.0	6	17.6	1	5.3
5	To adapt the paths for cyclist	19	16.4	0	0.0	6	24.0	3	12.0	9	26.5	1	5.3
6	To extend the network of public transport during high season	17	14.7	0	0.0	1	4.0	4	16.0	10	29.4	1	5.3
7	To restore mountain paths and trails	14	12.1	2	15.4	2	8.0	3	12.0	5	14.7	2	10.5
8	To reduce the price of public transport and to subsidise it for the local population	12	10.3	0	0.0	1	4.0	5	20.0	3	8.8	3	15.8
9	To improve the connection with the city of Valencia	12	10.3	0	0.0	3	12.0	2	8.0	4	11.8	3	15.8
10	To extend the inter-municipal mobility	11	9.5	0	0.0	5	20.0	0	0.0	4	11.8	2	10.5

4.3.9 Management of Cultural Heritage

See Table 4.9.

Table 4.9 Management of cultural heritage (results of the Public Participation Plan_phase 2)

		TOTAL		GÁTOVA		MARINES		SERRA		NÁQUERA		OLOCAU	
		No. votes	%	No. votes	%	No. votes	%	No. votes	%	No. votes	%	No. votes	%
1	Do not know/no answer	28	24.1	1	7.7	7	28.0	2	8.0	11	32.4	5	26.3
2	To rehabilitate existing heritage and include it in the heritage network	26	22.4	2	15.4	7	28.0	8	32.0	5	14.7	4	21.1
3	To use the proper means for rehabilitation (i.e. not using poor-quality materials like in Gátova's washing site. Not worsening their image like in Gátova's bridge) and a greater investment (i.e. for the castle. Church. Square and "Pla de l'Olla" in Marines; the castle. "Torre Vigía" and roman bridge in Serra; the bridge in Gátova …)	22	19.0	5	38.5	5	20.0	12	48.0	0	0.0	0	0.0
4	To promote the restored heritage	14	12.1	2	15.4	3	12.0	3	12.0	4	11.8	2	10.5
5	To promote local festivities and bullfighting	14	12.1	4	30.8	0	0.0	5	20.0	5	14.7	0	0.0
6	To inform the population regularly about what has been and will be done	13	11.2	2	15.4	3	12.0	2	8.0	4	11.8	2	10.5
7	To maintain the accesses to rehabilitated heritage	11	9.5	0	0.0	2	8.0	1	4.0	6	17.6	2	10.5
8	To restore the hydraulic heritage (deposits. Irrigation channels. Fountains. Water mills. Washers …)	11	9.5	4	30.8	2	8.0	1	4.0	0	0.0	4	21.1
9	To improve the access and marketing of the heritage as it is mostly unknown	8	6.9	0	0.0	2	8.0	2	8.0	3	8.8	1	5.3
10	To improve the signposting and information regarding how to visit the heritage sites (maps. Flyers. Brochures …)	8	6.9	1	7.7	2	8.0	2	8.0	3	8.8	0	0.0

4.3.10 Evolution of the Tourism Sector

See Table 4.10.

Table 4.10 Evolution of the tourism sector (results of the Public Participation Plan_phase 2)

		TOTALES		GÁTOVA		MARINES		SERRA		NÁQUERA		OLOCAU	
		No. votes	%	No. votes	%	No. votes	%	No. votes	%	No. votes	%	No. votes	%
1	To support rural tourism: lodges, farm school, equestrian centre, campsites open round the year …	45	38.8	5	38.5	13	52.0	10	40.0	8	23.5	9	47.4
2	To promote Sierra Calderona	21	18.1	3	23.1	5	20.0	4	16.0	8	23.5	1	5.3
3	To organise guided excursions	21	18.1	2	15.4	5	20.0	3	12.0	11	32.4	0	0.0
4	To implement adventure tourism	17	14.7	1	7.7	3	12.0	3	12.0	7	20.6	3	15.8
5	To facilitate family tourism as an initiation to mountain tourism	15	12.9	1	7.7	3	12.0	3	12.0	7	20.6	1	5.3
6	To improve routes (paths and resting points) for hikers, cyclist and horse riders	15	12.9	3	23.1	4	16.0	0	0.0	2	5.9	6	31.6
7	To promote the area in the city of Valencia	13	11.2	3	23.1	4	16.0	2	8.0	1	2.9	3	15.8
8	To promote rural, cultural and family tourism in the schools	12	10.3	2	15.4	1	4.0	4	16.0	4	11.8	1	5.3
9	To improve signposting	12	10.3	0	0.0	1	4.0	3	12.0	5	14.7	3	15.8
10	To create new itineraries for cyclist, hikers and horse riders	11	9.5	3	23.1	2	8.0	1	4.0	2	5.9	3	15.8

4.3.11 Landscape

See Table 4.11.

Table 4.11 Landscape (results of the Public Participation Plan_phase 2)

		TOTALES		GÁTOVA		MARINES		SERRA		NÁQUERA		OLOCAU	
		No. votes	%	No. votes	%	No. votes	%	No. votes	%	No. votes	%	No. votes	%
1	To improve the cleaning and maintenance works in forests and mountains	20	17.2	3	23.1	7	28.0	4	16.0	6	17.6	0	0.0
2	Do not know/no answer	19	16.4	2	15.4	5	20.0	2	8.0	6	17.6	2	10.5
3	To improve the conditions of the ravines	18	15.5	3	23.1	5	20.0	1	4.0	4	11.8	5	26.3
4	To clean and improve the existing paths	18	15.5	3	23.1	5	20.0	2	8.0	5	14.7	3	15.8
5	Nothing. The landscape is fine as it is	15	12.9	0	0.0	5	20.0	2	8.0	7	20.6	1	5.3
6	To improve the maintenance of parks, roads, urban edges and streets	15	12.9	2	15.4	6	24.0	2	8.0	1	2.9	4	21.1
7	Others	15	12.9	1	7.7	2	8.0	7	28.0	3	8.8	2	10.5
8	To give an identity to the "mountain villages"	14	12.1	1	7.7	3	12.0	5	20.0	2	5.9	3	15.8
9	To visually improve the accesses to urban centres and housing estates	14	12.1	0	0.0	2	8.0	4	16.0	6	17.6	2	10.5
10	To restore the forest and agriculture activity as the cultivated fields improve the visual quality of the landscape and a sustainable forest activity helps prevent fires	14	12.1	1	7.7	6	24.0	2	8.0	3	8.8	2	10.5

4.3.12 Evolution of Urban Land

See Table 4.12.

Table 4.12 Evolution of urban land (results of the Public Participation Plan_phase 2)

		TOTALES		GÁTOVA		MARINES		SERRA		NÁQUERA		OLOCAU	
		No. votes	%	No. votes	%	No. votes	%	No. votes	%	No. votes	%	No. votes	%
1	To consolidate properly the housing estates located in the plains and foothills of Sierra Calderona by revitalising their services	53	45.7	2	15.4	14	56.0	11	44.0	15	44.1	11	57.9
2	To stop urban sprawl. Adding services and densifying the existing low-density settlements. While revitalising the urban centres through a strategic offer of services and through the improvements of the quality of the housing stock	48	41.4	10	76.9	8	32.0	9	36.0	15	44.1	5	26.3
3	To create a built continuous of low-density housing estates in plains and Sierra Calderona foothills	7	6.0	0	0.0	0	0.0	3	12.0	2	5.9	1	5.3

4.3.13 Evolution of the Low-Density Housing System

See Table 4.13.

Table 4.13 Evolution of the low-density housing system (results of the Public Participation Plan_phase 2)

		TOTALES		GÁTOVA		MARINES		SERRA		NÁQUERA		OLOCAU	
		No. votes	%	No. votes	%	No. votes	%	No. votes	%	No. votes	%	No. votes	%
1	To improve the edges and interiors of these settlements. The neighbours should develop these improvements	39	33.6	3	23.1	10	40.0	9	36.0	9	26.5	6	31.6
2	Unify scattered settlements with nearby housing estates in order to improve their poor conditions. This will imply extending the built surface	8	6.9	1	7.7	2	8.0	1	4.0	2	5.9	2	10.5
3	To create a low-density built continuous in the plains and foothills of Sierra Calderona	7	6.0	0	0.0	0	0.0	3	12.0	2	5.9	1	5.3

4.3.14 Sustainability

See Table 4.14.

Table 4.14 Sustainability (results of the Public Participation Plan_phase 2)

		TOTAL		GÁTOVA		MARINES		SERRA		NÁQUERA		OLOCAU	
		No. votes	%	No. votes	%	No. votes	%	No. votes	%	No. votes	%	No. votes	%
1	Do not know/no answer	49	42.2	5	38.5	13	52.0	11	44.0	11	32.4	8	42.1
2	To facilitate recycling: not enough garbage containers. Recycling points within the urban centres. Improve the cleanness of containers …	14	12.1	0	0.0	1	4.0	1	4.0	9	26.5	3	15.8
3	To improve public transport and bicycle lanes: within and between municipalities	13	11.2	1	7.7	3	12.0	4	16.0	2	5.9	3	15.8
4	To install solar power farms	10	8.6	1	7.7	1	4.0	2	8.0	5	14.7	1	5.3
5	To install solar panels in buildings	9	7.8	1	7.7	2	8.0	1	4.0	4	11.8	1	5.3
6	To renovate the water. Electricity and gas installations as they are old and have leaks	7	6.0	1	7.7	1	4.0	2	8.0	3	8.8	0	0.0
7	To collect and reuse rain water	7	6.0	0	0.0	1	4.0	2	8.0	2	5.9	2	10.5
8	To secure grants to fund renewal energy plants	7	6.0	0	0.0	0	0.0	1	4.0	5	14.7	1	5.3
9	To optimise garbage collection	7	6.0	2	15.4	0	0.0	1	4.0	2	5.9	2	10.5
10	To reuse water after having been treated in a treatment plant	5	4.3	1	7.7	1	4.0	1	4.0	1	2.9	1	5.3

4.3.15 Socio-Demography

See Table 4.15.

Table 4.15 Socio-demography (results of the Public Participation Plan_phase 2)

		TOTAL		GÁTOVA		MARINES		SERRA		NÁQUERA		OLOCAU	
		No. votes	%	No. votes	%	No. votes	%	No. votes	%	No. votes	%	No. votes	%
1	Do not know/no answer	60	51.7	5	38.5	13	52.0	14	56.0	20	58.8	7	36.8
2	The population is not enough: facilitate affordable housing for young population. Create new jobs. Improve health and education services. Improve the communication between urban centres and housing estates. Create new industry. Promote agriculture....	39	33.6	5	38.5	6	24.0	9	36.0	6	17.6	12	63.2
3	To stop the emigration from the Sierra Calderona caused by the lack of employment	6	5.2	2	15.4	0	0.0	2	8.0	1	2.9	1	5.3
4	None	5	4.3	1	7.7	1	4.0	1	4.0	2	5.9	0	0.0
5	To improve services during high season (when the population increases)	4	3.4	0	0.0	0	0.0	0	0.0	4	11.8	0	0.0
6	To control the seasonal population: increasing services....	2	1.7	0	0.0	1	4.0	0	0.0	1	2.9	0	0.0
7	To stimulate employment within the Natural Park facilitating the access to housing to its employees	2	1.7	0	0.0	1	4.0	0	0.0	1	2.9	0	0.0
8	To safeguard the existing population: affordable housing, etc.	2	1.7	0	0.0	0	0.0	1	4.0	1	2.9	0	0.0
9	The population is already enough and varied	1	0.9	0	0.0	1	4.0	0	0.0	0	0.0	0	0.0
10	There is too much population in the area: too many housing estates and too many people	1	0.9	0	0.0	0	0.0	0	0.0	1	2.9	0	0.0

4.3.16 Economic Development

See Table 4.16.

Table 4.16 Economic development (results of the Public Participation Plan_phase 2)

		TOTAL		GÁTOVA		MARINES		SERRA		NÁQUERA		OLOCAU	
		No. votes	%	No. votes	%	No. votes	%	No. votes	%	No. votes	%	No. votes	%
1	To invest on quality mountain tourism: guided tours, information, sport activities ...	26	22.4	3	23.1	6	24.0	6	24.0	7	20.6	4	21.1
2	To promote adventure tourism for children and the youth: hiking, horse riding, orienteering, cycling, school camps ...	24	20.7	5	38.5	7	28.0	6	24.0	5	14.7	1	5.3
3	To improve sport tourism linked to the urban centres: itineraries for cyclist, lunch breaks ...	21	18.1	4	30.8	5	20.0	7	28.0	5	14.7	0	0.0
4	To restore organic agriculture lifting restrains for farmers	20	17.2	1	7.7	7	28.0	5	20.0	6	17.6	1	5.3
5	To promote tourism in Sierra Calderona from the city of Valencia (posters. flyers ...)	19	16.4	2	15.4	3	12.0	5	20.0	7	20.6	2	10.5
6	Do not know/no answer	17	14.7	0	0.0	3	12.0	3	12.0	7	20.6	3	15.8
7	To promote sustainable industry linked to the resources available in the Natural Park	14	12.1	2	15.4	5	20.0	3	12.0	3	8.8	1	5.3
8	Others	12	10.3	1	7.7	4	16.0	2	8.0	3	8.8	1	5.3
9	To improve and extend the offer of services	10	8.6	2	15.4	1	4.0	3	12.0	2	5.9	2	10.5
10	To promote the urban centres as access gates to Sierra Calderona	9	7.8	1	7.7	1	4.0	3	12.0	3	8.8	1	5.3

4.3.17 Management of Municipal Treasuries

See Table 4.17.

Table 4.17 Management of municipal treasuries (results of the Public Participation Plan_phase 2)

		TOTALES		GÁTOVA		MARINES		SERRA		NÁQUERA		OLOCAU	
		No. votes	%	No. votes	%	No. votes	%	No. votes	%	No. votes	%	No. votes	%
AIMS AND STRATEGIES TO IMPROVE THE ACCOUNTING POSITION													
1	To increase transparency in the management and control of public funds: embezzlement. Avoid corruption. Control better public contracts and budgets	52	44.8	5	38.5	7	28.0	14	56.0	21	61.8	5	26.3
2	Do not know/no answer	29	25.0	4	30.8	6	24.0	4	16.0	8	23.5	6	31.6
3	Do not increase taxes (It is not the only solution)	18	15.5	1	7.7	4	16.0	2	8.0	9	26.5	2	10.5
4	To better manage the available resources and existing heritage	13	11.2	0	0.0	1	4.0	3	12.0	8	23.5	1	5.3
5	To reduce the spending of festivities. Improve public lighting. Garbage collection and salaries for municipal employees and authorities avoiding at the same time double jobs	12	10.3	0	0.0	5	20.0	3	12.0	3	8.8	1	5.3
6	To demand exemplariness to the authorities in public spending and audits	12	10.3	2	15.4	0	0.0	5	20.0	1	2.9	3	15.8
7	Others	9	7.8	1	7.7	3	12.0	0	0.0	2	5.9	3	15.8
8	To develop a long-term project on sustainable development and growth	6	5.2	0	0.0	1	4.0	3	12.0	0	0.0	2	10.5
9	To invest the grants given by the regional government for the management of the Natural Park	4	3.4	0	0.0	2	8.0	0	0.0	1	2.9	1	5.3
10	To penalise malpractices	3	2.6	0	0.0	1	4.0	1	4.0	1	2.9	0	0.0

4.3.18 External Funding

See Table 4.18.

Table 4.18 External funding (results of the Public Participation Plan_phase 2)

		TOTALES		GÁTOVA		MARINES		SERRA		NÁQUERA		OLOCAU	
		No. votes	%	No. votes	%	No. votes	%	No. votes	%	No. votes	%	No. votes	%
AIMS AND STRATEGIES TO IMPROVE FUNDING													
1	Do not know/no answer	50	43.1	6	46.2	9	36.0	9	36.0	17	50.0	8	42.1
2	To increase public subsidies (local and national)	18	15.5	2	15.4	3	12.0	5	20.0	5	14.7	3	15.8
3	Private funding	10	8.6	0	0.0	1	4.0	4	16.0	3	8.8	1	5.3
4	Public funding: to combine political powers/responsibilities, i.e. municipalities (urban land) and regional government (Natural Park)	10	8.6	2	15.4	3	12.0	3	12.0	1	2.9	1	5.3
5	To create specific subsidies for agriculture and farmers. Develop designation of origin for local products. Improve the visual condition of the landscape (accesses. Urban edges) …	9	7.8	2	15.4	2	8.0	1	4.0	1	2.9	3	15.8
6	Others	8	6.9	0	0.0	4	16.0	1	4.0	2	5.9	1	5.3
7	To remove administration obstacles (i.e. currently business licences are approved after 3 months)	6	5.2	1	7.7	2	8.0	1	4.0	0	0.0	2	10.5
8	To allow compatible activities: integrated hotel-campsite. Guided visits. Ecclesiastic buildings used as accommodation facilities …	6	5.2	0	0.0	0	0.0	3	12.0	2	5.9	1	5.3
9	To link the protection of the Natural Park with subsidies	6	5.2	0	0.0	2	8.0	1	4.0	1	2.9	2	10.5
10	More efficient public spending	5	4.3	0	0.0	1	4.0	1	4.0	2	5.9	1	5.3

MIX
Papier aus verantwortungsvollen Quellen
Paper from responsible sources
FSC® C105338

If you have any concerns about our products,
you can contact us on
ProductSafety@springernature.com

In case Publisher is established outside the EU,
the EU authorized representative is:
Springer Nature Customer Service Center GmbH
Europaplatz 3, 69115 Heidelberg, Germany

Printed by Libri Plureos GmbH
in Hamburg, Germany